Educational Technology

Educational Technology

A Definition with Commentary

ALAN JANUSZEWSKI • MICHAEL MOLENDA

LEA Lawrence Erlbaum Associates
Taylor & Francis Group

New York   London

Lawrence Erlbaum Associates
Taylor & Francis Group
270 Madison Avenue
New York, NY 10016

Lawrence Erlbaum Associates
Taylor & Francis Group
2 Park Square
Milton Park, Abingdon
Oxon OX14 4RN

© 2008 by Taylor & Francis Group, LLC
Lawrence Erlbaum Associates is an imprint of Taylor & Francis Group, an Informa business

Printed in the United States of America on acid-free paper
10 9 8 7 6 5 4 3 2 1

International Standard Book Number-13: 978-0-8058-5861-7 (Softcover) 978-0-8058-5860-0 (Hardcover)

Visit the Taylor & Francis Web site at
http://www.taylorandfrancis.com

and the LEA and Routledge Web site at
http://www.routledge.com

CONTENTS

DEDICATION

This work is dedicated to the membership of the Association for Educational Communications and Technology, the other professionals who are practicing in our field, and our students.

PREFACE

This book presents a definition of the field of study and practice known as "educational technology" or "instructional technology." While recognizing that *educational* and *instructional* have different connotations, the authors intend that this definition encompass both terms. It could be argued that either term is broader and more inclusive in some sense, but the current Definition and Terminology Committee chooses to focus on the sense in which *education* is the broader term, incorporating both purposive and spontaneous learning, both teacher led and learner initiated, taking place in settings both formal and informal.

This project was sponsored by the Association for Educational Communications and Technology (AECT) and reflects the collaborative efforts of all the members of the AECT Definition and Terminology Committee. The definition statement itself was discussed and agreed to by the full committee during the annual conventions of 2002 and 2003, with final modifications made at the meeting of the Professors of Instructional Development and Technology in Spring of 2004, and was subsequently approved by the Board of Directors of AECT in the summer of 2004. Most of the chapters were written or coauthored by committee members, based on the understandings worked out during committee deliberations. All of the chapters were reviewed by outstandingly qualified members of the educational technology community and were improved immeasurably by their constructive criticism during the rounds of writing and reviewing during 2005 and 2006.

AECT has an international membership, and the study and practice of educational technology flourish in every part of the world. Although it is hoped that the ideas in this volume will be found relevant by a broad array of readers, especially those who are at the early stages of study, both domestic and foreign, we acknowledge that the text is centered in the experiences of the authors and the historical evolution of educational technology in the United States.

This volume begins with the statement of the definition itself (chapter 1), followed by chapters of commentary upon each of the key terms and concepts in the definition (chapters 2–9). Chapter 10 provides a historical context for the current definition by reviewing the salient elements of prior AECT definitions. Chapter 11 discusses the status of ethical considerations in the field, and chapter 12 concludes by discussing the ramifications of the current definition for academic programs in educational technology.

As has been the case with each of the preceding definition projects, this definition aims to reflect contemporary thinking and to stimulate a conversation about the many meanings of educational technology.

ACKNOWLEDGMENTS

Authors of the Definition

Members of the Definition and Terminology Committee of the Association for Educational Communications and Technology (AECT) that produced the one-sentence definition and the elaboration of it that constitute chapter 1 are:

Anthony Karl Betrus
Robert Maribe Branch
Philip Doughty
Michael Molenda
Robert Pearson
Kay A. Persichitte
Landra L. Rezabek
Rhonda S. Robinson
Chuck Stoddard
Alan Januszewski, Chair

Reviewers

The chapters of this book were subjected to several rounds of review, which gave invaluable guidance to authors in refining their chapters. The editors would like to thank these AECT members who contributed their time and expertise by reviewing various portions and versions of manuscripts of the chapters that comprise this book:

Anthony Karl Betrus
M. J. Bishop
Robert Maribe Branch
Buddy Burniske
Ward Cates
Kathy Cennamo
J. Ana Donaldson
Philip Doughty
Judy Duffield
Donald P. Ely
Tom Hergert
Mary Herring
Janette Hill

J. Randall Koetting
Barbara Lockee
Delia Neuman
Kay A. Persichitte
Robert Reiser
Rita Richey
Landra L. Rezabek
Rhonda S. Robinson
Edd Schneider
Barbara Seels
Kenneth Silber
J. Michael Spector
William Sugar
Connie Tibbits
Andrew R. J. Yeaman

Illustrations

All the illustrations—the figures and tables—appearing in this book were created by Ms. Jung Won Hur, doctoral candidate in Instructional Systems Technology, Indiana University, based on artwork submitted by the various authors of the chapters.

Institutional Support

The editors would like to thank the officers and staff of AECT and members of the AECT Board of Directors for their support and patience over the course of this project. In addition, all the contributors owe thanks to their schools, colleges, universities, and businesses that support them and provide a nurturing environment for doing work such as this.

1

DEFINITION

*Definition and Terminology Committee of the Association
for Educational Communications and Technology*

The Definition

CONCEPTIONS OF EDUCATIONAL TECHNOLOGY have been evolving as long as the field has, and they continue to evolve. Therefore, today's conception is a temporary one, a snapshot in time. In today's conception, educational technology can be defined as an abstract concept or as a field of practice. First, the definition of the *concept*:

> Educational technology is the study and ethical practice of facilitating learning and improving performance by creating, using, and managing appropriate technological processes and resources.

Elements of the Definition

Each of the key terms used in the definition will be discussed as to its intended meaning in the context of the definition.

Study. The theoretical understanding of, as well as the practice of, educational technology, requires continual knowledge construction and refinement through research and reflective practice, which are encompassed in the term *study.* That is, *study* refers to information gathering and analysis beyond the traditional conceptions of research. It is intended to include quantitative and qualitative research as well as other forms of disciplined inquiry such as theorizing, philosophical analysis, historical investigations, development projects, fault analyses, system analyses, and evaluations. Research has traditionally been both a generator of new ideas and an evaluative process to help improve practice. Research can be conducted based upon a variety of methodological constructs as well as several contrasting theoretical constructs. The research in educational technology has grown from investigations attempting to "prove" that media and technology are effective tools for instruction, to investigations formulated to examine the appropriate applications of processes and technologies to the improvement of learning.

Important to the newest research in educational technology is the use of authentic environments and the voices of practitioners and users as well as researchers. Inherent in the word *research* is the iterative process it encompasses. Research seeks to resolve problems by investigating solutions, and those attempts lead to new practice and therefore new problems and questions. Certainly, the ideas of reflective practice and inquiry based upon authentic settings are valuable perspectives on research. Reflective practitioners consider the problems in their environments (e.g., a learning problem of their students) and attempt to resolve the problems by changes in practice, based upon both research results and professional experience. Reflection on this process leads to changes in the considered solution and further attempts to identify and solve problems in the environment, a cyclical process of practice/reflection that can lead to improved practice (Schön, 1990).

Current inquiry problem areas are often determined by the influx of new technologies into educational practice. The history of the field has recorded the many research programs initiated in response to new technologies, investigating how to best design, develop, use, and manage the products of the new technology. However, more recently, the inquiry programs in educational technology have been influenced by growth and change in major theoretical positions in learning theory, information management, and other allied fields. For example, the theoretical lenses of cognitive and constructivist learning theories have changed the emphasis in the field from teaching to learning. Attention to learners' perspectives, preferences, and ownership of the learning process has grown. These theoretical shifts have changed the orientation of the field dramatically, from a field driven by the design of instruction to be "delivered" in a variety of formats to a field which seeks to create learning environments in which learners can explore—often assisted by electronic support systems—in order to arrive at meaningful understanding. The research emphasis has shifted toward observing learners' active participation and construction of their own path toward learning. In other words, interest is moving away from the design of prespecified instructional routines and toward the design of environments to facilitate learning.

Ethical practice. Educational technology has long had an ethical stance and a code of ethical practice expectations. The AECT Ethics Committee has been active in defining the field's ethical standards and in providing case examples from which to discuss and understand the implications of ethical concerns for practice. In fact, the recent emphasis in society on the ethical use of media and on respect for intellectual property has been addressed by this AECT committee for the educational technology field. The evolution and promulgation of AECT's ethical principles are discussed in depth in chapter 11.

There has been an increase in concerns and attention to the ethical issues within educational technology. Ethics are not merely "rules and expectations" but are a basis for practice. In fact, ethical practice is less a series of expectations, boundaries, and new laws than it is an approach or construct from which to work. The current definition considers ethical practice as essential to our professional success, for without the ethical considerations being addressed, success is not possible.

From the perspective of critical theory, professionals in educational technology must question even basic assumptions such as the efficacy of traditional constructs such as the systems approach and technologies of instruction, as well as the power position of those designing and developing the technological solutions. Contemporary ethics oblige educational technologists to consider their learners, the environments for learning, and the needs and the "good" of society as they develop their practices. Considering who is included, who is empowered, and who has authority are new issues in the design and development of learning solutions, but an ethical stance insists that educational technologists question their practice areas in these ways as well as in the more traditional constructs of efficiency or effectiveness.

The AECT Code of Professional Ethics includes principles "intended to aid members individually and collectively in maintaining a high level of professional conduct" (Welliver, 2001). AECT's code is divided into three categories: commitment to the individual, such as the protection of rights of access to materials and efforts to protect the health and safety of professionals; commitment to society, such as truthful public statements regarding educational matters or fair and equitable practices with those rendering service to the profession; and commitment to the profession, such as improving professional knowledge and skill and giving accurate credit to work and ideas published. Each of the three principal areas has several listed commitments that help inform educational technology professionals regarding their appropriate actions, regardless of their contexts or roles. Consideration is provided for those serving as researchers, professors, consultants, designers, and learning resource directors, for example, to help shape their own professional behaviors and ethical conducts.

Facilitating. The shift in views of learning and instruction reflected in cognitive and constructivist learning theories has engendered a rethinking of assumptions about the connection between instruction and learning. Earlier definitions in this field implied a more direct cause and effect relationship between instructional interventions and learning. For example, the first formal AECT definition (Ely, 1963) referred to "the design and use of messages which control the learning process." Later definitions were less explicit but continued to imply a relatively direct connection between well-designed, well-delivered instruction and effective learning. With the recent paradigm shift in learning theories has come a greater recognition of the learner's role as a constructor as opposed to a recipient of knowledge. With this recognition of learner ownership and responsibility has come a role for technology that is more facilitative than controlling.

In addition, when learning goals in schools, colleges, and other organizations shift toward deep rather than shallow learning, the learning environments must become more immersive and more authentic. In these environments, the key role of technology is not so much to present information and provide drill and practice (to *control* learning) but to provide the problem space and the tools to explore it (to *support* learning). In such cases, the immersive environments and cognitive tools educational technologists help design and use are created to guide learners, to make learning opportunities available, and to assist learners in finding the answers to their questions. Even in cases in which a more expository strategy is justified, where presentation and drill and practice are appropriately emphasized, the learner must still attend to, process, and take meaning from the activities. The learner is still in control, not the instructional program. Therefore, educational technology claims to *facilitate learning* rather than to cause or control learning; that is, it can help create an environment in which learning more easily could occur.

Facilitating includes the design of the environment, the organizing of resources, and the providing of tools. The learning events can take place in face-to-face settings or in virtual environments, as in microworlds or distance learning.

Learning. The term *learning* does not connote today what it connoted 40 years ago when the first AECT definition was developed. There is a heightened awareness of the difference between the mere retention of information for testing purposes and the acquisition of knowledge, skills, and attitudes used beyond the classroom walls.

One of the critical elements of instructional design is to identify the learning tasks to be pursued and to choose assessment methods to measure their

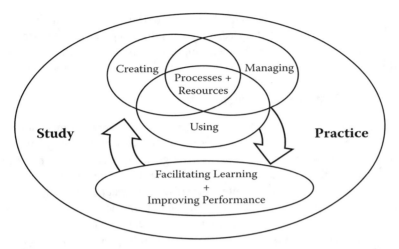

Figure 1.1. A visual summary of key elements of the current definition.

attainment. Learning tasks can be categorized according to various taxonomies. A straightforward one is suggested by Perkins (1992). The simplest type of learning is *retention* of information. In schools and colleges, learning may be assessed by means of paper-and-pencil tests that require demonstration of such retention. Computer-based instruction units (as in "integrated learning systems") may incorporate multiple-choice, matching, or short-answer tests comparable to paper-and-pencil tests.

The learning goal may include *understanding* as well as retention. Assessments that require paraphrasing or problem solving may tap the understanding dimension. Such forms of assessment are more challenging to the designer, mainly because they are more labor intensive to comstruct and evaluate.

Learning goals may be more ambitious, such that the knowledge and skills are applied in *active use*. To assess this level of learning requires real or simulated problem situations, something that is obviously challenging to arrange. Some would characterize these differences in types of learning simply as *surface* versus *deep* learning (Weigel, 2001).

Such types or levels of learning have long been acknowledged, but there has been a growing demand in schools, higher education, and corporate training for more attention to the *active use* level. It is increasingly perceived that time and money spent on inculcating and assessing "inert knowledge" (Whitehead, 1929) is essentially wasted. If learners do not use the knowledge, skills, and attitudes outside the classroom, what is the point of teaching them? So today, when educators talk about the pursuit of learning, they usually mean productive, active use, or deep learning. Pursuing deep learning implies different instructional and assessment approaches than surface learning, so this shift in connotation has profound implications for what processes and resources are "appropriate."

Chapter 2 explores *facilitating learning* in some depth.

Improving. For a field to have any claim on public support it must be able to make a credible case for offering some public benefit. It must provide a superior way to accomplish some worthy goal. For example, for chefs to claim to be culinary professionals they must be able to prepare food in ways that are somehow better than non-specialists—more appealing, safer, more nutritious, prepared more efficiently, or the like. In the case of educational technology, to *improve performance* most often entails a claim of effectiveness: that the processes lead predictably to quality products, and that the products lead predictably to effective learning, changes in capabilities that carry over into real-world application.

Effectiveness often implies efficiency, that is, that results are accomplished with the least wasted time, effort, and expense. But what is efficient depends on the goals being pursued. If you want to drive from San Francisco to Los Angeles in the shortest time, Interstate Highway 5 is likely to be efficient. However, if your real goal is to see the ocean views along the way, State Highway 1, which winds along the coastline, would be more efficient. Similarly, designers might well disagree on methods if they do not have the same learning goals in mind. To a great extent, the systematic instructional development movement has been motivated by concerns of efficiency, defined as helping learners reach predetermined goals that are measured by objective assessments.

The concept of efficiency is viewed differently in the constructivist learning approach. In this approach, designers place greater emphasis on the appeal of the instruction and on the extent to which learners are empowered to choose their own goals and their own learning paths. They would more likely measure success in terms of knowledge that is deeply understood, experienced, and able to be applied to real-world problems as opposed to less authentic or embedded measures of learning, such as objective tests. Such designs, however, would still need to be planned for learning to occur within a particular time frame with some goals in mind and resources for meeting those goals. Among parties who have managed to agree on goals, efficiency in reaching those goals surely would be regarded as a plus.

With high expectations for learning, and high stakes for successful achievement becoming ever more important in society, *other things being equal,* faster is better than slower and cheaper is better than more expensive.

Performance. First, in the context of this definition, *performance* refers to the learner's ability to use and apply the new capabilities gained. Historically, educational technology has always had a special commitment to results, exemplified by programmed instruction, the first process to be labeled "educational technology." Programmed instruction materials were judged by the extent to which users were able to perform the "terminal objective" after instruction. Terminal objectives were stated in terms of the actual conditions for which people were being trained or educated, and they were assessed according to how well learners functioned under these conditions. Thus, the reference to *improving performance* reinforces the newer connotation of learning: not just inert knowledge but usable capability.

Second, in addition to helping individual learners become better performers, the tools and ideas of educational technology can help teachers and designers to be better performers and they can help organizations reach their goals more efficaciously. That is, educational technology claims to have the power to increase productivity at the individual and the organizational levels.

The use of *improving performance* in this definition is not meant to imply that educational technology encompasses all forms of performance improvement. As is advocated in the related field of human performance technology (HPT), there are many different sorts of interventions that may be used in the workplace to improve performance, such as tools, incentives, organizational change, cognitive support, and job redesign, in addition to instruction (Pershing, 2006). Since it encompasses all these sorts of interventions, HPT is a broader concept than educational technology.

The definition mentions three major functions that are integral to the concept of educational technology—creating, using, and managing. These functions can be viewed as separate sets of activities that might be carried out by different people at different times. They can also be viewed as phases of the larger process of instructional development. Advocates of a systems approach to instructional development would go further to specify that these functions be accompanied by evaluation processes at each phase. Monitoring decisions and taking corrective actions at each phase are critical attributes of the systems approach. Examples of such evaluation activities are mentioned under the headings of creating, using, and managing below.

Chapter 3 discusses the implications of *improving performance* in the context of educational technology.

Creating. Creation refers to the research, theory, and practice involved in the generation of instructional materials, learning environments, and large teaching learning systems in many different settings, formal and nonformal.

The educational technology field has witnessed an evolution in media formats and in theoretical underpinnings for the materials and systems that have been created—from silent films to programmed instruction to multimedia packages to Web-based microworlds.

Creating can include a variety of activities, depending on the design approach that is used. Design approaches can evolve from different developer mindsets: aesthetic, scientific, engineering, psychological, procedural, or systemic, each of which can be employed to produce the necessary materials and conditions for effective learning.

A systems approach, for example, might entail procedures for *analyzing* an instructional problem, *designing* and *developing* a solution, *evaluating* and revising decisions made at each step, and then *implementing* a solution. Assessing results and taking corrective action along the way is referred to as *formative evaluation*, while assessing the impact of the project at the end is referred to as *summative evaluation*. Different sorts of evaluative questions are asked at different stages. At the *front-end analysis* stage, is there a performance problem and does it entail instructional needs? In *learner analysis,* what are the characteristics of the learners? In *task analysis,* what capabilities must the learners master? At the design stage, what are the learning objectives? Is the blueprint aligned with those objectives? Do instructional materials instantiate the principles of *message design*? At the development stage, does the prototype actually guide learners toward the objectives? At the implementation stage, is the new solution being used and used properly? What is its impact on the original problem?

Design and development processes are influenced by the varied analog and digital technologies used to create instructional materials and learning environments. Designing for teacher-led classroom instruction, for example, may follow a different path than designing for a computer-based simulation game. What is created may be not only the materials for instruction and the surrounding learning environments, but also such supporting tools as databases for knowledge management, online databases for problem exploration, automated help systems, and portfolios for displaying and assessing learning.

Chapter 4 provides elaboration of the concepts and processes associated with *creating.*

Using. This element refers to the theories and practices related to bringing learners into contact with learning conditions and resources. As such, it is Action Central, where the solution meets the problem. Using begins with the *selection* of appropriate processes and resources—methods and materials, in other words—whether that selection is done by the learner or by an instructor. Wise selection is based on materials evaluation, to determine if existing resources are suitable for this audience and purpose. Then the learner's encounter with the learning resources takes place within some environment following some procedures, often under the guidance of an instructor, the planning and conduct of which can fit under the label of utilization. If the resources involve unfamiliar media or methods, their usability may be tested before use.

In some cases, there is a conscious effort to bring an instructional innovation to the attention of instructors or to market it. This *diffusion* process can be another phase of using. When teachers incorporate new resources into their curricular plans, this is referred to as *integration*; when such integration takes place on a larger scale, incorporating the innovation into the organizational structure, it is referred to as *institutionalization*.

In a systems approach, the design team would also take responsibility for *change management*, taking steps at each phase of development to ensure that stakeholders and end users accept, support, and use the final product.

Chapter 5 is devoted to further discussion of "using."

Managing. One of the earliest responsibilities of professionals in the field of educational technology has been management; in the early years, this took the form of directing the operations of audiovisual centers. As media production and instructional development processes became more complicated and larger scale, they had to master project management skills as well. As distance education programs based on information and communications technologies (ICT) developed, educational technologists found themselves involved in delivery system management. In all of these managerial functions, there are subfunctions of personnel management and information management, referring to the issues of organizing the work of people and planning and controlling the storage and processing of information in the course of managing projects or organizations. Prudent management also requires program evaluation. In the systems approach, this entails quality control measures to monitor results and quality assurance measures to enable continuous improvement of the management processes.

People who carry out management functions may be seen as exercising *leadership*, combining management expertise with support of ethical practice in all phases of educational technology practice.

Chapter 6 explores the dimensions of "managing."

Appropriate. The term appropriate is meant to apply to both processes and resources, denoting suitability for and compatibility with their intended purposes.

The term *appropriate technology* is widely used internationally in the field of community development to refer to a tool or practice that is the simplest and most benign solution to a problem. The concept grew out of the environmental movement of the 1970s, sparked by the book, *Small is Beautiful* (Schumacher, 1975), in which the term was coined. In this sense, appropriate technologies are those that are connected with the local users and cultures and are sustainable within the local economic circumstances. Sustainability is particularly critical in settings like developing countries, to ensure that the solution uses resources carefully, minimizes damage to the environment, and will be available to future generations.

AECT's professional standards have longed recognized that appropriateness has an ethical dimension. A number of provisions in the AECT Code of Ethics (Welliver, 2001) are relevant. Section 1.7 is the broadest and perhaps most directly relevant item, specifying the requirement to "promote current and sound professional practices in the use of technology in education." Section 1.5 requires "sound professional procedures for evaluation and selection of materials and equipment." Section 1.6 requires researchers and practitioners to protect individuals "from conditions harmful to health and safety." Section 1.8 requires the avoidance of content that promotes gender, ethnic, racial, or religious stereotypes, and it encourages the "development of programs and media that emphasize the diversity of our society as a multicultural community." Further, Section 3 of AECT's Code calls for providing "opportunities for culturally and intellectually diverse points of view" and avoiding "commercial exploitation," as well as following copyright laws and conducting research and practice using procedures guided by professional groups and institutional review boards.

Of course, a practice or resource is appropriate only if it is likely to yield results. This implies a criterion of effectiveness or usefulness for the intended purpose. For example, a particular computer-based simulation game might be selected by a social studies teacher if past experience indicated that it stimulated the sort of pertinent discussion that she intended. It would be judged appropriate in terms of usefulness.

"Appropriateness" has sometimes been used as a rubric for attempts to censor books or other instructional materials. Challenges may be based on claims that the material is sexually explicit, contains offensive language, or is otherwise unsuited to a particular age group. That is not the connotation or the context intended in this definition.

In summary, the selection of methods and media should be made on the basis of "best practices" applicable to a given situation, as specified in Section 1.7 of the Code of Ethics. This implies that educational technology professionals keep themselves updated on the knowledge base of the field and use that knowledge base in making decisions. Random choices, which might be acceptable for those outside the profession, do not meet the criterion of "appropriate." Informed, professionally sound choices help learners learn productively while making wise use of the time and resources of the organization, including the time and effort of educational technologists themselves.

Technological. In terms of lexicography, it is undesirable to use the word *technological* in a definition of educational technology. In this case, the use is justified because *technological* is a shorthand term that describes an approach to human activity based on the definition of technology as "the systematic application of scientific or other organized knowledge to practical tasks" (Galbraith, 1967, p. 12). It is a way of thinking that is neatly summarized in one word. It would be more awkward to paraphrase the concept of *technological* within the new definition than to simply use the shorthand term.

The term modifies both processes and resources. First, it modifies processes. There are *nontechnological* processes that could be used in planning and implementing instruction, such as the everyday decision-making processes of teachers, which may be significantly different from those advocated in this field. The field advocates the use of processes that have some claim of worthy results, based on research or at least reflective development. Without the *technological* modifier, any sorts of models, protocols, or formulations could be included in the ambit of educational technology, blurring the boundaries with the field of curriculum and instruction or education in general.

Second, the term also modifies resources, the hardware and software entailed in teaching—still pictures, videos, audiocassettes, satellite uplinks, computer programs, DVD disks and players, and the like. These are the most publicly visible aspects of educational technology. To ignore them in this definition would be to create a greater communication gap between specialists and nonspecialist readers.

The values associated with *appropriate* and *technological* as well as other values of educational technology are discussed in greater depth in chapter 9.

Processes. A *process* can be defined as a series of activities directed toward a specified result. Educational technologists often employ specialized processes to design, develop, and produce learning resources, subsumed into a larger process of instructional development. From the 1960s through the 1990s, a central concern of the field was the pursuit of a systems approach to instructional development. To many, the systems approach was and is central to the identity of the field.

A paradigm shift occurred in the decade since the prior AECT definition (Seels & Richey, 1994) involving postmodern and constructivist influences among others. To simplify, the focus moved from what the instructor is doing to what the learner is doing. In this view, individuals construct their own knowledge and gain ownership based on their struggles to make sense of their experience. To the extent that the teaching-learning experience is abstracted from real-world application and to the extent that it is controlled and possessed by the teacher, it diminishes the likelihood of learner engagement, mastery, and transfer of the skill. This sensibility came into some conflict with the plan-and-control sensibility often associated with systematic instructional development, a conflict whose resolution is still being negotiated. Such conflicts are not unique to educational technology. For example, the field of software engineering struggles with prescriptive "waterfall" models versus more free-form "agile" approaches.

In the context of the definition, *processes* also include those of using and managing resources as well as those of creating them.

Chapter 7 discusses *processes* in greater detail.

Resources. The many resources for learning are central to the identity of the field. The pool of resources has expanded with technological innovations and with the development of new understandings regarding how these technological tools might help guide learners. Resources are people, tools, technologies, and materials designed to help learners. Resources can include high-tech ICT systems, community resources such as libraries, zoos, museums, and people with special knowledge or expertise. They include digital media, such as CD-ROMs, Web sites, WebQuests, and electronic performance support systems (EPSS). And they include analog media, such as books and other print materials, video recordings, and other traditional audiovisual materials. Teachers discover new tools and create new resources, learners can collect and locate their own resources, and educational technology specialists add to the growing list of possible resources as well.

Chapter 8 elaborates on the *resources* that are considered central to educational technology.

Conclusion

What is proposed here is a revised definition of the concept of educational technology, built upon AECT's most recent prior definition of instructional technology (Seels & Richey, 1994). It is a tentative definition, subject to further reconsideration over time. Educational technology is viewed as a construct that is larger than instructional technology, as education is more general than instruction. Further, educational or instructional technology can be seen as discrete elements within performance technology, the holistic approach to improving performance in the workplace through many different means, including training.

The *concept* of educational technology must be distinguished from the *field* and the *profession* of educational technology. The validity of each can be judged separately from the others and can be judged by different criteria.

This definition differs from previous ones in several regards. First, the term *study* instead of *research* implies a broader view of the many forms of inquiry, including reflective practice. Second, it makes an explicit commitment to *ethical* practice.

Third, the object of educational technology is cast as "*facilitating* learning," a claim more modest than that of controlling or causing learning.

Fourth, it is intentional that learning is placed at the center of the definition, to highlight the centrality of learning to educational technology. It is the goal of promoting learning that is distinctive about the field, compared to other fields with which it might be conflated, such as information technology or performance technology.

Fifth, "improving performance" implies, first, a quality criterion, a goal of facilitating learning better than is done with approaches other than educational technology. In addition, it refers to a goal of guiding learners to not just inert knowledge but active, ready to use knowledge, skills, and attitudes.

Sixth, it describes the major functions of the field (creation, use, and management) in broader, less technical terms than previous definitions in order to reflect an eclectic view of the design process.

Seventh, the definition specifies that the tools and methods of the field be "appropriate," meaning suited to the people and conditions to which they are applied. Finally, it makes the attribute of "technological" explicit, with the rationale that tools and methods that are not technological fall outside the boundaries of the field.

The terms *improving* and *appropriate* are explicitly included in the definition in order to recognize the centrality of such values to the core meaning of educational technology. If the work of the field is not done "better" by professionals than it is done by amateurs, the field has no justification for public recognition or support. It must represent some specialized expertise that is applied with professional soundness.

References

Ely, D. P. (1963). The changing role of the audiovisual process: A definition and glossary of related terms. *Audiovisual Communication Review, 11*(1), Supplement 6.

Galbraith, J. K. (1967). *The new industrial state.* Boston: Houghton Mifflin.

Perkins, D. N. (1992). Technology meets constructivism: Do they make a marriage? In T. M. Duffy, & D. H. Jonassen (Eds.), *Constructivism and the technology of education: A conversation.* Hillsdale, NJ: Lawrence Erlbaum Associates.

Pershing, J. A. (Ed.). (2006). *Handbook of human performance technology* (3rd ed.). San Francisco: Pfeiffer.

Schön, D. A. (1990). *Educating the reflective practitioner.* San Francisco: Jossey-Bass.

Schumacher, E. F. (1975). *Small is beautiful: Economics as if people mattered.* New York: Harper & Row.

Seels, B., & Richey, R. (1994). *Instructional technology: The definition and domains of the field.* Washington, DC: Association for Educational Communications and Technology.

Weigel, V. B. (2001). *Deep learning for a digital age: Technology's untapped potential to enrich higher education.* San Francisco: Jossey-Bass.

Welliver, P. W. (Ed.). (2001). *A code of professional ethics: A guide to professional conduct in the field of educational communications and technology.* Bloomington, IN: Association for Educational Communications and Technology.

Whitehead, A. N. (1929). *The aims of education and other essays.* New York: The Free Press.

2

FACILITATING LEARNING

Rhonda Robinson
Northern Illinois University

Michael Molenda
Indiana University

Landra Rezabek
University of Wyoming

Introduction

Educational technology is the study and ethical practice of *facilitating learning* and improving performance by creating, using, and managing appropriate technological processes and resources.

Focus on Learning

THE DEFINITION BEGINS WITH the proposition that "educational technology is the study and ethical practice of *facilitating learning . . .*" indicating that helping people to learn is the primary and essential purpose of educational technology. All of the AECT definitions since 1963 have referred to learning as the end product of educational technology. However, the definitions have differed regarding the strength of the connection between technological interventions and changes in learner capability.

Prior focus on messages and control. The 1963 definition centered the field on the "design and use of messages which *control* the learning process" (Ely, 1963, p. 18). In this version, the focus is on messages, specifically, messages that *control* learning. The 1963 definition makes the strongest connection between learning and educational technology interventions. Januszewski (2001) proposed that the word *control* had two connotations, which were derived from the dominant theories at that time: the behaviorist learning-theory notion that consequences of behaviors determined whether or not they were learned and the communication-theory notion that processes were regulated by feedback (pp. 42–43).

Prior claim of management of learning. Aside from the official definitions, the notion of control or management has long had strong support within the field. For example, Hoban (1965) observed that "the central problem of education is not learning but the management of learning, and that the teaching-learning relationship is subsumed under the management of learning" (p. 124). Later, in outlining the parameters for research in educational technology, Schwen (1977) proposed that inquiry should center on "the management-of-learning problem." Heinich (1984) also emphasized technology's commanding role: "The basic premise of instructional technology is that all instructional contingencies can be managed through space and time" (p. 68).

Prior focus on processes. Various definitions proposed in the 1970s focused on instruction, problem solving, and systematic design, with little mention of learning processes or outcomes. The Commission on Instructional Technology (1970), for instance, used the expression to "bring about more effective instruction" (p. 19) rather than mentioning learning, using theory from communications and systems as its base. In the Silber (1970) definition, the focus was on solving educational problems. Learners, and their learning improvement, were not mentioned explicitly in the definition. And in another definition of that period, the field was described as a study of the systematic means by which educational ends are achieved (Seels & Richey, 1994, p. 19).

The AECT (1977) and Seels and Richey (1994) definitions focused more on the processes that constitute the work activities of educational technology and then name human learning as the end purpose of those processes without specifying either "controlling" or "facilitating" learning. The 1977 definition returned to the idea of "involving" people and other resources to analyze problems and implement solutions to those problems "involved in all aspects of human learning." While this definition seems to focus on problem solution, which may or may not be learning, the complex nature of this definition (16 pages of it) and the many elements of learning resources and organizational structures, in some ways, may foreshadow the current definition terms. Facilitating learning does involve a complex organization of processes and resources including people, materials, settings, and so on. But facilitating learning puts the emphasis on the learners and their interests and abilities (or disabilities), rather than on an outside entity identifying and defining the "problem" to be solved. In this view, learners have more responsibility for actually defining what the learning problem will be as well as controlling their own internal mental processes.

The 1994 definition again defined the field primarily in terms of its work activities. These work activities yield "processes and resources for learning" but the center of the definition seems to be on the work activities rather than on the learner or learning.

An earlier definition foreshadowing the current one. Given the commonness of the notion of management and control in the 1970s, it is somewhat surprising that the 1972 definition comes close to the current one: "Educational technology is a field involved in the *facilitation* of human learning . . ." (Ely, 1972, p. 36). The authors of the 1972 definition consciously chose the term *facilitation*, as did the current authors, in order to loosen the connotation that either messages or methods *determine* learning outcomes. *Facilitate* is meant to convey the contemporary view that learning is controlled internally, not externally, and that an external agent can, at best, influence the process.

To summarize, all of these definitions in one way or another specify that learning is the purpose toward which educational technology is aimed. The current definition, like the 1972 one, explicitly adopts the term *facilitate* to avoid connotations of management or control. This is meant to reflect current views about how learning occurs. This term suggests synonyms such as promote, assist, and support, which is what external agents—such as teachers—can do, while learners themselves actually manage and control their own learning.

Chapter Purposes

Facilitating learning appears to be a simple, nonthreatening phrase. Its denotation is clear enough. Buts its connotations are associated with years of research, debate, divergent philosophies, and unresolved issues. The goal in this chapter is to present a framework for thinking about the variables involved in facilitating learning through the lenses of divergent scholarly perspectives. Therefore, this chapter presents multiple perspectives on the teaching-learning process, trying to provide a balanced overview of the differences in terminology and consequences of these perspectives for educational technology. It also discusses informal and formal learning activities and instructional methods, and considers the assessment and evaluation of learners whose learning has been facilitated using these activities.

From Learning Theory to Instructional Theory

Learning theories attempt to *describe* how humans learn. They provide an account of what are the key elements in the process of gaining new knowledge and capabilities and how those elements interact. For example, behaviorism focuses on the observable events that precede and follow certain behaviors; cognitivism focuses on inferred mental conditions—the chain of internal activities associated with learning. Learning theories are useful to the extent that they allow us to articulate issues sensibly and to conduct inquiry to test hypotheses that flow from the theory.

It is quite another question to construct *instructional* theories, which attempt to *prescribe* teaching methods, to create the best conditions to help learners to acquire new knowledge and capabilities. The descriptive-prescriptive distinction is discussed at some length in Reigeluth (1983), with Reigeluth, Gropper, and Landa providing logical analysis and examples to illustrate the distinction (pp. 21–23, 50–52, 59–66). They make the point that practical "implications" do not flow directly or easily from descriptive abstractions. As one philosopher of education (Phillips, 1994) points out,

> [A] defect of the 'isms' approach was that it was based on the untenable conception of 'implication.' In order to draw implications from an abstract or theoretical premise, other premises are required which link the first premise to the practical domain of interest. . . . The point is that these matters cannot be decided by deducing them in a simple way from some abstract philosophical position. (p. 3864)

Unfortunately, many learning theorists themselves set a bad example by leaping to conclusions about the instructional implications of their theories. It is no wonder that many other adherents of learning theories, convinced of their descriptive accuracy, quickly rush to spell out practical implications, which they assume to have as much prescriptive as descriptive accuracy. This conflation of learning theory and instructional theory leads to barren arguments about the merits of one theory or the other. Champions of a particular learning theory, which may have a strong grounding in research and is therefore a quite useful *description* of how people learn, sometimes forcefully argue that their *prescriptive* instructional implications must be equally true whether or not they have been tested and upheld empirically.

At this time, it is conventional to group the various theories of learning into three broad categories: behaviorism, cognitivism, and constructivism (e.g., see Ertmer & Newby, 1993). Each of these bodies of theory, as well as others, has its adherents. Each, some would claim, has suffered from overly enthusiastic advocacy of particular instructional solutions prematurely derived from a descriptive learning theory. The most recent victim of this confusion is constructivism. As Kirschner, Sweller, and Clark (2006) point out, "The constructivist description of learning is accurate, but the instructional consequences suggested by constructivists do not necessarily follow" (p. 78). Or, as the criticism was framed by Bransford, A. L. Brown, and Cocking (2000),

> A common misconception regarding "constructivist" theories of knowing (that existing knowledge is used to build new knowledge) is that teachers should never tell students anything directly but, instead, should always allow them to construct knowledge for themselves. This perspective confuses a theory of pedagogy (teaching) with a theory of knowing. (p. 11)

To avoid a lengthy, hair-splitting descriptive-prescriptive analysis, we will simply refer to each body of thought as a "perspective," not distinguishing rigorously between the descriptive learning theories and the prescriptive instructional theories within each body of thought. The goal is to represent each perspective roughly as it appears in the literature of educational technology.

Perspectives Have Consequences

How one creates, uses, and manages learning resources depends greatly on one's beliefs about how people learn. For example, a teacher inspired by the behaviorist perspective would be expected to determine what the learner already knows, select an appropriate goal for that learner, provide prompts to guide them toward desired behaviors, and arrange reinforcers for those desired behaviors. On the other hand, a teacher inspired by Montessori's (2004) developmental perspective would be expected to determine a child's developmental status, select an appropriate work activity, model that activity, and step back to observe and support the child's efforts to master the new task.

One's view of how learning takes place can also affect decision making about educational policies. If one considers learning to be under the control of teachers—believing that teaching equals learning—it is entirely reasonable to support policies that make teachers directly accountable for student test results. The teacher is the worker and student learning is the product produced. The assumption is that if teachers "work harder" students will learn better. A variation of this viewpoint is that of the student as customer, a metaphor that has become quite popular in higher education and corporate training, often called "learner-centered teaching." The student is seen as the recipient of services provided by the teacher, akin to getting a haircut. In this view, teaching is something done *to* learners, so, obviously, the service provider is the one accountable for the results.

However, if one views learning as being primarily under the control of learners (a constructivist view), teachers and students are seen more as collaborators in a common enterprise. They are coproducers of students' learning accomplishments. Nothing happens until the students do their part of the coproduction. In this view, a more appropriate model is psychotherapy rather than hair cutting. The student is not a customer but a worker doing the hardest part of constructing new knowledge, skills, and attitudes. This view would imply educational policies focused on student motivation to achieve. Teachers would be accountable for doing *their part* of the job professionally but would not be expected to take full responsibility for what students do and do not learn. The issue of motivation and who has control of it is discussed near the end of this chapter and in chapter 3.

Learning Defined and Viewed From Different Perspectives

Learning can be defined as "a persisting change in human performance or performance potential . . . . as a result of the learner's experience and interaction with the world" (Driscoll, 2005, p. 9). Different theories of learning regard different elements of the process as being of paramount importance, and they use a different vocabulary to describe the underlying processes that they believe are occurring within the learner. In the remainder of this chapter, the behaviorist, cognitivist, and constructivist perspectives are each discussed briefly regarding their main elements, emphases, and relationship to educational technology concerns. To these three categories is added the category of "eclectic," reflecting the widely accepted view that theory and practice can be enlightened by viewing problems through different lenses or even combining lenses.

Behaviorism

The name "behaviorism" refers collectively to several quite diverse bodies of thought in psychology and philosophy. This discussion will focus on radical behaviorism because its operationalization, operant conditioning, has had the greatest practical impact on theory and practice in educational technology (Burton, Moore, & Magliaro, 2004). Operant conditioning involves the contingent relationships among the stimuli that precede a response, the response itself, and the stimuli that follow a response, that is, the consequences of the behavior (p. 10). B. F. Skinner (Ferster & Skinner, 1957) discovered that by manipulating these three variables, he could elicit quite complex new behaviors from laboratory animals. Other researchers found that humans, too, responded in similar ways to certain types of consequences or reinforcers.

Behaviorism in Educational Technology. Prompted by his own experiences with schools as a parent, Skinner (1954) became interested in the possibility of applying operant conditioning to academic learning. His analysis of the problems of group-based traditional instruction and his invention of a mechanical device for interactive learning, referred to as a "teaching machine," gained national attention. The pedagogical organization of stimuli, responses, and reinforcers in teaching machines became known as programmed instruction, and programmed instruction lessons in book format were published in great profusion in the 1960s. By the mid-1960s, Skinner (1965; 1968) viewed programmed instruction as a practical application of scientific knowledge to the practical tasks of education and so he referred to his instructional strategies as a "technology of teaching." Other authors converted this term to *educational technology*; an early example is *Educational technology: Readings in programmed instruction* (DeCecco, 1964).

Teaching machines and programmed instruction. Between 1960 and 1970, the research focus of what had been the audiovisual education field shifted sharply toward work on teaching machines and programmed instruction, prompting the change of the name of the field to educational technology. Torkelson (1977) examined the contents of articles published in *AV Communication Review* between 1953 and 1977 and found that the topics of teaching machines and programmed instruction dominated the journal in the 1960s. In fact, between 1963 and 1967, these topics represented a plurality of all articles published.

Programmed tutoring. Programmed tutoring was developed to overcome some of the weaknesses of programmed self-instructional materials, specifically, their being limited to "knowledge of correct response" as a reinforcer and their totally expository strategy. In Ellson's (Ellson, Barner, Engle, & Kempwerth, 1965) programmed tutoring, a live person, usually a peer learner, followed instructions in leading the tutee through practice exercises, giving social reinforcers (a nod, a smile, an affirming phrase) when correct and hints toward a solution ("brightening") when incorrect. The brightening technique was meant to make the experience more of a discovery activity, in which learners figured out the answers rather than being told them. A meta-analysis of programmed and structured tutoring programs showed tutees scoring around the 75th percentile compared to the 50th percentile for conventional instruction (Cohen, Kulik, J. A., & Kulik, C. C., 1982); this difference is one of the largest ever recorded in research comparing methods.

Direct Instruction. Direct instruction (DI) is an empirically based, scripted method for small group instruction; it provides fast paced, constant interaction between students and the teacher (Englemann, 1980). Although it is not consciously derived from behaviorism, its procedure visibly applies behaviorist prescriptions, particularly continuous learner responses to teacher prompts followed by reinforcement or remediation, as appropriate. A large-scale comparison of 20 different instructional models used with at-risk children showed DI to be the most effective in terms of basic skills, cognitive skills, and self-concept (Watkins, 1988). After more than a quarter century of implementation, DI established a solid record of demonstrated success (Adams & Engelmann, 1996). Further, it was found to be one of three comprehensive school reform models "to have clearly established, across varying contexts and varying study designs, that their effects are relatively robust and . . . can be expected to improve students' test scores" (Borman, Hewes, Overman, & Brown, S., 2002, p. 37).

Personalized System of Instruction (PSI). F. S. Keller's (1968) Personalized System of Instruction (PSI), or "Keller Plan," is a method for organizing all the material of a whole course or curriculum. The subject matter is divided into sequential units (could be chapters of a textbook or specially created modules) that are studied independently by learners, progressing at their own pace. At the end of a unit, learners have to pass a competency test before being allowed to go forward to the next unit. Immediately after the test, they receive coaching from a proctor to correct any mistakes. This procedure protects students from accumulating ignorance and falling further and further behind if they miss a key point (Keller, F. S., 1968). The self-pacing and immediate remediation are the elements that lend a degree of personalization. During the period it was being tested at many colleges and universities, the 1960s and 1970s, it was the most instructionally powerful innovation evaluated up to that time (Kulik, J. A., Kulik, C. C., & Cohen, 1979; Keller, F. S., 1977).

Behaviorism's major impact on educational technology has been on the soft technology side, contributing several templates or frameworks for instruction— such as programmed instruction, programmed tutoring, Direct Instruction, and PSI (Lockee, Moore, & Burton, 2004). As hard technology advanced, these frameworks were incorporated in mechanical, electro-mechanical, and ultimately, digital formats, such as computer-assisted instruction (CAI) and online distance education.

Computer-assisted instruction (CAI). Experiments in CAI began just at the time that programmed instruction was at its peak, so many of the early CAI programs followed a drill and practice or tutorial format resembling programmed instruction: small units of information followed by a question and the student's response. A correct response was confirmed, while an incorrect response might branch the learner to a remedial sequence or an easier question. Beginning in the mid-1960s, the CAI research and development program at Stanford University, later the Computer Curriculum Corporation, created successful drill and practice materials in mathematics and reading, later adding foreign languages (Saettler, 1990, p. 308).

More innovative and more learner-centered programs were developed in the TICCIT project at Brigham Young University in the 1970s. These sophisticated programs yielded successful programs in mathematics and English composition. However, both the Stanford and TICCIT programs failed to gain major adoption in their intended sectors, K–12 and community college education (Saettler, 1990, p. 310).

The PLATO project at University of Illinois began in 1961, aiming to produce cost-efficient instruction using networked inexpensive terminals and a simplified programming language for instruction, TUTOR. Most of the early programs were basically drill and practice with some degree of branching, but a wide variety of subject matter was developed at the college level. Over time, terminals at outlying universities were connected to the central mainframe in a timesharing system, growing to hundreds of sites and thousands of hours of material available across the college curriculum. As software development continued, many innovative display systems evolved, including a graphical Web browser. With experience and with more capable hardware, more varied sorts of instructional strategies became possible, including laboratory and discovery oriented methods.

The PLATO system pioneered online forums and message boards, e-mail, chat rooms, instant messaging, remote screen sharing, and multiplayer games, leading to the emergence of what was perhaps the world's first online community (Woolley, 1994). It continued to grow and evolve right through the early 2000s, sparking the expansion of local CAI development and finding a niche in military and vocational education.

Behaviorism and Facilitating Learning How has behaviorism contributed to facilitating learning? For one thing, the behaviorism-based technologies demonstrated that it is possible to achieve dramatic achievement test gains through careful control of the contingencies among stimuli, responses, and consequences, as claimed. Thorough analysis of learning tasks, precise specification of objectives, subdivision of the content into small steps, eliciting active responses, and giving feedback to those responses constitute a successful formula, at least for certain types of learning goals. In addition, the planning process required to produce lessons of this sort gave birth to the larger planning methodology now known as instructional systems design (Magliaro, Lockee, & Burton, 2005).

Programmed instruction demonstrated that individual learners could work effectively at their own pace without the guidance of a live teacher, freeing instruction from the teacher-centered, group-based paradigm. In doing so, it also made the learner an active participant in the learning process, not active in the sense that learners had control of the process, but in the sense that they needed to respond overtly and thoughtfully at frequent intervals, requiring them to stay engaged with the material.

Last but not least, behaviorism, because it does not focus on internal cognitive processes, is not limited to use in the cognitive domain. The behaviors that are taught and learned may combine cognitive, affective, and motor dimensions. Behaviorist approaches have been applied effectively to athletic skills and attitudes as well as to intellectual skills.

However, despite the impressive track record of behaviorally based technologies of instruction in experiments and field trials, their reception in public education has been lukewarm at best. Adoption, where it has taken place, has been slow and piecemeal. This might be attributed both to the nature of academic learning and the nature of educational organizations. First, the learning outcomes in most of these projects are measured in terms of test scores. What some people understood in the 1960s and what more people understood 40 years later is that what students regurgitate on tests tends to be forgotten or ignored as they walk out the classroom door. Early skeptics were concerned whether the new knowledge gained through programmed instruction would be transferred to real-world problems or to future lessons. If students are gaining "inert knowledge," what is the advantage if it is learned 25% faster or better? Educators also questioned whether students in these treatments were gaining the skills, such as metacognitive ability, and attitudes, such as ownership of their learning, needed to help them become self-initiating lifelong learners.

Second, the organizational structures of schools and colleges are not conducive to innovations that require radical change in those structures, such as those proposed in programmed instruction, direct instruction, and PSI. To make sense economically, the costs of any technology must be self-liquidating, as they are in business and other sectors of the market economy. In order to become self-liquidating, technological interventions must replace costly human labor to some extent. This conflicts with the interests of those now doing the labor.

As Heinich (1984) pointed out a generation ago, technologies threaten power relationships within the organization and "as technology becomes more sophisticated and more pervasive in effect, consideration of its use must be raised to higher and higher levels of decision making" (p. 73). As Shrock (1990) put it,

> We can anticipate that teachers comfortable with their traditional role in the classroom will suppress any technology that threatens that role. Unfortunately, the traditional role preferred by most teachers—teacher centered, large group, expository, text supported teaching—is largely incompatible with the recommendations of instructional technologists (and the results of educational research). (p. 25)

Of course, it is not just resistance by teachers that impedes the acceptance of methods that would require rather major restructuring. Schools are complex enterprises, with many different power centers and constituencies, each having expectations and interests at stake. So it is not surprising that the behaviorism-based innovations—as well as other technology-based innovations—have been considered unaffordable or have tended to be resisted in terms of large-scale adoption, at least in most school systems in the United States.

Cognitivism

Like behaviorism, *cognitivism* is a label for a variety of diverse theories in psychology that endeavor to explain internal mental functions through scientific methods. From this perspective, learners use their memory and thought processes to generate strategies as well as store and manipulate mental representations and ideas. Theories that would later become very influential were being developed in the 1920s and 1930s by Jean Piaget in Switzerland and Lev Vygotsky in Russia, but these did not have significant impact on American educational psychology until translations were widely circulated in the 1960s. Cognitive theories gained momentum in the United States with the publication of Jerome Bruner's (1960) *The Process of Education*, the dissemination of Piaget's and Vygotsky's works, and the emergence of information-processing theory in the late 1960s. By 1970, when the journal *Cognitive Psychology* was begun, the cognitive perspective had gained not only legitimacy but also dominance.

Piaget's theory. Jean Piaget, a biologist, became deeply interested in the thought processes of doing science, especially in the development of thinking, which he called "genetic epistemology." Through interviews with children, he developed the theory that young children build up classification systems and try to fit the objects and events of their everyday experiences into the existing framework (he called this *assimilation*). When they encountered contradictions—things that just did not fit—they modified their mental structures (he called this *accommodation*). As he continued his investigation of children, he noted that there were periods where assimilation dominated, periods where accommodation dominated, and periods of relative equilibrium, and that these periods were similar across many different children, leading him to conclude that there were fixed stages of cognitive development.

Information processing theory. Another branch of cognitivism, information processing theory, uses the computer as a metaphor and views learning as a series of transformations of information through various (hypothesized) mental processes. It focuses on how information is *stored in memory*. In this theory, information is thought to be processed in a serial, discontinuous manner as it moves from one stage to the next, from sensory memory, where external stimuli are detected and taken into the nervous system, to short-term memory, to long-term memory (Atkinson & Shiffrin, 1968).

Schema theory. An approach that is more congruent with Piaget's theories, schema theory, suggests that material stored in long-term memory is arranged in organized structures that are amenable to change and that store knowledge in a more abstract form than our specific, concrete experiences. Ausubel's (1963) subsumption theory proposes that meaningful verbal learning involves superordinate, representational, and combinatorial processes that occur during the reception of information. A primary process is subsumption, in which new material is integrated with relevant ideas in the existing cognitive structure.

Cognitive load theory combines notions from information processing and schema theories, proposing that novices become experts as they expand and enhance their mental schemata. However, for schema acquisition to occur successfully the cognitive load should be controlled while processing is taking place in working memory because working memory has a finite capacity (Sweller, 1988).

Neuroscience. The neuroscience approach has become feasible only with the development of imaging technologies that allow observation of neurological activities. It attempts to understand mental processes by more or less direct observation of the physical functioning of the brain and nervous system. Leamnson (2000) provides an accessible account of the biological basis of learning, referring to the functioning of neurons, dendrites, and axons. Learning consists essentially of creating and stabilizing synaptic connections among neurons. Within the brain, the frontal lobes are the major site of organizing thoughts, and the frontal lobes communicate with the limbic system, site of emotion. Leamnson sees the challenge of education being to arouse emotions that inspire learners to focus on the learning tasks (p. 39). Winn (2004) suggests that the information-processing view of cognitivism has been losing favor in light of new evidence, particularly evidence from neuroscience.

In summary, cognitivism differs from behaviorism in its belief that the internal mental processes can and must be understood in order to have an adequate theory of human learning. There are differing hypotheses about how those internal processes operate.

Cognitivism in Educational Technology. Cognitivist instructional theories focus more on the presentation side of the learning equation—the organization of content so that it makes sense to the learner and is easy to remember. The goal is to activate the learner's thought processes so that new material can be processed in a way that it expands the learner's mental schemata.

Audiovisual media. Audiovisual technology, which could stimulate multiple senses, provided new tools to surmount the limitations of the textbook and teacher talk. Since the early days of the visual instruction movement, represented by C. F. Hoban, C. F. Hoban, Jr., and Zisman (1937), the field struggled against empty verbalism or rote memorization. Dale (1946), an early advocate of rich learning environments, expanded the notion of visual instruction by proposing in his Cone of Experience that learning experiences could be arrayed in a spectrum from concrete to abstract, each with its proper place in the tool kit. The prescriptions given in this era tended to be drawn from Gestalt psychology, which attempted to describe how humans and other primates perceived stimuli and used cognitive processes to understand and solve problems. The Gestaltists insisted that an understanding of human psychology required tools beyond those of scientific observation; they sought a unified study of psychology, rejecting the mind-body dichotomy and dealing with thoughts and feelings, aimed at understanding the human experiences of insight, creativity, and morality.

The Gestalt perspective, with its original emphasis on sensory perception and how humans construct meaning from bits and pieces of auditory and visual information, had great appeal to those in audiovisual education.

Visual learning. Educational technology's long and deep interest in message design, based on the principles of visual perception, fits into this agenda. A wide variety of theories, some derived from the Gestalt paradigm and some fitting under the conventional cognitivist umbrella, have been proposed to explain how humans construct and interpret visuals, according to Anglin, Vaez, and Cunningham (2004). In addition, a wide variety of classification schemes have been proposed for the various purposes that instructional visuals can serve. For example, Alesandrini (1984) proposes three broad categories: representational (pictures that resemble the thing or idea pictured), analogical (showing known objects and implying a similarity to the unknown concept), and arbitrary (charts or diagrams that attempt to organize thinking about a concept but do not physically resemble it). Others propose categories focusing on more specific mental functions, such as decorative, representational, mnemonic, organizational, relational, transformational, and interpretive (Carney & Levin, 2002; Lohr, 2003; Clark, R., & Lyons, 2004).

Regardless of these disagreements, researchers have identified a body of principles and generalizations about the juxtaposition of visuals and text that have informed the practice of message design—the layout of image and text to help learners to focus on important features and to understand and remember key ideas (Fleming & Levie, 1993; Lohr, 2003). Usability testing on Web pages is reconfirming the message design principles discovered in the predigital era.

Auditory learning. Learning based on hearing, too, has been examined through the lens of cognitive theories regarding the processing, storing, and retrieving of auditory information (Barron, 2004). Barron's review of research on auditory, visual, and verbal processing suggests that these sensory modalities are processed differently in the brain (p. 957). Many variables affect the productive use of audio materials in instruction, including cognitive load. The situation becomes even more complex when considering the combination of audio, visual, and verbal information in multimedia learning. Moore, Burton, and Myers (2004) attempt to summarize the rather disparate findings of research on multiple-channel presentations by observing that

> The human information processing system appears to function as multiple-channel system until the system capacity overloads. When the system capacity is reached, the processing system seems to revert to a single-channel system. (p. 998)

Overall, they do not consider the research on multiple-channel communication to offer reliable guidance for practice for instructional designers (p. 998), nor is it clear that the cognitivist information-processing model is the most fruitful one for continuing research in this area.

Digital multimedia. In more recent times, the computer captured the attention of cognitivists. First, the digital format can present multimedia displays more easily and more cheaply than was possible with earlier analog equipment. Learner use of multiple sensory modalities as presented in computer multimedia more closely resembles the natural human cognitive system. Second, computers can transform information from one symbol system to another. For example, you can input mathematical data and the computer can transform those data into graphs. In addition, the hypertext capability of computers allows the linking of ideas, both by authors and by learners. Kozma and Johnston (1991), looking at computer capabilities even before the spread of the World Wide Web, speculated about ways in which computers could advance the cognitivists' agenda:

- "From reception to engagement," moving from passive reception of lectures to more active involvement in immersive environments.
- "From the classroom to the real world," suggesting that technology can bring problems and resources from the real world into the classroom, and can allow students' learning to be focused outside of their classroom environment through resources and people they have access to through the Web.
- "From text to multiple representations," enabling the use of mathematical, graphical, auditory, visual, and other systems instead of just verbal symbols.

- "From coverage to mastery," using simulations, games, and drill-and-practice programs that encourage repeated practice of basic skills until they are automatized.
- "From isolation to interconnection," transforming the learner experience from a solitary one to a collaborative one.
- "From products to processes," helping students to engage in the work processes—and the ways of thinking—in their chosen field.
- "From mechanics to understanding in the laboratory," enabling students to use computer simulations that allow them to explore more hypotheses and cover more different processes in less time and at less expense. (pp. 16–18)

Cognitivism and Facilitating Learning. How has cognitivism contributed to facilitating learning? To begin with, we must acknowledge a limitation of cognitivist theory; it is meant to apply to learning in the cognitive domain— knowledge, understanding, application, evaluation, and metacognition. It has much less to say about motor skills or attitudes except as regards the cognitive elements of those skills.

Cognitivism's emphasis on careful arrangement of the content to make it meaningful, comprehensible, memorable, and appealing draws attention to message design issues. Cognitivist prescriptions include showing learners how the new knowledge is structured (e.g., advanced organizers), calling their attention to the salient features by stating objectives, chunking the material into digestible units, laying out text for easy comprehension, and complementing the text with helpful visuals (Silber, K. H., & Foshay, 2006, p. 374).

Both information-processing theory and schema theory suggest that the sequence of mental steps is an important part of facilitating learning, so instructional theorists have proposed a number of lesson frameworks or templates for arranging the steps of a learning event (Molenda & Russell, 2006, pp. 351–360). An example of such a lesson framework is Gagne's (Gagne & Medsker, 1996, p. 140) Events of Instruction, which recommends a specific sequence of events for a successful lesson: (a) Gain the learners' attention by telling them or dramatizing the reason for mastering this skill; (b) tell them clearly what they are expected to be able to do after the learning session; (c) remind them of what they already know and how the current lesson builds on that; (d) demonstrate the new skill or present the new information; (e) guide the learners in mastering the content by suggesting mnemonic devices, asking questions, or giving hints; (f) provide opportunities to practice the new knowledge or skill; (g) during the practice, confirm correct responses or desired performance and give feedback to help learners overcome errors; (h) test the learners' mastery, preferably by having them use the new knowledge, skills, and attitudes in real or simulated problem situations; and (i) help the learners transfer their new skills by giving them on-the-job practice or simulated practice involving varied problems.

Conducting a lesson in this sequence exemplifies an expository or deductive approach: telling the learners "the point"—the concept, rule, or procedure they are supposed to master—and then letting them apply "the point" in some practice setting. Sometimes a discovery or inductive approach may be specified, putting practice and feedback (steps f and g) before stating objectives, review of prior learning, presentation, and learning guidance (steps b, c, d, and e).

Another lesson framework based on cognitivist instructional theories is offered by Foshay, K. H. Silber, and Stelnicki (2003) in the form of "a cognitive training model." They recommend 17 specific tactics organized around five strategic phases: (1) gaining and focusing attention, (2) linking to prior knowledge, (3) organizing the content, (4) assimilating the new knowledge, and (5) strengthening retention and transfer of the new knowledge (p. 29). Examples of the tactics recommended by Foshay et al. are shown in Table 2.1.

Their five stages overlap with Gagne's (Gagne & Medsker, 1996) Events of Instruction, but there are some differences in content and emphasis. The cognitive training model puts special emphasis on the tasks of organizing and linking the new information; it integrates motivational elements from J. M. Keller's (1987) ARCS model; and it provides specific guidance for organizing information, in terms of chunking, layout, and use of illustrations.

Table 2.1. Selected examples of instructional tactics recommended in the Cognitive Training Model.

Learning Stage	Supporting Instructional Tactics
1. Select information to attend to	E.g., tell learners "what's in it for me."
2. Link new information to existing knowledge	E.g., compare and contrast new information and existing knowledge.
3. Organize the information	E.g., employ "chunking"—organize and limit the information according to information processing limits.
4. Assimilate new information with existing knowledge	E.g., demonstrate real-life examples of how the new knowledge is applied.
5. Retain and transfer knowledge	E.g., give practice in real or simulated setting.

Note: Adapted from Figure 2.2 in *Writing training materials that work,* by W. R. Foshay, K. H. Silber, and M. B. Stelnicki. San Francisco: Jossey-Bass/Pfeiffer, 2003.

Constructivism

The most talked about learning perspective of the past decade is labeled *constructivism*. It is difficult to characterize the claims of *constructivism* because there are a number of claimants embracing a diversity of views. The label itself is most closely identified with the self-educated philosopher, logician, linguist, and cognitive theorist, Ernst von Glasersfeld (1984), beginning with his treatise, *An introduction to radical constructivism*. Von Glasersfeld (1992) attempted to construct an epistemology, a theory of knowing, in which the "experiential world is constituted and structured by the knower's own ways and means of perceiving and conceiving, and in this elementary sense it is always and irrevocably subjective."

The Problem of Defining Constructivism.   However, the authors who were probably most influential in introducing *constructivism* to the educational technology audience in North America—Bednar, Cunningham, Duffy, and Perry (1991)—did not refer to von Glasersfeld as a source. Their primary source for a "new epistemology" was Lakoff (1987) and his work in sociolinguistics (although Lakoff used the label *experientialism*, not *constructivism*, for his theory of language acquisition). In discussing instructional applications of constructivism, these authors gave the examples of situated cognition, anchored instruction, cognitive flexibility, problem based learning, cognitive apprenticeship, and everyday cognition (although none of these theories are based on either von Glasersfeld's or Lakoff's epistemology). After the introduction of Bednar et al., the most visible advocates for *constructivism* in educational technology—Duffy, Cunningham, and Jonassen (e.g., Jonassen, 1991; Duffy & Jonassen, 1992; Duffy & Cunningham, 1996) used *constructivism* as an umbrella term for a wide range of ideas drawn primarily from recent developments in cognitive psychology (which were not necessarily dependent on a "new epistemology"). Piaget and Vygotsky are also usually cited as formative influences on the development of this perspective.

Vygotsky observed that mental abilities developed through social interactions of the child with parents, but also other adults. Through these interactions, children learn the habits of mind of their culture—speech patterns, written language, and other symbolic knowledge that influence how they construct knowledge in their own minds. Because of the importance of social and cultural influences in his theory, it is termed a *sociocultural* approach to learning and the branch that follows this theory is often termed *social constructivism*.

Philosopher D. C. Phillips (1995) pointed out the semantic morass that had come to hinder discourse about "constructivism":

> The rampant sectarianism, coupled with the array of other literatures that contain pertinent material, makes it difficult to give even a cursory introductory account of constructivism, for members of the various sects will object that their own views are nothing like this! (p. 5)

Phillips (1995) examined a number of authors or groups of authors, holding widely divergent and sometimes conflicting views, who are most closely associated with the various sects of constructivism: Ernst von Glasersfeld, Immanuel Kant, the feminist epistemologists, Thomas S. Kuhn, Jean Piaget, Lev Vygotsky, and John Dewey (pp. 6–7).

An analysis of "constructivist didactics" by Terhart (2003) attempted to parse out which elements of constructivist didactic theory are dependent on a new paradigm and which are consistent with evolution of thought within cognitivism. He concluded that it is difficult to distinguish *moderate* constructivist principles of instruction, which are the ones most frequently encountered in education literature, from cognitivist principles. On the other hand, *radical* constructivism "would ultimately render didactic thought and activity in specific subjects impossible as well as morally illegitimate" (p. 33). Terhart concludes,

> . . . [moderate] constructivist didactics really does not have any genuine new ideas to offer to the praxis of teaching. Rather, it recommends the well-known teaching methods and arrangement of self-directed learning, discovery learning, practical learning, co-operative learning in groups. I think that the 'new' constructivist didactics in the end is merely *an assembly of long-known teaching methods (albeit not practiced!).* (p. 42)

In view of these many differing and sometimes conflicting streams of thought, Driscoll (2005) concludes, "There is no single constructivist theory of instruction" (p. 386). She cites as constructivism's common denominator the assumption "that knowledge is constructed by learners as they attempt to make sense of their experiences" (p. 387). This overlaps with the assumptions of cognitivists. Where constructivists (some of them) seem to differ from cognitivists, according to Driscoll, is that they argue, that "knowledge constructions do not necessarily bear any correspondence to external reality" (p. 388). This aligns with von Glasersfeld's (1992) "irrevocably subjective" stance.

A possible solution to this labeling problem is to follow the advice of Terhart (2003) and use the label *moderate constructivist* to refer to constructivist theories and strategies that accept the assumptions of cognitivists and the label *radical constructivist* to refer to constructivist theories and strategies that depend on the subjectivist epistemology of von Glasersfeld. In the remainder of this chapter, we are discussing the moderate constructivist perspective unless otherwise indicated.

Setting aside the semantic issues, it is quite clear that the constructivist perspective is the one that holds the "commanding heights" in educational technology research and development at the beginning of the 21st century. The American Psychological Association's (1995) *Learner-centered psychological principles*, the most authoritative recent position paper on learning, features constructivist ideas as its driving force.

Constructivist Prescriptions. Prescriptive principles derived from constructivism include, according to Driscoll (2005): "1. Embed learning in complex, realistic, and relevant environments. 2. Provide for social negotiation as an integral part of learning. 3. Support multiple perspectives and the use of multiple modes of representation. 4. Encourage ownership in learning. 5. Nurture self-awareness of the knowledge construction process" (pp. 394–395). What sorts of instructional strategies are derived from these principles? We will focus on those mentioned in the early article by Bednar et al. (1991)—situated cognition (which is associated with cognitive apprenticeship), anchored instruction, and problem-based learning—plus collaborative learning.

Situated cognition. The theory of situated cognition emphasizes the notion that all human thoughts are conceived within a specific context—a time, a place, and a social setting. J. S. Brown, Collins, and Duguid (1989) point out that academic learning is situated in the classroom environment and therefore tends to become "inert knowledge," not transferred to life outside the classroom. This theory puts the social aspect at the center of the learning process, viewing expertise as developing within a community of practice.

Cognitive apprenticeship, which embodies the first two principles cited by Driscoll (2005), provides a theoretical framework for the process of helping novices become experts through one-to-one guidance. It takes a method traditionally applied in trades and crafts and applies it to learning in the cognitive domain. Dennen (2004) views cognitive apprenticeship as being grounded in "scaffolding, modeling, mentoring, and coaching . . . all methods of teaching and learning that draw on social constructivist learning theory" (p. 813).

Anchored instruction. The Cognition and Technology Group at Vanderbilt (CTGV) introduced anchored instruction as a strategy in the 1990s to incorporate the insights of situated cognition into classroom instruction. CTGV developed interactive videodiscs that allowed students and teachers to plunge into complex, realistic problems requiring the use of mathematics and science principles to solve. The video materials served as anchors or macrocontexts for a series of learning episodes. As explained by CTGV (1993), "The design of these anchors was quite different from the design of videos that were typically used in education . . . our goal was to create interesting, realistic contexts that encouraged the active construction of knowledge by learners. Our anchors were stories rather than lectures and were designed to be explored by students and teachers" (p. 52). These video materials have been often cited as examples for multimedia design and production within constructivist frameworks.

Problem-based learning. Problem-based strategies embody Driscoll's (2005) first principle, complex and realistic environments, and usually all of the other principles as well. They have been used in medical education for several decades. Since the 1990s, computer-based simulations, sometimes being self-contained ecological systems known as microworlds, have been used to immerse learners in problem spaces. These immersive environments overlap considerably with anchored instruction, but claim to emphasize first-hand involvement in, rather than observation of, problem situations. They also often entail collaborative group work, thus also embodying Driscoll's second principle of social negotiation. The group members are encouraged to reflect on their learning, thus embodying the principle of self-awareness of the knowledge construction process.

Moderate constructivists tend to recommend immersing learners in simplified versions of the problem to begin with, moving toward more complex versions as learners master the knowledge and skills needed to cope with growing complexity, as in Reigeluth's (1979) elaboration theory and Merrill's (2002) pebble-in-the-pond strategy. Radical constructivists tend to value the authenticity of the experience, not being as concerned about complexity or cognitive load.

Collaborative learning. Driscoll's (2005) second principle, social negotiation (derived from Vygotsky's theories of the sociocultural nature of knowledge), is represented in collaborative learning, which is incorporated in most of the constructivist instructional strategies discussed earlier. Computer-supported collaborative learning (CSCL) is currently the most prominent format. Roschelle and Pea (2002) speculate that wireless handheld devices will allow CSCL to evolve in new directions from those possible in traditional computer labs.

Collaborative learning is not achieved only through CSCL, of course. Educators and teachers at all levels have been using and continue to use collaboration as a strategy for learners. Classroom teachers especially have been urged to employ engaged learning activities, based upon constructivist principles, within small-group authentically based inquiries, in order to improve communication skills, problem solving and creative thinking skills, and cooperation and team learning abilities in students. These activities can be computer mediated or computer supported, or can involve the use of computer software for recording and reporting results of inquiry by students.

Constructivism in Educational Technology. The engaged learning principles as promoted by the North Central Regional Educational Laboratory (NCREL) (Tinzmann, Rasmussen, & Foertsch, 1999) include many of the components of constructivism and the use of educational technology as a tool for achieving learning. The description of engaged learning includes:

> Students are explorers, teachers, cognitive apprentices, producers of knowledge, and directors and managers of their own learning. Teachers are facilitators, guides, and colearners; they seek professional growth, design curriculum, and conduct research. Learning tasks are authentic, challenging, and multidisciplinary. Assessment is authentic, based on performance, seamless and ongoing, and generates new learning. (p. 1)

Engaged learning, as developed by teachers through the use of technology, is worthwhile when it helps students reach important district, state, or national standards. Many teachers have learned through their initial education, staff development, or inservice education to plan for student activities that represent engaged learning, are authentic, are worthwhile, and involve constructivist principles while employing educational technologies as tools for learning. Advocates of constructivism have repeatedly encouraged such development through texts and articles for educators, based upon constructivist ideals.

These advocates also frequently point out the needed changes in the methods by which learning is assessed. Assessment in these classrooms must also be authentic and focused on performance, use complex and meaningful activities, be based upon construction of knowledge rather than repetition of facts, and be conducted through observation, presentation, and other realistic, real-world-based activities (Jonassen, Howland, Moore, & Marra, 2003).

Constructivism and Facilitating Learning.   How has constructivism contributed to facilitating learning? First, the strong advocacy advanced by its adherents has captured the attention of educational technologists. Since the late 1980s, the conversation within educational technology has revolved around the claims of constructivism, debating their merits and imagining their implications.

At the very least, a host of earlier innovations, such as anchored instruction, problem-based learning (PBL), and collaborative learning, have been explored as instantiations of constructivist theory. Constructivism has infused these explorations with a sense of mission.

Cautions emerging from research.   The profusion of research and development has provided results that allow some conclusions to be drawn regarding the efficacy of these methods for different audiences and learning goals. One of the clearest syntheses of this research is offered by Kirschner, Sweller, and R. E. Clark (2006), who examine "minimal guidance." Problem-based or inquiry-based programs are often set up so that learners explore a problem space freely, with minimal guidance. Kirschner et al. find that, for learners who are at the novice or intermediate stage, such programs are less effective as well as less efficient than programs with strong instructional guidance. Further, minimally guided programs "may have negative results when students acquire misconceptions or incomplete or disorganized knowledge" (p. 84). They hypothesize that minimally guided learning environments subject learners to a heavy cognitive load that interferes with use of their cognitive processing abilities.

In medicine and science courses, the inquiry-based approach is often justified on the basis that it forces learners to "think like scientists." Kirschner et al. (2006) point out, "The way an expert works in his/her domain (epistemology) is not equivalent to the way one learns in that area (pedagogy)" (p. 78). So, the consistently poor results of these methods when applied to learners who are at the novice or intermediate stages should not be surprising. Going back to the original proposition of von Glasersfeld, a "new epistemology" does not necessarily equate with new or unique instructional prescriptions.

In summary, it is difficult to identify any particular learning theory or instructional strategy as unequivocally constructivist. But the instructional methods most often advocated under the guise of constructivism seem to be most suited to facilitating learning for advanced or complex learning goals being pursued by learners who already have a high level of skill in that domain.

An Eclectic Perspective

As discussed in chapter 5, an eclectic perspective, combining principles from different theories, may provide a synthesis that serves well in practice. In philosophy, blithely tacking together conflicting doctrines can produce incoherent theoretical structures, but in practical matters, eclecticism often makes sense. Educators can easily see that different theories of learning lead to instructional theories that offer guidance for different sorts of learning goals. The theories do not necessarily contradict each other; rather, they explain certain phenomena better than others. Ertmer and Newby (1993) suggest one such fairly simple formula for combining the theoretical perspectives discussed here: Employ the behaviorist perspective in situations in which learners have lower levels of task knowledge and for learning goals requiring lower cognitive processing; use the cognitivist perspective for middle levels of task knowledge and cognitive processing; and consider the constructivist perspective for situations in which learners have a higher level of prior knowledge and are working on higher level tasks, such as complex problem solving in ill-structured domains (pp. 68–69). While not all may agree with this as a recommendation, it illustrates the sort of synthesis that can flow from an eclectic approach.

Since the late 1990s, an umbrella under which different perspectives, especially cognitivist and constructivist, converge is *learner-centered* education. This concept gained wide credibility when it was endorsed by the APA Board of Educational Affairs (1995) in the form of 14 principles, shown in Table 2.2.

These principles addressed cognitive and metacognitive, affective and motivational, developmental, social, and individual differences factors. They were "learner-centered" in the sense that they attempt to derive instructional implications from research on the learning process and in the sense that they encourage adapting instruction to individual learners. The list is somewhat enigmatic in that it is a list of observations (descriptions) about the learning process, but the items are referred to as "principles," implying prescriptive advice. In any event, the APA's learner-centered principles have played a major role in shaping the discussion about how to facilitate learning early in the 21st century.

Formal and Informal Learning

Thus far we have assumed learning to be a formal, planned process such as is usually associated with schooling. It is interesting to note, however, that the definition of educational technology and its goal to facilitate learning is not necessarily limited to a formal process. The old AECT (1977) definition text included a definition of *learner* as an individual "engaged in acquiring new

Table 2.2. APA's learner-centered psychological principles.

1. **Nature of the learning process.** The learning of complex subject matter is most effective when it is an intentional process of constructing meaning from information and experience.

2. **Goals of the learning process.** The successful learner, over time and with support and instructional guidance, can create meaningful, coherent representations of knowledge.

3. **Construction of knowledge.** The successful learner can link new information with existing knowledge in meaningful ways.

4. **Strategic thinking.** The successful learner can create and use a repertoire of thinking and reasoning strategies to achieve complex learning goals.

5. **Thinking about thinking.** Higher order strategies for selecting and monitoring mental operations facilitate creative and critical thinking.

6. **Context of learning.** Learning is influenced by environmental factors, including culture, technology, and instructional practices.

7. **Motivational and emotional influences on learning.** What and how much is learned is influenced by the learner's motivation. Motivation to learn, in turn, is influenced by the individual's emotional states, beliefs, interests and goals, and habits of thinking.

8. **Intrinsic motivation to learn.** The learner's creativity, higher order thinking, and natural curiosity all contribute to motivation to learn. Intrinsic motivation is stimulated by tasks of optimal novelty and difficulty, relevant to personal interests, and providing for personal choice and control.

9. **Effects of motivation on effort.** Acquisition of complex knowledge and skills requires extended learner effort and guided practice. Without learners' motivation to learn, the willingness to exert this effort is unlikely without coercion.

10. **Developmental influences on learning.** As individuals develop, there are different opportunities and constraints for learning. Learning is most effective when differential development within and across physical, intellectual, emotional, and social domains is taken into account.

11. **Social influences on learning.** Learning is influenced by social interactions, interpersonal relations, and communication with others.

12. **Individual differences in learning.** Learners have different strategies, approaches, and capabilities for learning that are a function of prior experience and heredity.

13. **Learning and diversity.** Learning is most effective when differences in learners' linguistic, cultural, and social backgrounds are taken into account.

14. **Standards and assessment.** Setting appropriately high and challenging standards and assessing the learner as well as learning progress—including diagnostic, process, and outcome assessment—are integral parts of the learning process.

Note: Adapted from Learner-Centered Psychological Principles: A Framework for School Redesign and Reform. The full list of principles is available online at: http://www.apa.org/ed/lcp2/lcp14.html.

skills, attitudes or knowledge whether with a specified sequence of instruction or a random assortment of stimuli" (p. 209). So learning, it might follow, can be formal or informal, and a learning environment can include structured and unstructured settings.

It may be important to consider informal learning as a salient aspect for educational technologists as technologies and media continue to provide and expand learning opportunities for learners of all ages. It cannot be said that most learning occurs in schooling or training situations. Individuals are motivated to learn through the Web, through print materials, and through informal encounters with "experts" in the community. This informal learning is neither designed nor assessed by educators, but must be considered when we discuss the role of facilitating learning for learners of all ages and stations of life. The field may need to increase its awareness of these public resources and continue to consider their instructional potential for both motivating and providing learning opportunities.

In fact, even in formal learning settings, planned instruction is not the only, or even the most important, determinant of success or failure in learning. To simplify a complex situation somewhat, we can say that learning is most directly dependent on three factors: aptitude, effort, and instruction (Walberg, 1984). Those who come into the setting with a high level of native ability—aptitude— may succeed without even trying very hard or receiving quality instruction. Or those who exert tremendous effort may succeed even if they have limited aptitude and uninspired teaching. The investment of effort is assumed to be driven by the individual's motivation, which itself is a product of home and personal background, expectations, and interest in the subject matter.

Therefore, it is important to recognize that instruction, no matter how well designed and executed, is only one part of the learning equation, often overshadowed by learners' developmental abilities, their needs, and their interests. Instructional designers can influence effort through *motivational design*—making the materials as interesting and relevant as possible and arranging the total learning environment so that learners have an expectation of success and achieve satisfying results (Keller, J. M., 1987). However, the motivation that comes from beyond the classroom is largely beyond the instructional designer's span of control. Looking at the instructional setting as a total system and seeing how the various factors interact is discussed in greater depth in chapter 3.

Media Versus Methods

Some enthusiasts for using media to improve learning seem to assume that merely embedding the content into a newer media format will automatically improve its effectiveness. This assumption has been under attack since R. E. Clark (1983) declared that "The best current evidence is that media are mere vehicles that deliver instruction but do not influence student achievement any more than the truck that delivers our groceries causes changes in our nutrition" (p. 445). He based this conclusion on a meta-analysis of hundreds of research reports from studies in which instructional presentations in one media format were compared with presentations in a different format. R. E. Clark concludes, "It seems not to be media but variables such as instructional methods that foster learning" (p. 449).

A debate about "media versus methods" raged for a decade. The most effective counterargument was raised by Kozma (1991), who contended that the studies cited by R. E. Clark (1983) were based on a presentation paradigm— learners watching or listening to a presentation. Kozma agreed that, under such conditions, different media formats only made a difference in time and cost, not learning effectiveness. Kozma proposed that different results could be expected from a different instructional paradigm, one in which media are used as tools by learners, not as presentations. In other words, not learning *from* media (Clark's term), but learning *with* media (Kozma's term). In subsequent years, as the use of media more and more comes to mean digital media, educational technology looks forward to a new research agenda, studying the possibilities of this new paradigm.

Summary

The current definition of educational technology explicitly adopts the term *facilitating learning* in order to emphasize the understanding that learning is controlled and owned by learners. Teachers and designers can and do influence learning, but that influence is facilitative rather than causative. The term *facilitating learning* is posited as the purpose of the field, not as the result of processes that are the raison d'etre of the field.

Different theories of learning and instruction emphasize different variables in the learning process, so *facilitating* has different meanings for each theory. Understanding the implications of the different theories is impeded by the practice of conflating instructional theories with learning theories and even epistemologies. For the purposes of this chapter, the bodies of theory are viewed simply as different perspectives on teaching and learning. Behaviorism, cognitivism, and constructivism each have prompted interesting and successful applications of educational technology. Each has added to our overall understanding of how people learn and how instruction might be improved. It is possible to envision an eclectic umbrella under which various creative uses can be combined to provide rich environments for active learning.

Assessment and evaluation methods are an important link in the chain of successful implementation of any behaviorist, cognitivist, or constructivist instructional innovation. If the innovative program is striving toward the goal of deeper, higher level, metacognitive, or applied knowledge, its results will not be adequately captured by conventional paper-and-pencil tests.

Although most of the discussion in the chapter is framed in terms of formal instructional situations, the current definition is also intended to apply to informal learning. In fact, that is one of the reasons that the definition chooses the term *educational technology* rather than *instructional technology*, using the term with the broader connotation in order to capture both planned and spontaneous learning situations.

We conclude with some comments about the *values* underlying this whole chapter. In facilitating the process of learning, regardless of associated theoretical perspectives, the practice of educational technology actually helps or hinders the *people* who are in pursuit of learning. In other words, we do what we do as educational technologists not so much to facilitate learning in and of itself but to facilitate learning by the intended audience. This shift in emphasis from the process to the people indicates an increasing focus and awareness of students as the core of our activities as educational technologists. When the learner is the focus, as opposed to the hardware, the design, or the materials, then the idea of facilitating learning must also focus on the learner and their abilities and responsibilities. Learner-centered thinking reminds us that at its core, learning is still an idiosyncratic or at least not completely controllable activity. As instructors and designers, we take advantage of generalizations about people and the ways they may learn. In our efforts to facilitate learning truly, however, we must acknowledge the diversity of the individual. We may not be capable of always facilitating learning for that particular person, but we must not forget facilitating learning for each individual is the goal. Facilitation suggests that we attend more completely to the learner within the setting, consider the context and the environment, and make an attempt to relate our designs to the cultural and societal aspects of the setting as we design or create learning environments. The diversity of learners would be addressed and learning supported through our use of both hardware and software, and in fact, this becomes the goal of technology integration into learning environments.

References

Adams, G. L., & Engelmann, S. (1996). *Research on Direct Instruction: 25 years beyond DISTAR*. Seattle, WA: Education Achievement Systems.

Alesandrini, K. L. (1984). Pictures and adult learning. *Instructional Science, 13*, 63–77.

American Psychological Association Board of Educational Affairs. (1995). *Learner-centered psychological principles: A framework for school reform and redesign*. Washington, DC: American Psychological Association. Retrieved October 12, 2005, from http://www.apa.org/ed/lcpnewtext.html

Anglin, G. J., Vaez, H., & Cunningham, K. L. (2004). Visual representations and learning: The role of static and animated graphics. In D. H. Jonassen (Ed.), *Handbook of research on educational communications and technology* (2nd ed., pp. 865–916). Mahwah, NJ: Lawrence Erlbaum Associates.

Association for Educational Communications and Technology (AECT). (1977). *The definition of educational technology*. Washington, DC: Author.

Atkinson, R. C., & Shiffrin, R. M. (1968). Human memory: A proposed system and its control processes. In K. Spence, & J. Spence (Eds.), *The psychology of learning and motivation* (Vol. 2, pp. 486–522). New York: Academic Press.

Ausubel, D. (1963). *The psychology of meaningful verbal learning*. New York: Grune & Stratton.

Barron, A. E. (2004). Auditory instruction. In D. H. Jonassen (Ed.), *Handbook of research on educational communications and technology* (2nd ed., pp. 949–978). Mahwah, NJ: Lawrence Erlbaum Associates.

Bednar, A. K., Cunningham, D., Duffy, T. M., & Perry, J. D. (1991). Theory into practice: How do we link? In G. Anglin (Ed.), *Instructional technology: Past, present and future* (pp. 88–101). Denver, CO: Libraries Unlimited.

Borman, G. D., Hewes, G. M., Overman, L. T., & Brown, S. (2002). *Comprehensive school reform and student achievement: A meta-analysis* (Rep. No. 59). Baltimore, MD: Center for Research on the Education of Students Placed at Risk (CRESPAR), Johns Hopkins University.

Bransford, J. D., Brown, A. L., & Cocking, R. R. (Eds.). (2000). *How people learn: Brain, mind, experience, and school* (expanded edition). Washington, DC: National Academy Press.

Brown, J. S., Collins, A., & Duguid, P. (1989, January/February). Situated cognition and the culture of learning. *Educational Researcher, 18*, 32–42.

Bruner, J. S. (1960). *The process of education*. Cambridge, MA: Harvard University Press.

Burton, J. K., Moore, D. M., & Magliaro, S. G. (2004). Behaviorism and instructional technology. In D. H. Jonassen (Ed.), *Handbook of research on educational communications and technology* (2nd ed., pp. 3–36). Mahwah, NJ: Lawrence Erlbaum Associates.

Carney, R. N., & Levin, J. R. (2002). Pictorial illustrations still improve students' learning from text. *Educational Psychology Review, 14*(1), 5–26.

Clark, R., & Lyons, C. (2004). *Graphics for learning.* San Francisco: Pfeiffer.

Clark, R. E. (1983). Reconsidering research on learning from media. *Review of Educational Research, 53*(4), 445–459.

Cohen, P. A., Kulik, J. A., & Kulik, C. C. (1982, Summer). Educational outcomes of tutoring: A meta-analysis of findings. *American Educational Research Journal, 19*(2), 237–248.

Commission on Instructional Technology. (1970). *To improve learning: A report to the President and Congress of the United States.* Washington, DC: U.S. Government Printing Office.

Cognition and Technology Group at Vanderbilt (CTGV). (1993). Anchored instruction and situated cognition revisited. *Educational Technology, 33*(3), 52–70.

Dale, E. (1946). *Audio-visual methods in teaching.* New York: The Dryden Press.

DeCecco, J. P. (1964). *Educational technology: Readings in programmed instruction.* New York: Holt, Rinehart, and Winston.

Dennen, V. P. (2004). Cognitive apprenticeship in educational practice: Research on scaffolding, modeling, mentoring, and coaching as instructional strategies. In D. H. Jonassen (Ed.), *Handbook of research on educational communications and technology* (2nd ed., pp. 3–36). Mahwah, NJ: Lawrence Erlbaum Associates.

Driscoll, M. P. (2005). *Psychology of learning for instruction* (3rd ed.). Boston: Allyn & Bacon.

Duffy, T. M., & Cunningham, D. J. (1996). Constructivism: Implications for the design and delivery of instruction. In D. H. Jonassen (Ed.), *Handbook of research for educational communications and technology* (pp. 170–198). New York: Macmillan Library Reference U.S.A.

Duffy, T. M., & Jonassen, D. H. (Eds.). (1992). *Constructivism and the technology of instruction: A conversation.* Hillsdale, NJ: Lawrence Erlbaum Associates.

Ellson, D. G., Barner, L., Engle, T., & Kempwerth, L. (1965, Fall). Programmed tutoring: A teaching aid and a research tool. *Reading Research Quarterly, I,* 71–127.

Ely, D. P. (Ed.). (1963). The changing role of the audiovisual process in education: A definition and glossary of related terms. *AV Communication Review, 11*(1), Supplement 6.

Ely, D. P. (Ed.). (1972, October). The field of educational technology: A statement of definition. *Audiovisual Instruction, 17*(8), 36–43.

Englemann, S. (1980). *Direct Instruction.* Englewood Cliffs, NJ: Educational Technology Publications.

Ertmer, P. A., & Newby, T. J. (1993). Behaviorism, cognitivism, constructivism: Comparing critical features from an instructional design perspective. *Performance Improvement Quarterly, 6*(4), 50–72.

Ferster, C. B., & Skinner, B. F. (1957). *Schedules of reinforcement.* New York: Appleton-Century-Crofts.

Fleming, M. L., & Levie, W. H. (1993). *Instructional message design: Principles from the cognitive and behavioral sciences* (2nd ed.). Hillsdale, NJ: Educational Technology Publications.

Foshay, W. R., Silber, K. H., & Stelnicki, M. B. (2003). *Writing training materials that work.* San Francisco: Jossey-Bass.

Gagne, R. M., & Medsker, K. L. (1996). *The conditions of learning: Training applications.* Fort Worth, TX: Harcourt Brace College Publishers.

Heinich, R. (1984). The proper study of instructional technology. *Educational Communication and Technology Journal, 32*(2), 67–87.

Hoban, C. F. (1965). From theory to policy decisions. *AV Communication Review, 13*, 121–139.

Hoban, C. F., Hoban, C. F., Jr., & Zisman, S. B. (1937). *Visualizing the curriculum.* New York: The Cordon Co.

Januszewski, A. (2001). *Educational technology: The development of a concept.* Englewood, CO: Libraries Unlimited.

Jonassen, D. H. (1991). Objectivism versus constructivism: do we need a new philosophical paradigm? *Educational Technology Research and Development, 39*(3), 5–14.

Jonassen, D. H., Howland, J., Moore, J., & Marra, R. M. (2003). *Learning to solve problems with technology: A constructivist perspective* (2nd ed.). Columbus, OH: Merrill/Prentice-Hall.

Keller, F. S. (1968). Goodbye teacher . . . *Journal of Applied Behavior Analysis, 1,* 78–79.

Keller, F. S. (1977). *Summers and sabbaticals: Selected papers on psychology and education.* Champaign, IL: Research Press Company.

Keller, J. M. (1987). Development and use of the ARCS model of instructional design. *Journal of Instructional Development, 10*(3), 2–10.

Kirschner, P. A., Sweller, J., & Clark, R. E. (2006). Why minimal guidance during instruction des not work: An analysis of the failure of constructivist, discovery, problem-based, experiential, and inquiry-based teaching. *Educational Psychologist, 41*(2), 75–86.

Kozma, R. B. (1991, Summer). Learning with media. *Review of Educational Research, 61*(2), 179–211.

Kozma, R. B., & Johnston, J. (1991, January/February). The technological revolution comes to the classroom. *Change, 23*(1), 10–23.

Kulik, J. A., Kulik, C. C., & Cohen, P. A. (1979). A meta-analysis of outcome studies of Keller's personalized system of instruction. *American Psychologist, 34*, 307–318.

Lakoff, G. (1987). *Women, fire, and dangerous things.* Chicago: University of Chicago Press.

Leamnson, R. (2000, November/December). Learning as biological brain change. *Change, 32*(6), 35–40.

Lockee, B., Moore, D., & Burton, J. (2004). Foundations of programmed instruction. In D. H. Jonassen (Ed.), *Handbook of research on educational communications and technology* (2nd ed., pp. 545–569). Mahwah, NJ: Lawrence Erlbaum Associates.

Lohr, L. (2003). *Creating graphics for learning and performance: Lessons in visual literacy.* Upper Saddle River, NJ: Merrill Prentice Hall.

Magliaro, S., Lockee, B., & Burton, J. (2005). Direct instruction revisited: A key model for instructional technology. *Educational Technology Research and Development, 53*(4), 41–55.

Merrill, M. D. (2002). A pebble-in-the-pond model for instructional design. *Performance Improvement, 41*(7), 39–44.

Molenda, M., & Russell, J. D. (2006). Instruction as an intervention. In J. A. Pershing (Ed.), *Handbook of human performance technology* (3rd ed., pp. 335–369). San Francisco: Pfeiffer.

Montessori, M. (2004). *The Montessori method: The origins of an educational innovation: Including an abridged and annotated edition of Maria Montessori's The Montessori Method* (G. L. Gutek, Ed.). Lanham, MD: Rowman & Littlefield.

Moore, D. M., Burton, J. K., & Myers, R. J. (2004). Multiple-channel communication: The theoretical and research foundations of multimedia. In D. H. Jonassen (Ed.), *Handbook of research on educational communications and technology* (2nd ed., pp. 979–1005). Mahwah, NJ: Lawrence Erlbaum Associates.

Phillips, D. C. (1994). Philosophy of education: Historical overview. In T. Husen, & T. N. Postlethwaite (Eds.), *International encyclopedia of education* (2nd ed., pp. 4447–4456). Oxford, UK: Pergamon.

Phillips, D. C. (1995, October). The good, the bad, and the ugly: The many faces of constructivism. *Educational Researcher, 24*(7), 5–12.

Reigeluth, C. M. (1979). In search of a better way to organize instruction: The elaboration theory. *Journal of Instructional Development, 2*(3), 8–15.

Reigeluth, C. M. (Ed.). (1983). *Instructional-design theories and models.* Hillsdale, NJ: Lawrence Erlbaum Associates.

Roschelle, J., & Pea, R. (2002). A walk on the WILD side: How wireless handhelds may change computer-supported collaborative learning. *International Journal of Cognition and Technology, 1*(1), 145–168.

Saettler, P. (1990) *The evolution of American educational technology.* Englewood, CO: Libraries Unlimited.

Schwen, T. M. (1977). Professional scholarship in educational technology: Criteria for judging inquiry. *Educational Communication and Technology Journal, 25*(1), 5–24.

Seels, B. B., & Richey, R. C. (1994). *Instructional technology: The definition and domains of the field.* Washington, DC: Association for Educational Communications and Technology (AECT).

Shrock, S. A. (1990). School reform and restructuring: Does performance technology have a role? *Performance Improvement Quarterly, 3*(4), 12–33.

Silber, K. (1970). What field are we in, anyhow? *Audiovisual Instruction, 15*(5), 21–24.

Silber, K. H., & Foshay, W. R. (2006). Designing instructional strategies. In J. A. Pershing (Ed.), *Handbook of human performance technology* (3rd ed., pp. 370–413). San Francisco: Pfeiffer.

Skinner, B. F. (1954). The science of learning and the art of teaching. *Harvard Educational Review, 24,* 86–97.

Skinner, B. F. (1965). The technology of teaching. *Proceedings of the Royal Society, 162*(Series B), 427–443.

Skinner, B. F. (1968). *The technology of teaching.* New York: Appleton-Century-Crofts.

Sweller, J. (1988). Cognitive load during problem solving: Effects on learning. *Cognitive Science, 12,* 257–285.

Terhart, E. (2003). Constructivism and teaching: a new paradigm in general didactics? *Journal of Curriculum Studies, 35*(1), 25–44.

Tinzmann, M. B., Rasmussen, C., & Foertsch, M. (1999). *Engaged and worthwhile learning.* Retrieved November 20, 2005, from http://www.ncrtec.org/pd/lwtres/ewl.pdf

Torkelson, G. M. (1977). AVCR—One quarter century: Evolution of theory and research. *AV Communication Review, 25*(4), 317–358.

von Glasersfeld, E. (1984). An introduction to radical constructivism. In P. Watzlawick (Ed.), *The invented reality* (pp. 17–40). New York: W.W. Norton.

von Glasersfeld, E. (1992, August). *Aspects of radical constructivism and its educational recommendations.* Paper presented at ICMe-7, Working Group #4, Quebec, Canada.

Walberg, H. J. (1984, May). Improving the productivity of America's schools. *Educational Leadership, 41*(8), 19–27.

Watkins, C. L. (1988). Project follow through: A story of the identification and neglect of effective instruction. *Youth Policy, 10*(7), 7–11.

Winn, W. (2004). Cognitive perspectives in psychology. In D. H. Jonassen (Ed.), *Handbook of research on educational communications and technology* (2nd ed., pp. 79–112). Mahwah, NJ: Lawrence Erlbaum Associates.

Woolley, D. R. (1994). *The emergence of online community.* Retrieved October 17, 2005, from http://www.thinkofit.com/plato/dwplato.htm

3

IMPROVING PERFORMANCE

Michael Molenda
James A. Pershing
Indiana University

Introduction

Educational technology is the study and ethical practice of facilitating learning and *improving performance* by creating, using, and managing appropriate technological processes and resources.

*T*HE TERM IMPROVING PERFORMANCE represents educational technology's claim of offering the societal benefit of accomplishing a worthy goal in a superior fashion. What is that goal? Beyond just facilitating learning, educational technology claims to improve the performance of individual learners, of teachers and designers, and of organizations. This chapter discusses each of those goals in turn.

Please note that this chapter is *not* about *performance improvement* as it is conceived in business management theory or the field of human performance technology (HPT). In those venues, people view *performance improvement* as a process of using *all* available means to solve performance problems in organizations. Those means may include interventions such as personnel selection, incentive programs, and organizational redesign in addition to training. This book and this chapter, on the other hand, are about *educational* interventions only. Therefore, this chapter deals only with the ways in which technology can enhance educational interventions in ways that improve human performance. At the end of the chapter, we discuss the broader theory of HPT and show how educational technology and HPT interface with each other to form a powerful integrated concept.

Improving Individual Learner Performance

Educational technology extends individual learning into improved performance in several ways. First, the learning experiences are made more valuable by being focused on worthwhile goals, not just passing tests. Second, through technology the experiences can lead to deeper levels of understanding, beyond rote memory. Then they are made more valuable by being designed in ways that make the new knowledge and skill transferable. That is, the new learning is applicable to real-life situations, not simply left behind in the classroom. Through these means, learners become doers, with knowledge better connected to performance beyond the classroom setting.

More Valuable Learning

The Problem of Superficial Testing.   In formal education, learning outcomes tend to be measured in terms of paper-and-pencil test results, whether teacher made or standardized. The formats of these achievement tests tend to be those that are most easily and reliably scored—true/false, multiple-choice, matching, and other such close-ended formats. A limitation of such instruments is that they are useful primarily for cognitive skills alone and especially cognitive skills of the lower levels—knowledge and comprehension as opposed to application, evaluation, and problem solving. Surveys of evaluation practices in corporate training indicate that in that sector, too, outcomes are most often measured by paper-and-pencil instruments rather than more authentic measures (Sugrue, 2003, p. 18). A problem arises if instructors then "teach to the test," and they are often under considerable pressure to do so. If the test requires only lower level skills, instructors may teach only these skills.

Such narrowing and lowering of goals may have been taking place in the public schools of the United States since the national implementation of high-stakes testing in the years after 2001. According to Nichols and Berliner (2005), news sources reported that,

> Teachers are forced to cut creative elements of their curriculum like art, creative writing, and hands-on activities to prepare students for the standardized tests. In some cases, when standardized tests focus on math and

reading skills, teachers abandon traditional subjects like social studies and science to drill students on test-taking skills. (p. iii)

In a national survey, teachers confirmed that the pressure of doing well on a standardized test seriously compromises their instructional practice (Pedulla et al., 2003).

Multiple intelligences. Meanwhile, more diverse types of knowledge, skills, and attitudes may be valuable for individual learners and for society. Howard Gardner (Gardner & Hatch, 1989), for example, suggests that there might be seven different types of intelligence, of which only two—linguistic and logical mathematical—are typically addressed in formal education. The other intelligences—musical, spatial, bodily kinesthetic, interpersonal, and intrapersonal—are addressed to some extent in the curricula of schools and colleges and to a greater extent in schools experimenting with curricula based on Gardner's theory (Gardner & Hatch, 1989, p. 7). However, they usually are not addressed in the high-stakes tests that actually drive day-to-day teaching priorities. Consequently, references to learning outcomes in formal education tend to be equated with narrow, limited, and low-level knowledge.

Domains and levels of objectives. The best-known taxonomy of domains and levels of learning objectives is known as Bloom's taxonomy. In its original form (Bloom, Englehart, Furst, Hill, & Krathwohl, 1956), it proposed that educational objectives could be classified broadly into three domains— (a) cognitive, (b) affective, and (c) psychomotor. Each of these, in turn, could be subdivided into several levels, reflecting simpler and more complex skills within each domain.

The cognitive domain was viewed as basically hierarchical—from simple to complex—beginning with knowledge and proceeding to comprehension, application, analysis, synthesis, and evaluation. More recently, a team representing the original authors and publisher (Anderson & Krathwohl, 2001) suggested a revision of the cognitive categories into a two-dimensional matrix, reflecting current research and terminology. They renamed the categories as (a) remember, (b) understand, (c) apply, (d) analyze, (e) evaluate, and (f) create. On the second dimension, each of these levels may be applied to facts, concepts, procedures, or metacognitive knowledge.

The affective domain, dealing with attitudes and feelings, is organized according to the level of internalization of the attitude, starting with receiving and proceeding to the more deeply internalized levels of responding, valuing, organization, and characterization (Krathwohl, Bloom, & Masia, 1964).

The classification of objectives in the psychomotor domain is especially challenging since these tasks involve combinations of physical and mental skills. Simpson (1972) proposed that psychomotor skills can be organized according to their complexity, beginning with guided responses and proceeding to habitual mechanical skills, then to fluent combinations of skills, and eventually to the ability to adapt and originate new physical skills.

Romiszowski (1981) proposed that a major dimension of learned skills was missing from the traditional taxonomies—the interpersonal domain, one of the neglected domains later identified by Gardner and Hatch (1989). Romiszowski contended that not only were interpersonal skills not represented, but also they were very frequently the subject of training and education. In the school setting, teachers often aim to help students work better in groups as well as to interact productively with their peers in general. In the corporate world, supervisory and management training often dwells on human relationships. For example, the American Management Association (AMA, n.d.) offered over two dozen courses in this domain, related to assertiveness, leadership, communicating, managing emotions, listening, and negotiating. This "missing" domain has not yet been fleshed out in terms of an authoritative taxonomy but is recognized in textbooks on instructional design (Morrison, Ross, & Kemp, 2004) and instructional media utilization (Heinich, Molenda, & Russell, 1985).

During the programmed instruction era of the 1960s, Mager (1962) insisted that in order to be useful, objectives must not only clearly specify the domain and level of the skill but also the conditions under which the skill would be performed and the criterion or level of mastery required. The notion of precisely stated performance objectives was absorbed into the emerging doctrine of the systems approach to instructional design (ID). Systems approach models place a heavy emphasis on specifying learning objectives precisely, since a clear path of action cannot be chosen until the goal is set. On one hand, the practice of specifying objectives precisely can enrich education by offering a broad menu of targets at which to aim. However, on the other hand, it can lead to narrow and often low-level objectives being implemented. This latter tendency was noted in the programmed instruction era, when authors of programmed materials often found it convenient to achieve precision by specifying behaviors that were easy to observe and measure "answer correctly 90% of the questions on the post-test," or "list five reasons."

On the more positive side, many contemporary instructional design textbooks reflect quite a sophisticated view of types and levels of learning. Taking Morrison et al. (2004) as a sample of what is advocated in systematic ID models, we find that they refer to the cognitive, affective, psychomotor, and interpersonal domains, and within those domains describe multiple types and levels of skill. For each level in each domain, they provide a list of verbs representing indicators of each level. Although this elaboration of types and levels of learning does not necessarily match the breadth of Gardner's (Gardner & Hatch, 1989) typology, it does provide a broad array of learning objectives. Therefore, one of the ways in which educational technology seeks to improve performance is through instructional design practices that lead planners to think about a wide range of learning outcomes and clarify what types of learning, at what levels, are desired. If such advice is followed, learners are more likely to experience learning activities and assessment methods that are appropriate for the wide range of human learning needs, not just those that are emphasized on standardized tests.

Surface Versus Deep Learning.   Settling for verbal recall as the goal of instruction was a major problem that Edgar Dale (1946) was combating in the first modern textbook on audiovisual education. Dale contrasted "bookish learning" with "real learning," by which he meant learning that was permanent, laden with emotional overtones, and ready to be applied to real-world problems. Therefore, this issue has a venerable and central place in the tradition of educational technology. Dale's position is echoed by many other contemporary educators. It is at the heart of cognitivists' "meaningful learning," and much of the rhetoric of constructivism is aimed at replacing rote learning with learning that is situated in applied contexts.

The difference between rote knowledge and applicable knowledge is qualitative, according to the findings of neuroscience: "Overall, neuroscience research confirms the important role that experience plays in building the structure of the mind by modifying the structures of the brain . . . " (Bransford, Brown, & Cocking, 1999). Weigel (2002) suggests the terms *surface learning* and *deep learning* to characterize these contrasting goals. Surface learning is represented in mere memorization of facts, treating material as unrelated bits of information, and carrying out procedures routinely without thought or strategy (p. 6). In deep learning, learners relate ideas to previous knowledge, look for underlying patterns, examine claims critically, and reflect on their own understandings (p. 6).

Weigel (2002) and others propose that the venue in which deep learning can best take place is an inquiry-oriented community of learners. They suggest that such communities could be created through information technology. Using the workplace team as a paradigm, educators using local and Web-based networked computers, set up learning communities to allow learners to collaborate on realistic tasks. As they work in such problem-based and task-based environments, they develop deep learning by proposing solutions, testing them, debating them with others, and arriving at a group synthesis.

Transfer of Learning in Formal Education. Technology can help learners not only to master higher-level skills, but also to apply new knowledge to novel situations, especially those outside the classroom—referred to as transfer of learning. Research on situated cognition suggested that what is learned in the classroom context tends to be confined to that setting unless learners have opportunities to practice the new skill in contexts that resemble the real world. Hard technology in the form of computer-based simulations offers a way to be immersed virtually in environments that would be impractical or even impossible to duplicate in reality.

Computer-based microworlds immerse learners in problems that are embedded in the complexities of reality. Some examples developed recently at the Center for the Study of Problem Solving (CSPS) include computer-based simulations that allow learners to step into the shoes of a homeless single mother, design a new highway interchange, develop a new food product in an agribusiness lab, or play the role of a peacekeeper in a war-torn nation (CSPS, n.d.). Such immersive virtual environments add to the student experience by pushing academic learning into the realm of application.

Transfer of Training in Corporate Settings. In corporate training, there is a long-standing concern for the ability of trainees to put their newly acquired knowledge and skills to work in their everyday jobs, expressed in the term *transfer of training* (Baldwin & Ford, 1988). The systems approach to instructional design helps planners to focus on transfer of training, not just by activities that happen after instruction, but also those that happen before and during instruction,

- Before training: focus on transfer goals in the needs analysis; involve supervisors and trainees at the needs analysis stage; ask supervisors and trainees to develop a transfer plan together as a prerequisite for participation.
- During training: focus on application-oriented activities; incorporate visualization experiences into instruction; have participants develop individual transfer plans.
- After training: follow up with reaction surveys; observe and validate changed work behavior directly or through supervisors; conduct follow-up refresher or problem-solving workshops (Broad & Newstrom, 1992).

Therefore, individual learner performance in the classroom and in the workplace can be enhanced through soft technology, a systematic approach to ID, and through hard technology, the creation and use of immersive environments in which learners can practice and apply knowledge and skills in realistic settings.

Improving Performance of Teachers and Designers

Educational technology can improve the performance not only of learners but also of those who design and deliver instruction. It can reduce learning time and increase learning effectiveness, both of which enhance the productivity of instructors and designers. Equally importantly, educational technology can help create instruction that is more appealing and respectful of human values, thus aligning instructors and designers with their highest professional commitments.

Reducing Instructional Time

Early in the evolution of modern educational technology as behavioral psychologists were translating laboratory findings into real-world applications, they quickly came to appreciate the importance of articulating the goal of any instructional intervention. It is axiomatic in operant conditioning that the process starts by specifying the desired behavior. The formula for behavior modification is to specify the behavioral goal, observe the learner's practice, and provide appropriate consequences for performance. Carried over into corporate training, precise performance objectives became the starting point of any design project (Mager, 1962). This, in turn, required close analysis of purported training needs to discriminate between objectives that were "nice to know" and those that were "need to know."

Procedures for needs analysis and task analysis were refined to relentlessly weed out unnecessary training activity. In fact, many of the early triumphs of systematic instructional design were attributable to the reduction of learner time spent in unnecessary training. As Robert Mager (1977) put it in his keynote address at the ASTD national conference, "Since the objectives for this type of instruction are usually derived from a task or goal analysis, the instruction is more tightly tuned to the needs of the corporation than was previously the case" (p. 13). He went on to cite specific cases of dramatic reductions in instructional time: a broadcasting corporation's course on transmitter maintenance reduced from four weeks to an average of two weeks, self-paced, per person; an army typewriter-repair course reduced in length by 35%; an airline's flight crew training reduced from 15 days to an average of 8; and the U.S. Air Force reducing instructional time between 10 and 25% per course over a range of over 1,000 courses. These time reduction achievements obviously yield great benefits to the organization, enhancing its performance, but they can be seen as enhancements to the performance of those who plan and deliver instruction—designers and teachers. The same number of staff can produce more and better instruction, instruction that is targeted to organizational needs.

Creating More Cost-Beneficial Instruction

Systematic instructional design allows ordinary planners to achieve extraordinary results. For novices, it can replace intuition and trial-and-error approaches with approaches that have been tested and refined. Beginning instructional designers can attain expert status more rapidly.

Instructional design can lead more reliably to effective learning, especially if the procedures include careful attention to selection of powerful instructional strategies. It can also arrive at that goal more efficiently. In the corporate setting, when trainees return to the job sooner as more skilled performers, the training function contributes to profits. When training is a profit center rather than a cost center, the instructional designer becomes a hero. Here we are discussing the benefits of increased productivity for teachers and designers; in the later section on "improving performance of organizations," we will discuss the benefits for the organizations themselves.

In formal education, the growing demand for learner-centered, active learning means advance planning of new sorts of learning environments. The development of such environments requires a different approach than ordinary day-to-day ad hoc teaching. Educators who can apply a disciplined approach to instructional design are more valued professionals.

Creating More Humane Instruction

More Appealing Instruction. Instructional design theory aims at creating instruction that is appealing as well as being effective and efficient (Reigeluth, 1983, p. 20). Making this one of the major criteria for good instruction is justified by the expectation that learners are more likely to want to continue learning when the experience is appealing. If nothing else, being appealing can at least increase time on task, which is consistently associated with improved learning.

What is appealing? This will vary from case to case, but in general instruction that has appeal has one or more of these qualities:

- Provides a challenge, evokes high expectations
- Has relevance and authenticity in terms of learners' past experiences and future needs
- Employs humor or a playful element
- Holds attention through novelty
- Is engaging intellectually and emotionally
- Connects with learners' own interests, goals
- Uses multiple forms of representation (e.g., audio and visual).

Keller (1987) referred to his ARCS model as a method for improving the "motivational appeal" of instructional materials (p. 2), meaning materials that attract attention, are relevant to the learner, inspire the learner's confidence, and provide satisfaction (p. 3).

Educational technology has a long history of concern for appealing instruction. Comenius (1592–1670), one of the major precursors of the field, created an impressive body of work about pedagogy, particularly advocating the use of sensory stimuli to enrich instruction. He opposed the punitive character of schools of his time, proposing instead to introduce children "to knowledge of the prime things that are in the world, by sport and merry pastime" (Comenius, 1657/1967). In the 19th century and early 20th century, Johann Herbart in Germany and William James and John Dewey in the United States developed educational theories that placed "interest" at the heart of the process.

The original rationale behind the audiovisual movement of the early 1900s was to escape the empty verbalism of lecture- and reading-based instruction by the use of films, audiovisual media, and other sensory experiences. For Dale (1946), the ideal was "rich experiences," involving the senses in ways that are engaging and fresh: "The richest experiences are almost always personal adventures, in which the outcome has the appeal of the unpredictable" (p. 22).

Research by Csikszentmihalyi (1988) and others suggested a high correlation among positive emotional states, engagement, concentration, and enjoyment. Many of the instructional innovations inspired by cognitivist and constructivist theories—such as problem-based learning, cognitive apprenticeship, immersion in microworlds—have been designed to arouse interest as a key component in motivating learners to become deeply engaged with the material (Schiefele, 1991).

Respectful of Human Values. Humanism and technology are not contradictory concepts. Classrooms can be inhumane with or without technology, and technology can be used in ways that liberate people or constrain them. Many of the innovations advocated in educational technology have focused on advancing human values.

Programmed instruction, structured tutoring, Direct Instruction, and other design formats that sprang from behaviorist roots—which are often perceived as quite mechanistic—actually aimed to liberate learners from the tedium of large-group, passive instruction (Skinner, 1968). Being modular, lessons in these formats could be prescribed according to individual needs. Being paced according to individual progress, each learner received a customized program. Being mastery based, learners' confidence was built through experiencing success. Being based on operant conditioning, learners were constantly receiving feedback on their performance; in structured tutoring and Direct Instruction much of the feedback takes the form of social reinforcers (e.g., smiles and compliments).

More recently, constructivist and postmodernist theories make a strong claim to place humane values as the highest priority. The methods favored by constructivism place special emphasis on emotional and motivational features, and they often depend on technology-based experiences to attain these features. Immersive environments, such as computer-based microworlds and simulation games, provide a venue for "serious play" (Rieber, Smith, & Noah, 1998). Discovery activities based on exploration of Web resources are also favored. Besides stimulating curiosity, they put learners in control of the action, allowing them to determine the nature and sequence of the experience. Such environments require that individuals take ownership of their learning, which in part is intended to nurture lifelong interest in learning. Reflection activities during and after instruction are meant to help learners to become more conscious of the strategies they have followed so that they can grow in their ability to take control of their own learning processes.

Improving Performance of Organizations

Previous definitions have focused on the role of technology in improving individual learning to the exclusion of its role in improving the performance of organizations. Historically, technology has been adopted by organizations as a way to improve productivity—to reduce costs and/or increase output. This economic motive is certainly a major one for training programs in business and industry, but it has been less prominent in schools and universities. Given the enormous public benefit that could be achieved by increasing the productivity of public educational institutions, we will review the issues of efficiency and effectiveness and some possible roles for technology in improving productivity in education.

Promoting Efficiency and Effectiveness

Efficiency in education is a delicate subject. It is easy to agree that human endeavors ought to be pursued efficiently, but it is more difficult to agree about the extension of this idea to education. The problem is posed clearly by Monk (2003):

> Educators often feel ambivalent about the pursuit of efficiency in education. On the one hand, there is a basic belief that efficiency is a good and worthy goal; on the other hand, there is [a] sense of worry that efforts to improve efficiency will ultimately undermine what lies at the heart of high-quality education. Part of the difficulty stems from a misunderstanding about the meaning of efficiency as well as from the legacy of past, sometimes misguided, efforts to improve the efficiency of educational systems. (p. 700)

The pursuit of *effective* results is less controversial, but the concept of effectiveness is often intertwined with that of efficiency. We can begin to sort out these issues by examining the meanings of both concepts. Since both concepts are derived from economics, we begin with their meaning in economics.

Efficiency Defined. Economic *efficiency* is the production of goods and services in the least costly way. Its focus is on how an organization transforms inputs to outputs (McConnell & Brue, 2002). In the context of education and training, efficiency could be viewed as the design, development, and conduct of instruction in ways that use the least resources for the same or better results. Preserving and not wasting resources is necessary when resources are scarce, and in educational institutions, resources are typically limited. All organizations are better off when they leverage their available resources. By leveraging available resources, educational institutions benefit by being able to conduct more instruction with the same resources or the same instruction using fewer resources (thereby releasing funds for other functions of the organization). Further, if the institutions have rivals providing the same services, efficiency makes them more competitive.

Effectiveness Defined. Economic *effectiveness* is the production of goods and services that are valued by society and its members (Heilbroner & Thurow, 1998). In short, someone is willing to pay for them. In the context of education, effectiveness has to do with the degree to which learners attain worthy learning goals; that is, the school, college, or training center prepares learners with knowledge, skills, and attitudes that are desired by their stakeholders.

From an economic perspective, efficiency is concerned with supply side factors while effectiveness focuses on demand side factors (Nas, 1996; Brinkerhoff & Dressler, 1990). From a systems perspective, efficiency is concerned with inputs and how they are processed while effectiveness is concerned with outputs. Often, efficiency is characterized as doing things right, and effectiveness is doing the right things (a formulation attributed to Peter F. Drucker). In the short term, effectiveness—doing the right thing—is more important than efficiency—doing things the right way. In the longer term, effectiveness and efficiency must go hand in hand. We need both. Instruction that is efficient is pointless if it misses the mark of producing desired knowledge, skills, or attitudes. Similarly, instruction that produces desired learning results but consumes excessive resources, is not timely, or does not affect the right people is also unproductive. It wastes scarce resources.

Productivity Defined. In simplest economic terms, productivity is output divided by input. An operation is productive to the extent that it is both efficient and effective—it produces desired results with the least necessary cost. As we will discuss, in education "desired results" may mean different things to different people. That is why it is so important to be clear about measurement: how costs are defined and measured and how outcomes are defined and measured. There is virtually unanimous agreement among economists that education, both elementary/secondary and postsecondary, has been declining in productivity over the past decade—costs constantly rising without any noticeable improvement—or even decline—in the attainments of students.

What Inputs (for Efficiency) and Outcomes (for Effectiveness) to Measure? Judgments about efficiency and effectiveness, and therefore productivity, depend heavily on how costs and benefits—human and monetary—are calculated. However, there is not a consensus among economists as to what factors should go into the equation of what economists refer to as "the production function" in education (Hanushek, 1986, p. 1149). First, what factors should be considered as inputs? Second, what takes place during the throughput, or the processing step? In other words, how is learning "produced?" Third, what factors should be measured to determine the success of education? Although these issues are better understood today and although the statistical methodologies continue to advance, economists and educators still have not reached consensus on the answers (Schwartz & Stiefel, 2001).

Input measures.   Hanushek (1986) proposes that, for K–12 education, student achievement is a function of "the cumulative inputs of family, peers or other students, and schools and teachers. These inputs also interact with each other and with the innate abilities . . . of the student" (p. 1155). He breaks down "school and teacher" factors into teachers' educational level and experience, class size, facilities, instructional expenditures, and wealth of the community or school district.

These factors and the interactions among them are shown in Fig. 3.1 (and discussed in detail later in this chapter), which depicts the relationships according to research on factors associated with student academic learning. The noteworthy point seen in Fig. 3.1 is that some factors—such as aptitude, motivation, and instructional experiences—contribute more directly to

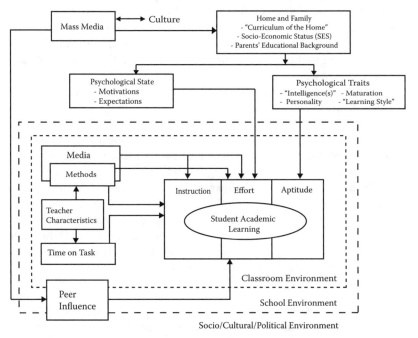

Figure 3.1.   Student Academic Learning Model. © M. Molenda, 2005. Used with permission.

learning than others, which are filtered through these more central factors. This helps to explain the failure of economic research and education research to find direct correlations between, for example, class size or teacher experience, and achievement test results (Hanushek, 1986, p. 1161, provided a meta-analysis of 147 such studies). Class size does not cause learning. It may affect learning *indirectly* by influencing what instructional strategies are chosen by the teacher or by coloring the motivational atmosphere in the classroom. The same applies to the factor of teacher experience. Having a lot of experience does not cause learning. It may affect learning *indirectly* by influencing the teacher's judgment in choosing instructional or motivational strategies.

Economic models for higher education differ from those for K–12 education because educational inputs and outputs are only a part of the total university enterprise: "Universities are a classic example of a multiple output firm, with outputs including research, housing, and entertainment (sports) in addition to education" (Bosworth, 2005, p. 70). Studies of instructional costs and benefits tend to be carried out at the departmental or course level. Such studies also tend to assume faculty expertise and student aptitude and motivation as constants, ignoring their contribution to the equation. Consequently, they focus on the factors of instructor time and hardware, software, and development costs. This conceptualization of the problem of improving efficiency lends itself well to the use of technology. The National Center for Academic Transformation (NCAT; http://www.theNCAT.org) sponsored a series of R&D projects to demonstrate that technology-assisted instruction can reduce the instructor time costs while maintaining quality (Twigg, 1999).

Beyond existing traditions in economics, questions plague the attempt to measure efficiency. Obviously, the instructor's planning and teaching time is an important input in the equation. But what about the learner's time? In cases where collaborative learning is emphasized, do you count the time spent by partners helping each other learn? In the case of peer tutoring, do you count the tutor's time? If so, what value do you put on such time? And how do you count the learning benefits accruing to peer learners? Obviously, the cost of purchasing textbooks and other instructional materials should be counted, but what about the development costs for locally produced materials and systems? What amortization schedule should be used for equipment and materials?

Throughputs, or the "production" process. Although it is not made explicit in economic models of education, the instructor seems to be assumed to be the party who does the "production." This is certainly the assumption when students are considered "customers." When using this metaphor, the instructor is clearly viewed as performing a service for a client. However, as discussed in chapter 2, the contemporary view of the learning process considers the learner the producer. There is no learning without the willing and active participation of the learner. Rather than receiving a service, the learner is actually creating the product—his or her own learning gains—sometimes in collaboration with an instructor and sometimes without.

The instructor's role is still large—providing the *conditions* (instructional and, especially, motivational) needed for successful learning—but not predominant. Thus, for an economic model to bear any resemblance to the reality of the situation, the learner must be viewed as at least the coproducer of learning gains. The throughput part of the model must include learners, and it must take into account their psychological traits (e.g., aptitude, developmental level, and personality) and psychological states (e.g., motivations and expectations), shown in Fig. 3.1.

Outcome measures. As thorny as the issues are for input and throughput variables, they are thornier for outcome measures. As Bosworth (2005) noted, "Medical care and education are two major examples of activities that raise challenging, and thus far unresolved, issues of how to measure output" (p. 68). What inputs cause learning and the factors involved in "producing" learning are empirical questions, which can be settled by research, but deciding on outcome measures is much more a matter of judgment, involving educational, social, and political values as well as economic analyses.

In the corporate setting, training outcomes are often viewed within Kirkpatrick's (1998) framework. He proposes that one can measure program success at any of four levels: (1) the reaction or satisfaction of learners, (2) the attainment of learning objectives, (3) the on-the-job behavior changes that follow instruction, or (4) organizational results, the overall impact of the instructional program on the organization's well-being. The selection of any of these targets could be justified, but the choice is often made out of convenience rather than out of careful consideration of hte demands of the situation.

For example, in public schools in the United States in 2006, the reality is that, as a matter of public policy, outcomes measured in terms of standardized test scores heavily outweigh all other benefits in the cost-benefit equation. This is defended in terms of needing some sort of objective measure of outcomes. Others would argue that this is too narrow a measure and that other outcomes should be counted, for example,

- Student achievement in learning domains not included in standardized testing, such as social development, civic virtues, creative arts, health and athletics, and love of learning

- Student achievement in basic skills that are not measured on standardized tests, such as enjoyment of reading, critical thinking in science, application of math to everyday life, and the like
- A healthy learning environment, where each student has the opportunity to develop toward leading a successful and productive life
- A productive work environment for teachers, in which their efforts are rewarded and they are motivated to stay and grow.

Because of their interests in efficiency and effectiveness, educational technologists have a special interest in making sure that both the processes and the outcomes are measured accurately. Thus, for example, when rich environments for active learning (REALs) are used to pursue deep learning and applied skills, it is paramount that the assessment be more than simple paper-and-pencil tests. Simulations and portfolios are much more likely to give an accurate gauge of the attainment of those higher-level skills. In other words, you cannot be sure about effectiveness unless you measure accurately what the outputs are.

It is entirely possible for one instructional system to be more cost *efficient* than another based on one set of outcomes, but less cost *effective* based on another set of outcomes. Monk (2003) referred to this problem as "the legacy of past, sometimes misguided, efforts to improve the efficiency of educational systems." Quality too often suffers when administrators focus narrowly on cutting costs. And the quality of outputs is often measured in intangibles, factors that are not as apparent as test scores.

For example, in teaching spelling, a structured tutoring program that has older students using flash cards to teach younger students to spell may result in 80% of the younger students spelling correctly 80% of weekly spelling test words 80% of the time. A computer-based program that teaches the same spelling words is purchased. Within a year, its costs are more than offset by replacing the hourly costs of the teacher aides that coordinated the peer-tutoring program. Further, the computer-based program results in 85% of the younger students spelling correctly 85% of the weekly spelling test words 85% of the time. This reduction in costs and improvement in outputs is technically more efficient. However, is it more effective? The answer is yes if the overall goal is improving spelling test scores of the younger students on weekly spelling tests. But what if there were unspoken goals?

In our hypothetical case, after one year the teachers begin to notice two phenomena. First, the younger students' spelling in their written work, that is, spelling in context, had become problematic. When the teachers investigate, they are reminded by the younger students that in the peer tutoring program the older students often presented words in example sentences and in contexts often individualized to the experiences of the younger students. Second, the teachers of the older students report a drop in their spelling ability. The older students report that by teaching the younger students spelling, their spelling skills were kept sharp by practice and thinking about ways to help the younger students devise ways to remember the spelling of troublesome words. So we have increased efficiency but decreased effectiveness if the goal is for all students to apply good spelling to all their work. In other words, it is more cost efficient but less cost effective.

This "efficiency without effectiveness" has been the historical problem. Callahan (1962) eloquently tells the story of the attempt to apply scientific management to American schools in the first decades of the 20th century and how quality, or effectiveness, was often sacrificed at the altar of business-like procedures. Such episodes lead educators to be suspicious of appeals to efficiency. They know intuitively that schools, colleges, and other learning institutions have numerous goals, many of them unstated or intangible, and they are concerned about what unintended consequences may develop.

There will always be debate, in businesses and educational institutions, about what goals are worth pursuing and what indicators should be used to measure progress toward those goals. Educational technologists, as much as any other stakeholders, should be part of that conversation. Taking a systems view, they can help their institutions define and achieve worthy goals (outputs) with means (instructional processes) that are as efficient *and* effective as possible. They can point to research indicating that technology-based instructional processes can contribute to educational productivity. For example,

- Ellson's (1986) meta-analysis of comparison studies, seeking experimental treatments that were more than *twice* as productive as the control treatment (defined as learning an equivalent amount in half the time or at half the expense). Among the 125 studies that met this criterion, about 70% constituted some variation on programmed instruction, structured tutoring, or "programmed teaching," such as Direct Instruction. In the latter instructional configuration, an instructor—who could be a student or a paraprofessional—conducts structured lessons following a template developed and pretested by a qualified design team, thus making economical use of division of labor.
- Levin, Glass, and Meister's (1984) computer modeling of the costs and benefits of four instructional treatments that made claims to cost effectiveness: lowering class size, tutoring programs, computer-assisted instruction (CAI), and increased instructional time.

Peer tutoring (soft technology) had by far the largest effect size, with CAI second. The other interventions yielded negligible benefits per dollar spent.

- In the first decade after Keller's (1987) invention of personalized system of instruction (PSI), described in chapter 2, some 75 comparison studies, mostly at the college level, had been published. A meta-analysis (Kulik, J. A., Kulik, C. L., & Smith, 1976) showed that the typical PSI student scored at the 75th percentile on a standardized test compared with the 50th percentile for the control treatment—one of the largest advantages for any experimental treatment in all of educational research.

Organizational Learning

The very survival of organizations is contingent on their abilities to learn and adapt to changing conditions. In contemporary management theory, organizational learning is regarded as more than just the sum of the knowledge and skills of an organization's individual members. In addition to this, organizations may have institutionalized processes for collecting, interpreting, storing, and disseminating knowledge. In the following sections we will discuss, first, individual learning *in* organizations, and, second, group learning *by* organizations.

Individual Learning in Organizations. As information and communication technologies (ICT) have grown in mass penetration and advanced in capability, more instructional functions can be mediated through technology. At the same time, economic pressures have motivated organizations to consider changing the way they conduct education and training.

ICT or "hard" technologies have proven capable of many economies related to education. In particular, they can deliver instructional materials cheaply over long distances, and they can do routine operations such as record-keeping less expensively and more reliably than human operators can. Perhaps more importantly from a learning standpoint, they can bring individuals and small groups together in conversation, thus enabling collaborative work as well as reflection on that work. By capitalizing on such advances in carrying out education and training, the productivity of the organization can improve: Learners spend less time in training and become expert performers more rapidly.

"Soft" technologies offer a new paradigm for organizing the work of education. This new paradigm starts with adopting some of the innovations of the industrial revolution—division of labor, specialization of function, and team organization. Corporations and distance education institutions have used this new work paradigm to create and offer online modules and courses at very competitive prices; the courses vary in instructional quality, but most are at least comparable to average residential courses; some are comparable to the best of traditional courses. Such new "technological" ways of working offer productivity improvements, sometimes dramatic.

Technology in business. For profit-making organizations, the role of technology has long been clear: technology is adopted primarily to replace costly human labor with cheaper means of production. Technologies that are more pervasive, such as information technology, tend to have even greater potential for transformational change. By the 1990s, corporations were experiencing competitive pressures not only from companies in their own country but also from companies in neighboring countries and countries many time zones and oceans distant. Globalization was gaining momentum. Consequently, pressure for cost cutting pushed American companies to find ways to do business with fewer employees. It was called "downsizing." Hence, businesses invested millions of dollars in computer systems, which they expected to recoup in the form of reduced costs of generating the products and services they sell. By the beginning of the 21st century, these investments were clearly paying off and many business processes had been transformed fundamentally.

Technology in K–12 education. What role technology should play in educational institutions has not been as clear. The administrative functions that schools and colleges share with businesses have been subjected to a good deal of automation—payrolls, recording of grades, enrollment figures, bus routes, financial records, and the like. However, the core function, providing education, has not been as radically affected.

A number of compelling cases of exemplary use of technology in schools have gained visibility from time to time, but few have persisted and expanded beyond the experimental stage. One prominent current example is Project CHILD, an elementary school model (described in chapter 5) that has been implemented and sustained in dozens of schools since 1995 (Butzin, 2005). This curricular plan exemplifies soft technology in the sense that it was systematically designed based on research and rigorous evaluation, and it also makes exemplary use of hard technology, employing computer-based activities as one of its pillars. Project CHILD has been recognized by a taxpayer group in Florida as an exemplary model of cost effectiveness (Florida TaxWatch, 2005). Unfortunately, for every school making exemplary use of technology to improve cost effectiveness, there are a hundred that do not.

There are many reasons that schools lag behind other sectors in their uses of technology in their core functions. First, the teaching-learning process is complex and highly intertwined with human feelings, such as altruism, submission, passionate interest in one's subject matter, and mutual trust and respect. It is not simple or easy to automate such a process, or even parts of the process. Second, key organizational decision makers have a stake in making and keeping the teaching-learning process labor intensive. As Heinich (1984) pointed out, this is reflected most clearly in the tendency of teachers' unions to protect jobs by opposing policies that might reduce the labor intensiveness of teaching (pp. 77–78). Third, most elementary and secondary schools in the United States are public institutions operated by local districts and funded largely by state appropriations. They have had, to a great extent, a monopoly position. There have been few competitors (nonpublic schools) within their local area and fewer from beyond. For most "customers," the only way to exercise choice is to physically uproot and relocate the whole family to a new location. So competitive pressure is largely lacking—or at least it has been in the past. Virtual schools may be changing the competitive environment.

Virtual schools. Distance education approaches first developed in higher education are now appearing at the elementary/secondary level in the form of virtual schools. For-profit ventures offer online courses aimed primarily at home schooling households. This puts competitive pressure on the public schools, which need to maintain their daily attendance rates in order to continue receiving state per-pupil allocations. Thus, public schools are pushing to implement online distance education programs. Online delivery is also an answer for hard-to-serve students, such as full-time workers, pregnant and young mothers, disciplinary force-outs, students with health problems, and others who are not well served by the regular schools.

Thus educational technology may help improve the organizational performance of schools by providing the communications capability (hard technology) and the courseware design (soft technology) to allow schools to expand their reach to changing audiences.

Technology in higher education. In higher education this issue has risen in visibility as distance education has migrated to an Internet-based platform. Educational institutions are able to reach distant audiences at little additional cost, compared with the costs of residential or television-based instruction. Many potential "customers" for higher education view educational services as a commodity that can be purchased from any one of many vendors, regardless of location. This is particularly true for nontraditional college students—adults with families and jobs. For such students, residential education involves many indirect costs—in terms of time, money, and aggravation—that can be avoided by working toward a degree online. This is not to say that the online option is necessarily superior in other ways, only that it can reduce cost and increase convenience. Experience to date suggests that it requires an exceptional degree of commitment for students to complete a program at a distance. In a relatively short period of time, a host of new distance education institutions, many of them for-profit, have sprung up and taken root. The largest, University of Phoenix, has become the largest private university in the United States, with over 200,000 students in its online and face-to-face courses. Although residential campuses still offer unique advantages and a ready supply of students, the competitive heat is rising.

It may not be competition, strictly speaking, that is driving interest in technology in higher education. Rather, administrators now have a concrete image of an alternative approach to education. They see that distance education institutions are able to offer education at much lower prices because of the way they are employing technology. Interestingly, it is not hard technology that gives such distance institutions an advantage (residential institutions have lots of hard technology, too) but rather soft technology. This was articulated clearly by Sir John Daniel, then Vice-Chancellor of the British Open University:

> The most important thing to understand about using distance education for university-level teaching and learning that is both intellectually powerful and competitively cost-effective is that you must concentrate on getting the soft technologies right. . . . These soft technologies are simply the working practices that underpin the rest of today's modern industrial and service economy: *division of labour, specialisation, teamwork and project management* [italics added]. (Daniel, 1999)

Division of labor and specialization refer to "unbundling" the various functions performed by instructors: instructional designer, developer, subject-matter expert, lecturer, discussion leader, evaluator, remediator, and adviser. By forming a team of specialists in these different functions each job can be done more expertly, a course can be designed, and the team can move on to the next course, thus industrializing the process. A well-designed course can be largely self-instructional, leaving the tutorial function to low-paid paraprofessionals working the phones in a cubicle somewhere. So far, this soft technology approach has been confined mainly to distance only operations, but administrators at traditional universities are taking note. There are examples of this approach being applied at traditional universities. One notable case is the

Math Emporium at Virginia Tech University (http://www.emporium.vt.edu), a large computer center encompassing a dozen core mathematics courses, all of which are available on demand in a self-instructional format.

Group Learning by Organizations. Argyris (1977) drew attention to the problem of people's ignoring or hiding errors in organizations. He proposed and later elaborated (Argyris & Schön, 1978) a distinction between single-loop learning—the detection of error in a particular case—and double-loop learning—when errors are detected and corrected in ways that alter the organization's future capabilities. Senge (1990) extended the concept of double-loop learning further, to generative learning—a posture of ongoing experimentation and feedback, critically examining the organization's actions and policies. The idea underlying these concepts is that organizations themselves can learn, that is, they can become smarter in dealing with the challenges they face.

If organizations do not actually have brains, how can they learn? Popper and Lipshitz (2000) propose that organizations can build organizational learning mechanisms (OLM), "institutionalized structural and procedural arrangements that allow organizations to learn non-vicariously, that is, to collect, analyse, store, disseminate, and use systematically information that is relevant to their and their members' performance" (p. 185).

Technology, both hard and soft, can contribute significantly to building OLMs. ICT can provide powerful means for storing, retrieving, and sharing knowledge. Audio and video conferences, Internet discussion forums, and groupware such as Lotus Notes enable a dynamic and growing organizational memory. Of course, the hard technology only works effectively when it is combined with the soft technologies of man-made policies and practices in a synergistic whole (Goodman & Darr, 1998).

The ultimate goal, proposed by Senge (1990) is the evolution of *learning organizations*—schools, colleges, and businesses "in which you cannot *not* learn because learning is so insinuated into the fabric of life" (p. 9). Learning organizations would be ideal environments for both individual learning *in* organizations and group learning *by* organizations.

A Systems Perspective on Organizational Performance

A powerful way to visualize the influences of technology within organizations is to adopt a systems view. Organizations of all types can be viewed as complex enterprises of interconnected parts that in ideal circumstances work in harmony to effectively convert numerous types of inputs to valued outputs: valued in the sense that individuals and other organizations are willing to use or support them. People are central to organizations. They work alone and in teams to create a work environment and culture that enables them to contribute to the generation of valued goods and services. The effectiveness of an organization as a whole depends to a great extent upon the effectiveness of the work that people perform individually and in teams as members of the organization's component parts.

Moreover, organizations do not exist in vacuums. They exist within a larger environment, or suprasystem, that places pressures, constraints, and expectations upon it. Other organizations provide its inputs and consume its outputs. The marketplace, natural forces, and governments regulate both directly and indirectly an organization's inputs, processes, and outputs. These forces, external to the organization, constitute its environment. An effective organization, through ongoing feedback from its external environment and back-and-forth feedback among its internal parts, continually calibrates and adjusts its inputs, processes, and outputs to achieve its overall goals and objectives in timely and cost effective ways.

Organizations, as complex systems, behave systemically. The parts are not independent or freestanding. As such, interventions must look beyond simple cause-and-effect relationships and recognize that a cause and its effect cannot be isolated or separated from its context. Systemic problem solving is a matter of holism over reductionism (Douglas & Wykowski, 1999; Hallbom & Hallbom, 2005).

Systems theory has been a key theory in educational technology since the 1960s, particularly through the early work of Bela Banathy (1968). It rose to greater prominence in the 1980s and 1990s as more and more American educators publicly acknowledged the need for systemic change. These calls ultimately led to the creation of the New American Schools Development Corporation (NASDC) as part of a national government initiative to develop new, whole-school designs for American schools, which functioned from 1992 through 1995.

The essence of the systems view is to step back and note the factors that surround and influence events in the classroom. Only by first seeing the classroom in its larger context can one restructure the environment to be more supportive of more powerful instructional strategies. The model shown in Fig. 3.1 is intended to provide this systemic perspective. The elements of the model and the interconnections among them are based on generalizations gleaned from meta-analyses of the educational research, especially those reported by Walberg (1984).

Direct Influences on Learning. The core of the model shows three influences that *directly* affect student academic learning. They are derived primarily from Walberg's (1984) overall conclusion that "the major causal influences flow from aptitudes, instruction, and the psychological environment to learning" (p. 21). The direct influences are:

- Aptitude—relatively permanent psychological traits, including intelligence(s), maturation level, personality, and "learning style" (which has been defined in many different ways)
- Effort—often characterized as amount of invested mental effort (AIME) or how hard the learner is working on learning tasks
- Instruction—the amount and quality of teaching-learning activities in which the learner is involved

The relative importance of these three factors is hotly debated among educators, under the rubric of the "nature-nurture" debate. Some psychologists have proposed that up to 90% of the variability in learning stems from aptitude factors; most would agree that aptitude is responsible for at least half of the variability. Effort may be the next most important. There is ample evidence that if students have high aptitude and/or motivation to invest a lot of mental effort, almost any instructional treatment will succeed.

However, to the extent that learners have lower aptitude or are less highly motivated, better designed instruction and longer engagement in it can improve the amount learned, retained, and applied.

Second-Level Influences on Learning.   Many of the forces that consistently show a causal relationship to learning actually impact learners indirectly, that is, they affect aptitude, effort, or instruction rather than affecting learning directly. As shown in Fig. 3.1, effort is especially affected by second-level influences. First, effort depends on the learner's psychological state, especially the motivations and expectations that are salient at the time of instruction. Second, effort can be affected by peer influences. Third, the media and methods selected in the instructional process can arouse effort.

Walberg found two aspects of instruction to be critical—time on task and the "quality" of the educational experience, which is represented by method and media in the diagram. The combination of methods and media provide the structure of the learning environment as well as the teaching-learning activities that are employed.

Walberg (1984) identified the classroom social setting as an important influence, defining it as "the cohesiveness, satisfaction, goal direction, and related social-psychological properties or climate of the classroom group perceived by students" (p. 24). This is indicated in Fig. 3.1 by the dotted line encompassing the classroom environment. Given the right climate, teachers are more likely to offer higher quality instruction and students are more likely to feel motivated to invest effort and activate their innate aptitudes.

Peer influences can act both inside and outside the classroom, hence this element is shown as straddling the boundary of the classroom in the diagram.

Third-Level Influences on Learning. Some of the other factors identified by Walberg (1984) as critical are represented in the diagram as third-level influences; that is, they do not influence learning directly, but indirectly, through some of the second-level forces. Chief among the third-level influences is home and family. This category includes a number of factors deemed very important by Walberg:

- A good home environment increases supervised homework and reduces the time spent watching television (p. 24). Since the time of Walberg's analysis recreational uses of the computer may be displacing television as the chief competitor for children's attention.
- The "curriculum of the home" promotes achievement in several ways, through informed parent-child conversations about school, encouraging leisure reading, deferring immediate gratifications in favor of longer-term goals, expressions of affection and interest in the child's activities, and other intangible psychological supports. Taken together, the home and family environment "is twice as predictive of academic learning as socio-economic status" (p. 25).

Mass media play a third-level role also, in that they help create a culture (just as they are also shaped by the culture) that may support or inhibit healthy psychological states, including motivation and expectations. They have an influence on peer groups' attitudes toward school also.

Surrounding all these influences—home and family, classroom, school, mass media, and peers—is the overall socio/cultural/political environment, both local and national. Within the United States, there are many subcultures, each of which exerts different influences on the forces within it, ultimately promoting or undermining the forces that affect academic achievement.

Only through a systemic lens of this sort can educators fully understand the interplay of the forces that actually impact the quality of learning. If schools or other organizations are to become learning communities, they must incorporate structures and policies that will be supportive of, not hostile to, the goal of facilitating learning. Educational technology, by nature devoted to a systemic view of problem situations, helps organizations improve performance by identifying the elements of the system, understanding the linkages among those elements, and treating root causes rather than mere symptoms.

Improving Performance of Organizations: Beyond Learning

Organizations can promote the productivity of the people within them by helping them gain new knowledge, skills, and attitudes, but they can also promote productivity by changing the conditions within the organization so that people can accomplish more, with or without additional instruction. For example, they can provide people with better tools, give them better working conditions, motivate them better, or provide job aids. Noninstructional interventions are often pursued under the label of *performance improvement* or *human performance improvement*. Those that entail changes in organizational structure are commonly seen as *organizational development* efforts. All of these would fall outside the field of educational technology. Those who advocate a systemic approach toward the total process of instructional and noninstructional performance improvement prefer the label of *HPT*.

Human Performance Technology (HPT)

Evolving since the 1970s as a separate field, HPT embraces the viewpoint that organizational effectiveness can be advanced by employing a wide range of interventions, including, but not limited to, instruction. Deficiencies in performance may be caused partly by ignorance, but more often there are problems of motivating people or giving them the tools needed to do the job, or even selecting people who are better suited to the demands of the job.

Therefore HPT pursues "... the systematic and systemic identification and removal of barriers to individual and organizational performance" (International Society for Performance Improvement, 2005). As a concept and field of practice it is comparable to educational technology. Like many instructional designers, performance technologists advocate systematic processes of analysis, selection, design, development, implementation, and evaluation to cost effectively influence human behavior and accomplishment (Harless, as cited in Geis, 1986). The difference is that performance technologists consider instruction to be just one of many possible interventions to improve performance in the workplace. This viewpoint was summarized in Pershing's (2006) definition of HPT as "the study and ethical practice of improving productivity in organizations by designing and developing effective interventions that are results-oriented, comprehensive, and systemic" (p. 6).

The systematic ID approach and the HPT approach are quite compatible with each other. A visual model that shows how the two concepts dovetail is shown in Fig. 3.2.

The strategic impact model (Molenda & Pershing, 2004) begins by emphasizing strategic alignment, showing how needs of the organization are

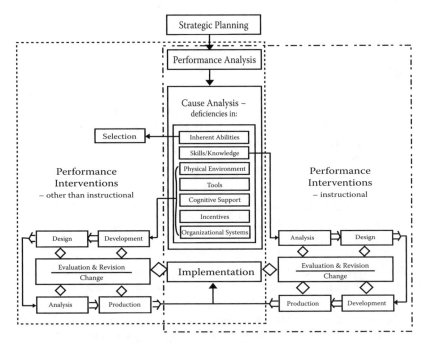

Figure 3.2. The Strategic Impact Model, showing the connection between instructional interventions (right side of model) and non-instructional performance interventions (left side of model). © J. Pershing & M. Molenda, 2003. Used with permission.

derived through strategic planning. Then performance analysis determines where there are deficiencies in the organization. Next, these deficiencies are examined as to their causes (cause analysis). Ignorance, or lack of skill/ knowledge, is only one of the possible classes of performance deficiencies, so instruction is only one of several possible solutions.

The steps in solving instructional problems are shown on the right side of the model. Other causes of deficiencies—low motivation, poor working conditions, lack of information, and poor organizational structure—can be addressed by other sorts of interventions, shown on the left side of the model.

All the interventions needed in a given case will pass through processes of analysis, design, development, and production (with evaluation and revision accompanying each of those stages) before they are brought together in a coordinated implementation. The model also represents the requirement of change management at each step along the way in order to increase the odds that the interventions will be accepted by the people in the system and incorporated into the organizational culture.

Summary

Educational technology can claim to improve the performance of individual learners, of teachers and designers, and of organizations as a whole.

To begin with, the educational experience is more likely to lead to improved performance because the instructional design doctrine of educational technology advocates the selection of objectives that fully represent the types and levels of capability to be learned. Further, educational technology has a commitment to promoting "deep learning," learning that is based on rich experience and that can be applied in real world contexts. Transfer of learning is promoted by learner immersion in microworlds, virtual environments in which learners have the opportunity to experience the consequences of decisions. In the corporate setting, the systems approach recommends activities before, during, and after training that make it more likely that workers will use their new skills on the job.

Teachers and instructional designers' performance is improved by the systems approach, which helps focus on high-value objectives, weeding out irrelevancies, thus reducing instructional time, which conserves the resources of educators. Systematic development processes also tend to yield more effective learning results, further enhancing productivity. Educational technologists are also sensitive to the need to make instruction appealing and humane. The innovations they advocate, from programmed instruction to constructivist learning environments have been tools to free learners from passive, lock-step teaching, to provide more exciting and involving learning experiences.

Productivity has been declining in the education sector. To improve productivity requires defining and improving both efficiency and effectiveness. Technology has the potential to improve both efficiency and effectiveness. ICT can reduce the time and cost of distributing materials as well as all sorts of administrative tasks. Soft technologies, especially modern work processes, can help improve organizational performance by unbundling the many functions associated with instruction and reorganizing those functions more rationally. Distance education universities have achieved enormous economies of scale this way, and some traditional universities have restructured programs to make them more learner centered and more efficient. To accomplish this restructuring, a systemic view is necessary, a view that is synonymous with educational technology.

Beyond improving learning, organizations can solve people problems that are larger than just those of lack of knowledge or skills. The umbrella of HPT provides a framework for combining instructional interventions with motivational, ergonomic, environmental, organizational, and other interventions into coordinated initiatives that can dramatically improve productivity.

References

American Management Association. (n.d.). *Communication and interpersonal skills.* Retrieved October 18, 2006, from http://www.amanet.org/seminars/category.cfm?cat=204

Anderson, L. W., & Krathwohl, D. R. (Eds.). (2001). *A taxonomy for learning, teaching, and assessing: A revision of Bloom's taxonomy of educational objectives.* New York: Longman.

Argyris, C. (1977). Double loop learning in organizations. *Harvard Business Review, 55*(5), 115–125.

Argyris, C., & Schön, D. (1978). *Organizational learning.* Reading, MA: Addison-Wesley.

Baldwin, T. T., & Ford, J. K. (1988). Transfer of training: A review and directions for future research. *Personnel Journal, 41,* 63–105.

Banathy, B. (1968). *Instructional systems.* Palo Alto, CA: Fearon.

Bloom, B. S., Englehart, M. D., Furst, E. J., Hill, W. H., & Krathwohl, D. R. (1956). *Taxonomy of educational objectives. Handbook I: Cognitive domain.* New York: Longmans, Green.

Bosworth, B. (2005). Productivity in education and the growing gap with service industries. In M. Devlin, R. C. Larson, & J. W. Meyerson (Eds.), *The Internet and the university: Forum 2004.* Boulder, CO: EDUCAUSE.

Bransford, J. D., Brown, A. L., & Cocking, R. R. (Eds.). (1999). *How people learn: Brain, mind, experience, and school.* Washington, DC: National Academy Press.

Brinkerhoff, R. O., & Dressler, D. E. (1990). *Productivity measurement: A guide for managers and evaluators.* Thousand Oaks, CA: Sage.

Broad, M. L., & Newstrom, J. W. (1992). *Transfer of training: Action-packed strategies to ensure high payoff from training investments.* Reading, MA: Addison-Wesley.

Butzin, S. M. (2005). *Joyful classrooms in an age of accountability.* Bloomington, IN: Phi Delta Kappa.

Callahan, R. E. (1962). *Education and the cult of efficiency.* Chicago: University of Chicago Press.

Center for the Study of Problem Solving (CSPS). (n.d.) University of Missouri at Columbia: http//csps.missouri.edu/pastprojects.php. Retrieved October 20, 2006.

Comenius, J. A. (1967). *Orbis sensualium pictus: Facsimile of the 3rd London edition 1672 with an introduction by James Bowen*. Sydney, Australia: Sydney University Press. (Original work published 1657)

Csikszentmihalyi, M. (1988). Motivation and creativity: Towards a synthesis of structural and energistic approaches to cognition. *New Ideas in Psychology, 6*, 159–176.

Dale, E. (1946). *Audio-visual methods in teaching*. New York: Dryden Press.

Daniel, J. (1999, April). *Technology is the answer: What was the question?* Paper presented at TechEd99, Ontario, Canada.

Douglas, N., & Wykowski, T. (1999). *Beyond reductionism: Gateways for learning and change*. Boca Raton, FL: St. Lucie Press.

Ellson, D. G. (1986). *Improving the productivity of teaching: 125 exhibits*. Bloomington, IN: Phi Delta Kappa.

Florida TaxWatch. (2005, March 2). *Education innovation creating powerful results: Program bringing higher achievement with lower costs*. Tallahassee, FL: Author. Retrieved May 16, 2006, from http://www.ifsi.org/Press%20release%20Taxwatch%203-2-05.htm

Gardner, H., & Hatch, T. (1989). Multiple intelligences go to school: Educational implications of the theory of multiple intelligences. *Educational Researcher, 18*(8), 4–9.

Geis, G. L. (1986). Human performance technology: An overview. In M. E. Smith (Ed.), *Introduction to performance technology* (Vol. 2). Washington, DC: National Society for Performance and Instruction.

Goodman, P. S., & Darr, E. D. (1998, November/December). Computer-aided systems and communities: Mechanisms for organizational learning in distributed environments. *MIS Quarterly*, 417–422.

Hallbom, T., & Hallbom, K. J. (2005). *The systemic nature of the mind and body and how it relates to health*. Retrieved May 26, 2005, from http://www.nlpca.com/articles/article2.htm

Hanushek, E. A. (1986, September). The economics of schooling: Production and efficiency in public schools. *Journal of Economic Literature, 24*, 1141–1177.

Heilbroner, R., & Thurow, L. (1998). *Economics explained: Everything you need to know about how the economy works and where it's going* (4th ed.) New York: Touchstone.

Heinich, R. (1984). The proper study of instructional technology. *Educational Communication and Technology Journal, 32*(2), 67–87.

Heinich, R., Molenda, M., & Russell, J. D. (1985). *Instructional media and the new technologies of instruction* (2nd ed.). New York: John Wiley & Sons.

International Society for Performance Improvement. (2005). *What is human performance technology?* Retrieved October 2, 2005, from http://www.ispi.org

Keller, J. M. (1987). Development and use of the ARCS model of instructional design. *Journal of instructional development, 10*(3), 2–10.

Kirkpatrick, D. L. (1998). *Evaluating training programs: The four levels* (2nd ed.). San Francisco: Berrett-Koehler.

Krathwohl, D. R., Bloom, B. S., & Masia, B. B. (1964). *Taxonomy of educational objectives. The classification of educational goals. Handbook II: Affective domain.* New York: David McKay Co.

Kulik, J. A., Kulik, C. C., & Smith, B. B. (1976). Research on the personalized system of instruction. *Programmed Learning and Educational Technology, 13*, 23–30.

Levin, H. M., Glass, G. V., & Meister, G. R. (1984). *Cost effectiveness of four educational interventions* (NIE Project Report No. 84-AS11). Stanford, CA: Institute for Research on Educational Finance and Governance.

Mager, R. F. (1962). *Preparing objectives for programmed instruction.* San Francisco: Fearon Publishers.

Mager, R. F. (1977, October). The 'winds of change.' *Training and Development Journal, 12*–20.

McConnell, C. R., & Brue, S. L. (2002). *Economics: Principles, problems, and policies* (15th ed.). New York: McGraw-Hill College.

Molenda, M., & Pershing, J. A. (2004, March/April). The strategic impact model: An integrative approach to performance improvement and instructional systems design. *TechTrends, 48*(2), 26–32.

Monk, D. H. (2003). Efficiency in education. In J. W. Guthrie (Ed.), *Encyclopedia of education* (2nd ed., pp. 700–704). New York: Macmillan Reference.

Morrison, G. R., Ross, S. M., & Kemp, J. E. (2004). *Designing effective instruction* (4th ed.). Hoboken, NJ: John Wiley & Sons.

Nas, T. F. (1996). *Cost-benefit analysis: Theory and application.* Thousand Oaks, CA: Sage.

Nichols, S. L., & Berliner, D. C. (2005, March). *The inevitable corruption of indicators and educators through high-stakes testing.* Tempe, AZ: Education Policies Laboratory, Arizona State University.

Pedulla, J. J., Abrams, L. M., Madaus, G. F., Russell, M. K., Ramos, M. A., & Miao, J. (2003, March). *Perceived effects of state-mandated testing programs on teaching and learning: Findings from a national survey of teachers.* Boston: National

Board on Educational Testing, Boston College. Retrieved October 1, 2005, from http://www.bc.edu/research/nbetpp/statements/nbr2.pdf

Pershing, J. A. (2006). Human performance technology fundamentals. In J. A. Pershing (Ed.), *The handbook of human performance technology: Principles, practices, and potential* (3rd ed., pp. 5–34). San Francisco: Pfeiffer.

Popper, M., & Lipshitz, R. (2000). Organizational learning: Mechanisms, culture, and feasibility. *Management Learning, 31*(2), 181–196.

Reigeluth, C. M. (Ed.). (1983). *Instructional-design theories and models.* Mahwah, NJ: Lawrence Erlbaum Associates.

Rieber, L. P., Smith, L., & Noah, D. (1998). The value of serious play. *Educational Technology, 38*(6), 29–37.

Romiszowski, A. J. (1981). *Designing instructional systems: Decision making in course planning and curriculum design.* New York: Nichols.

Schiefele, U. (1991). Interest, learning, and motivation. *Educational Psychologist, 26*(3), 299–323.

Schwartz, A. E., & Stiefel, L. (2001). Measuring school efficiency: Lessons from economics, implications for practice. In D. H. Monk, H. J. Walberg, & M. Wang (Eds.), *Improving educational productivity* (pp. 115–137). Greenwich, CT: Information Age Publishing.

Senge, P. M. (1990). *The fifth discipline: the art and practice of the learning organization.* New York: Doubleday/Currency.

Simpson, E. J. (1972). The classification of educational objectives in the psychomotor domain. In *The psychomotor domain: A resource book for media specialists* (pp. 43–56). Washington, DC: Gryphon House.

Skinner, B. F. (1968). Why teachers fail. In B. F. Skinner, *The technology of teaching* (pp. 93–113). New York: Appleton-Century-Crofts.

Sugrue, B. (2003). *State of the industry.* Alexandria, VA: American Society for Training and Development.

Twigg, C. A. (1999). *Improving learning and reducing costs: Redesigning large-enrollment courses.* Troy, NY: Center for Academic Transformation, Rensselaer Polytechnic Institute.

Walberg, H. J. (1984, May). Improving the productivity of America's schools. *Educational Leadership, 41*(8), 19–27.

Weigel, V. B. (2002). *Deep learning for a digital age: Technology's untapped potential to enrich higher education.* San Francisco: Jossey-Bass.

4

CREATING

Michael Molenda
Elizabeth Boling
Indiana University

Introduction

Educational technology is the study and ethical practice of facilitating learning and improving performance by *creating*, using, and managing appropriate technological processes and resources.

*T*HERE ARE NO PROCESSES or resources to use or manage unless someone first creates them. This chapter deals with the activities and theories related to the whole complex process involved in *creating* instructional materials, learning environments, and larger teaching-learning systems.

The previous AECT definition (Seels & Richey, 1994) used the terms *design*, *development*, and *evaluation* to refer to the function of creating resources for learning. In addition, the term *production* refers to the application of creative arts and crafts to generate the actual materials used by learners. The current definition avoids those terms in order to reserve them to be used as technical terms to describe certain steps in the larger process of creation. People have generated successful teaching-learning resources without consciously engaging in formal "design," "development," or "evaluation" activities. These terms tend to be associated with a particular approach—the systems approach. But design methodologies spring from many different approaches: aesthetic, scientific, engineering, psychological, procedural, or systemic, each of which can be employed to generate effective materials and conditions for learning.

It is the intent of this chapter to discuss many different ways of creating many different types of materials and systems for learning. The first half of the chapter shows how the meanings and the methods of creating have evolved as the spotlight has moved from one media form to another throughout the modern history of the field, with the different media bringing different research issues and theories into the field. The second half of the chapter deals with the "big ideas," including message design principles and instructional design models, that underlie the process of creating instructional media. The chapter concludes with a look at several contemporary issues related to creating.

Evolution of Practices and Theories for Creating

Overview: New Media Trigger Paradigm Shifts

The field that would become educational technology began as *visual education*, as educators explored the potentials of motion pictures and projected slides at the turn of the 20th century. As radio, sound film, and recorded audio developed, the field evolved into *audio*visual (AV) education around midcentury. Television in the 1950s added the new dimension of widespread broadcasting of AV programming. During this period, the field's design and production focus was on the creation of presentations that were attractive to the eye and ear. Educational film, radio, and television directors relied on creative imagination to capture the "wow" factor that viewers had come to expect from commercial versions of film, radio, and television.

The first great paradigm shift occurred in the 1950s and 1960s, as the new psychological technologies spawned by applied behaviorism—behavior management, teaching machines, and programmed instruction—confronted the AV paradigm. The focus shifted to what learners were *doing*, rather than what they were watching, and so the focus of design and production shifted from making AV presentations to creating learning environments in which learners had the opportunity to practice new skills under conditions of constant feedback. The field's new name, *educational technology*, reflected both the era's new hard technologies—teaching machines and AV hardware—and its soft technologies—the theory-guided programming inside the machine.

The second great paradigm shift occurred after the birth of microcomputers in the early 1980s. Similar to the programmed instruction movement of the 1960s, the information technology movement of the 1980s brought a whole new set of people, with a different mindset, into the domain of educational technology. Computer-assisted instruction (CAI) became the dominant paradigm.

Computer capabilities became networked through the Internet in the early 1990s, greatly magnifying the potential educational value of computers. Then in 1993 Mosaic's graphical user interface (GUI) and later Web browser software allowed the World Wide Web to become by far the most popular Internet protocol. Use of the Web grew exponentially for the rest of the decade. Because the Web made it easy for individuals to think and work collaboratively, and because it allowed anyone, anywhere to access interesting computer-based exploratory environments (e.g., simulations and games), traditional design processes came under challenge. Designers now were trying to design *experiences*, not just materials, and their tools came from the realms of computer programming and cognitive science. By the early 21st century the field was into its third paradigm shift—from CAI to Web-based learning environments—and facing the possibility of a fourth, ubiquitous learning through mobile media.

In the following sections, we will trace the evolution of the practices that constituted *creating* and the ideas that shaped practice as the spotlight moved from film, to radio and television, to AV materials, to programmed instruction, to CAI, to digital media, to Internet and Web-based learning, and to blended learning and mobile media.

Educational Film

The origins and early use of silent films in education in the 1910s and 1920s is discussed in chapter 8. During the 1930s, sound films competed with silent films but did not really become standard until after World War II.

Creating Educational Films. In the 1920s and 1930s, the creation of educational films was not explicitly guided by pedagogical theories or instructional design methodologies. Producers tended to choose subjects that were visual in nature, then to apply the methodology of one of the existing film genres—drama, travelogue, documentary, ethnography, historical reenactment, nature study, scientific experiment or demonstration, lecture, procedural guide, and the like—depending on what genre fit the subject matter. To begin the planning process, the overall approach to the film was described in a concise document known as a *treatment*, which could be considered an early version of rapid prototyping. The various gatekeepers on the project could decide at this point whether the approach was on target and within budget. Changes could be made at the treatment stage before time and funds were expended on production.

Subject matter experts and teaching or training specialists served as educational consultants, often sitting with the production staff in production meetings, discussing the content and the filmic techniques to be used prior to the development of a full *script*, the next major step in the planning process. Typically, they checked and approved the scripts as they evolved over time, a precursor of one aspect of formative evaluation. Once a script was in place, it was possible to arrange the actual shooting of scenes. Sometimes, especially for educational projects, a *storyboard* was created to allow discussion of and production of visual effects. After shooting came the editing of the scenes into a finished narrative or presentation.

During World War II, the need for "rapid mass training" of literally millions of combatants and industrial workers brought films to the forefront of military training. In the United States between 1941 and 1945, the Division of Visual Aids for Military Training produced over 400 sound films and over 400 silent filmstrips (Saettler, 1990, p. 181). The participation of Hollywood directors and actors lent an artistic and professional patina to these training films, but pedagogical design was slower in coming.

Underlying Research and Theory

Gestalt and cognitive theory.   During the war, as films were being produced and used in training, the U.S. Army commissioned a series of psychological studies, later published as *Experiments on Mass Communication* (Hovland, Lumsdaine, & Sheffield, 1949), which rigorously tested hypotheses about various filmic techniques and their instructional effectiveness. The hypotheses mainly revolved around the issues prominent in the Gestalt and cognitive psychology of that time: providing an introduction to provide a mental set for comprehending and remembering the film's message; pacing the presentation to suit the audience's cognitive ability; choosing words and images to illustrate points as clearly as possible; controlling the density of visual and aural messages for comprehension; avoiding distracting cues; and using repetition and summaries to enhance retention. Because of the concentration of time, money, effort, and research expended on these productions, a genre of *instructional film* came into its own. New filmic conventions were established, for example, showing procedural tasks from the performer's viewpoint rather than the viewer's and using a first-person stream-of-consciousness narration to model the thought process of the performer.

After the war, this line of research continued under U.S. Navy sponsorship at Pennsylvania State University, a research program known as "the Penn State studies" that yielded over a hundred publications (Hoban & Van

Ormer, 1970). Some of the experiments dealt with utilization techniques, but many explored presentation variables, such as camera angles, pacing, narration, music, and color (Saettler, 1990, p. 246).

Behaviorist theory. The U.S. Air Force also commissioned a series of studies in the early 1950s; these explored the possible interactions between film and programmed instruction techniques—examining the value of learner response during films and other types of instruction. In its later years, the Penn State team also turned to studying the potential of combining film or video with behaviorist principles. Some of their experimental lessons looked like a programmed instruction lesson filmed and projected on a screen, with the audience asked to watch presentations of information, then hear or read questions about the content, to which they responded by writing an answer on a worksheet or silently thinking the answer before being told the correct response.

Curriculum theory. In the postwar period, many companies competed to provide educational films for the school market. A development that was emblematic of their approach was the decision of the McGraw-Hill Book Company in 1947 to prepare a series of "textfilms." The explicit purpose of these films was to supplement the textbook by providing special visual materials that could not be duplicated in textbooks or in teachers' lectures (Saettler, 1990, p. 115). The textbook and films were accompanied by filmstrips and teachers' guides, which suggested how teachers could integrate all these materials into a coherent lesson plan. From that point onward, films and videos were designed mainly as supplementary materials rather than as replacements for traditional materials.

Although the formal research programs did not necessarily have a large practical impact on the design of educational films, they did bring new theoretical frameworks and vocabulary to the discourse about the creation of educational films, from psychological theories of perception, cognition, and operant conditioning.

Educational Radio and Television

As described in chapter 8, educational radio stations proliferated in the 1920s and 1930s. The first programs for schools in the United Kingdom were broadcast by the BBC in 1926. By the 1930s, radio programs tailored for school use were being broadcast by a number of city, state, and provincial authorities in the United States and Canada (as well as by the Canadian National Railways, CNR, system). Programs were produced on a broad range of subjects, from science and social studies to music and art.

Creating Educational Radio and Television. The programming tended to be "informally educative" (Levenson & Stasheff, 1952) rather than directly instructional. Both radio and television services had difficulty carving out a distinctly instructional role, and hence tended to play a peripheral role in schools and colleges. For one thing, the advantage of broadcasting is its coverage of a broad area, but that meant crossing school district and even state and provincial boundaries. It is difficult to create any lesson that would meet the content, scope, sequence, and timing demands of multiple school systems. For another thing, teachers, the gatekeepers of the classroom, were reluctant to turn over responsibility for core subject matter, sensing that it would threaten their authority.

After the popularization of video tape recording, later video cassette recording, educational television programming was increasingly created and used as off-the-shelf packaged units rather than being received through broadcasting. One of the leading creators and distributors of recorded television programming was the Agency for Instructional Technology (AIT), beginning in 1962 as National Center for School and College Television. During the 1970s and 1980s, AIT became a major producer of instructional television series, many of them award-winning, for the K–12 sector, developing an innovative consortium process for pooling the resources of state education departments that bought into projects on a case-by-case and step-by-step basis. After the demand for production of new series declined, AIT continued to be a major distributor of instructional television programs in cassette, CD, and DVD formats.

In business and industry, broadcast radio and television were not used as such, but after the popularization of videocassettes in the 1970s, many companies chose this format as a training tool. Until the late 1990s, nearly 70% of all U.S. companies used video recordings to some extent, either purchased off the shelf for generic objectives or locally produced for topics specific to the company (Bichelmeyer & Molenda, 2006, p. 7).

Radio and television had a lot in common in terms of design and production. They operated on the paradigm of the script, as with film, to create a self-contained package of information, usually intended to be communicated one way. As with educational films, radio and television programs tended to emulate the familiar genres: lecture, demonstration, voice-over visualization, interview, panel discussion, dramatization, field trip, or documentary (Wood & Wylie, 1977, p. 259). The production processes were comparable to those used in commercial radio and television: "We borrowed from commercial television certain ideas about what constitutes a program, and we have not shaken free from these concepts" (Suchman, 1966, p. 30). Generally speaking, the people who created educational programs had backgrounds in commercial radio and television. No other special expertise was considered necessary.

Underlying Research and Theory

Reflective practice. There was little attention to psychological research or theory on radio/TV program production until after World War II. However, there were some exemplary practices that evolved through reflective practice. For example, the CNR schools radio producers discovered in the 1920s that the incorporation of active audience participation vastly improved program usage (Buck, 2006). And in the 1930s, Cleveland schools radio programs were produced with questions, pauses for audience response, and answers to the questions. Such pseudointeractive radio programs were reinvented to teach mathematics and English in several countries in Latin America and Africa in recent years (Heinich, Molenda, & Russell, 1993).

At the Cleveland, Ohio Board of Education's radio station, WBOE, in the 1930s, they pretested programs by creating rough drafts and trying them out with student audiences. This practice foreshadowed the later notion of improving artifacts and validating their worth through formative and summative evaluation (Cambre, 1981).

Communication theory. During the later days of educational radio and the earlier days of educational television, communication theory was a dominant paradigm both in the physical and social sciences. Flowing from Shannon and Weaver's (Shannon, 1949) *information theory*, through Wiener's (1950) *cybernetics* and Berlo's (1960) *process of communication*, thinkers in educational technology were viewing teaching-learning problems as communication problems. The key variables were the natures, abilities, and intentions of senders and receivers; the capacities of different communication channels; the structure and content of the messages sent; the sorts of noise encountered in communicating; and the quality of feedback exchanged between receiver and sender. Improvement of communication depended on detecting where the weak points in the process were and ameliorating them—choosing a more visual medium, building more redundancy into the message, matching the receiver's language capability better, providing the sender with better feedback about the receiver's response, and the like.

This conceptual framework fit quite well with the producer's viewpoint because it addressed issues that were within producers' span of control. They were in a good position to think about the audience's needs and interests, to select the content and shape it into a message, and to choose a delivery system. Educators who used learning resources were not quite as satisfied with the communication paradigm because they were aware of the importance of what learners *did* with the messages after they were received. They saw communication as only one step in the process of instruction.

Research on presentation variables. By the time that intensive studies were being done on presentation variables after World War II, interest had shifted from radio and film to television. So, the guiding principles discovered through research were applied primarily to the production of broadcast television programs or shorter videotaped sequences. In addition to the military sponsored research, a great deal of university research was conducted, sparked by an infusion of federal grant money under Title VII of the National Defense Education Act of 1958. One of the most ambitious attempts to summarize this body of research was *Learning from Television* (Chu & Schramm, 1968). Only a small proportion of the studies cited in the monograph deal with "pedagogical variables" related to design and production, including such issues as humor, dramatic versus expository presentation, questions with pauses, problem-solving techniques, and lecture versus discussion format (pp. 28–37). Other chapters deal with learning from television in general, television in the context of the classroom, physical variables (e.g., screen size, viewing angles), utilization practices, attitudes toward instructional television, and lessons learned in developing countries.

The questions raised in many of the studies of this period were inspired by the practical concerns of production staff rather than by psychological or pedagogical theories. However, two streams of instructional theory stimulated considerable experimentation: programmed instruction and discovery learning.

Research on learner response. The Penn State studies and the Air Force studies discussed earlier attempted to create and test film and video materials that embodied programmed instruction features. Other research done with school and college audiences studied such issues as overt versus covert practice and the effect of knowledge of results, and found that it could work: "It has been established that television can be used in the 'lock-step' regulation of linear (Skinnerian) programming for groups of students. The television system submits cue frames, students make responses on printed answer sheets, after which the system provides knowledge of results" (*Television in Instruction*, 1970, p. 9).

For the most part, the findings were impractical to implement in a mass-media setting. The whole point of programmed instruction was to escape the whole-class arrangement and allow individuals to learn at their own pace, while the economics of broadcasting demanded large audiences over which to spread the considerable production costs.

Interestingly, the case of pseudointeractive radio mentioned earlier is a counterexample. First, while educational radio was abandoned in the United States, it rose in prominence in the less developed countries after educational television projects proved unsustainable in the 1970s. Second, it demonstrated that it was possible, with considerable trial and revision, to prepare programs that successfully incorporated student choral response and pseudoreinforcement of those responses (Friend, Searle, & Suppes, 1980).

Research on discovery learning. Around the same time as behavioral technology was having its largest impact on educational technology, the so-called Cognitive Revolution was gathering steam, led by Jerome Bruner (1960). A major theme of Bruner's was that learning is an active process in which learners construct new ideas based upon their existing knowledge. He argued that the function of school should be to provide conditions that will foster the discovery of relationships. This ideal suggested that television should be participative rather than passive. It should ask questions, pose challenging problems, and spark discussion and search for answers. In short, it should trigger inquiry (McBride, 1966). The discovery learning movement eventually led to the production of recorded series, especially in science and social studies, that portrayed problematic situations and invited learners to discuss them. This required a mindset change—viewing the visual presentation as part of a larger classroom activity rather than as a complete package in itself.

Research on children's attention and comprehension. Beginning in the late 1960s, the Children's Television Workshop (CTW) became the locus for major R&D activities related to the creation of educational television for children. The CTW developers were focused on the issues of how to capture and hold attention and then improve comprehension of televised material. They sought to teach basic cognitive skills and shape prosocial attitudes. Led by Keith Mielke, CTW pioneered in the systematic use of formative and summative evaluation to test the effects of various message design variables on attention and comprehension (Seels, Fullerton, Berry, & Horn, 2004, p. 257).

Over a period of several decades, CTW applied this R&D approach to the creation of a number of television series that were used in homes and schools, aimed at specific skills for different audiences: *Sesame Street*—cognitive and social development for preschoolers, *Electric Company*—reading skills for the early elementary school years, *3-2-1 Contact*—scientific interests and attitudes for the later elementary school years, and *Square One*—mathematics at the elementary school level (pp. 300–301). Later programs, such as *Blue's Clues,* continued the tradition of improving design through systematic testing, and they extended further into promoting participation and active problem solving.

Audiovisual (AV) Materials

Throughout the 20th century, a wide array of other types of auditory and visual materials were used for education and training. As described in chapter 8, lantern slides were in use by the end of the 19th century and silent films were in use by the 1910s. The phonograph, then sound films, added audio to visual media in the 1920s. By the post–World War II period, two-by-two-inch slides, 35mm filmstrips, and overhead transparencies were standard parts of school and college AV programs In the 1970s, the audiocassette format displaced reel-to-reel tape for amateur and educational recording. This format remained popular into the 21st century in many countries, although commercial distribution of popular music moved to the compact disc (CD) in the 1990s in more technologically advanced areas.

Creating AV Materials. Filmstrip and slide-set creation followed a process similar to that of filmmaking. The developer began by learning as much as possible about the topic, the audience, and the teaching objectives. This was done through reading and interviews with subject matter experts and other stakeholders, especially the client. The developer jotted ideas onto note cards, which were eventually arranged into logical clusters. As the structure of the filmstrip or slide set took shape, taking into account the "psychology" of the audience's needs, a script could be written (*Facts You Should Know,* 1965, p. 17).

With the script in hand, a visual storyboard could be constructed, consisting of thumbnail sketches of the visuals plus the accompanying text. Ideally, a rough draft of the visuals in slide format and a tape recording of the text could be presented to a representative sample of the target audience to test their reactions. After making revisions, the final script and storyboard could be converted to a finished product using professional performers and producers (*Facts You Should Know,* 1965, pp. 19–21).

Underlying Research and Theory. Research on the creation of AV materials has revolved around three major issues: the perception, interpretation, and retention of *visual* images; the perception, interpretation, and retention of *auditory* material; and the interaction of the visual and auditory mechanisms in *multimedia* formats.

Most of the basic research on visual and auditory perception has been done outside the field of educational technology. The research inside the field received a major stimulus by the founding of the journal, *Audio-Visual Communication Review,* in 1953 by the Department of Audio-Visual Instruction (DAVI), the predecessor of AECT. Then the National Defense Education Act in 1958 provided a flood of funding for AV research under Title VII.

Space here does not permit an adequate summary of the sorts of research done or its findings, but some of this work is alluded to later under the topic of message design. Dwyer (1972; 1978) provided early distillations of findings of research on improving visual learning, based primarily on the author's systematic experimental studies at Pennsylvania State University. A recent and authoritative synthesis of visual learning research was provided by Anglin, Vaez, and Cunningham (2004). A parallel review of research on auditory learning was provided by Barron (2004), and multimedia research was reviewed by Moore, Burton, and Myers (2004).

Programmed Instruction and Teaching Machines

The field, which in the 1950s and the 1960s was generally known as *educational media*, focused on the creation and use of auditory and visual materials to enhance instruction. The first great paradigm shift in the field's central interest occurred when teaching machines and programmed instruction burst upon the public consciousness. B. F. Skinner (1954) presented his first teaching machine based on operant conditioning principles, and major school demonstration projects were underway between 1957 and 1962.

Programmed instruction, whether presented in the format of a teaching machine or a book prescribed

> (a) an ordered sequence of stimulus items, (b) to each of which a student responds in some specified way, (c) his responses being reinforced by immediate knowledge of results, (d) so that he moves by small steps, (e) therefore making few errors and practicing mostly correct responses, from what he knows, by a process of successively closer approximations, toward what he is supposed to learn from the program. (Schramm, 1962, p. 2)

DAVI, AECT's predecessor organization, joined the new programmed instruction movement by publishing *Teaching machines and programmed learning: A source book* (Lumsdaine & Glaser, 1960). The 1959 DAVI convention program had no mention of programmed instruction, but there was a major session in 1960 on "Programmed instructional materials for use in teaching machines." This title gives a clue to the link between AV administrators and programmed instruction: the machines that were initially used to deliver the programmed lessons. When schools and colleges acquired teaching machines, someone had to take care of them—the AV coordinator! The primacy of the machine was indicated by the name that marked this special interest group at the next several DAVI conventions: the Teaching Machine Group.

Gradually, though, the emphasis shifted to designing and utilizing interactive self-instructional systems. The concept of "technology of teaching" was popularized by B. F. Skinner (1968) to describe his view of programmed instruction as a systematic application of the science of learning. This supplemented the notion promoted earlier by James D. Finn (1965) that instructional technology could be viewed as a *way of thinking* about instruction, not just a conglomeration of devices. Thereafter, technology had the dual meanings of "application of scientific thinking" and the various communications media and devices.

Creating Programmed Instruction. The process of creating the software for programmed instruction clearly was vastly different from that for AV materials. Now the critical steps were analyzing the task to be learned in order to break it down into a series of small steps, specifying the behavioral indicator of mastery of each step (performance objective), sequencing the activities into a hierarchical order, creating prompts for the desired responses, requiring learner response, and administering appropriate consequences (the possible contingencies: positive or negative reinforcement, punishment, or removal of reinforcement) for each response.

Underlying Research and Theory. Research on programmed instruction eventually falsified the sanctity of the specific prescription as given by Schramm (1962) previously: an ordered sequence of stimulus items, overt response, immediate knowledge of results, small steps, and mostly correct responses. Each of these elements was dispensable, yet programmed instruction lessons consistently led to better achievement when compared with so-called conventional instruction. What accounted for the improvement, if not the formulaic framework? Gradually, practitioners began to realize that it was the painstaking development *process*, which included frequent formative evaluation to ensure the learners were making correct responses. They discovered that "programming is a process" (Markle & Tiemann, 1967). Further, that process—of analyzing learners and learning tasks, specifying performance objectives, requiring active practice and feedback, and subjecting prototypes to testing and revision—was highly compatible with the analysis, design, develop, evaluate, and implement cycle proposed in systems approach models.

Computer-Assisted Instruction (CAI)

As discussed in chapter 2, CAI began just at the time that programmed instruction was at its peak, and so, many of the early CAI programs followed a drill and practice or tutorial format similar to teaching machines or programmed instruction books: small units of information followed by a question and the student's response. A correct response was confirmed while an incorrect response might branch the learner to a remedial sequence or an easier question. The design work therefore resembled that of programmed instruction, while the development-production work entailed skills in computer program writing.

Creating CAI. The PLATO project, begun in 1961, aimed to reduce costs by networking inexpensive terminals and offering programmers a simplified programming language for instruction, TUTOR. It became a locus for intensive R&D on the message design features of successful lessons as well as on authoring systems. The PLATO system pioneered many advanced functions (e.g., graphical interface, user discussion groups, e-mail, and instant messaging), and it continued to grow and evolve right through the early 2000s. This R&D program also led the way in developing creative approaches to CAI such as discovery learning and problem-based learning (PBL) through participation in laboratory experiments and other simulations. Like a lot of other CAI software, PLATO software eventually migrated to floppy disk format, then CD-ROM, then World Wide Web.

In the days of mainframe-based computing and into the early years of microcomputers (outside the PLATO environment), memory and display limitations dictated lesson designs similar to those for teaching machines and printed programmed instruction: frame-by-frame advancement through content presentation followed by questions to which the learner responded by means of an input device—keyboard, number pad, touch screen, or possibly a graphics tablet. The computer judged the correctness of the response and gave feedback to the learner, possibly branching to a remedial set of frames.

Underlying Research and Theory. The programmed instruction type lesson format lent itself to the same sort of design processes as used in programmed instruction (Burke, 1982). The product was a series of teaching frames and criterion (testing) frames. The development and production phases depended heavily on what sort of programming language or *author-ware* were being employed to input the lesson into the computer system. As with programmed instruction, evaluation and validation of the lesson was expected (but not always done).

The research paradigms were very much in the pattern of programmed instruction, as were the findings. This research also tended to be guided by the same theoretical constructs as in programmed instruction research, although CAI research more often included investigation of presentation variables and economic issues (since computing hardware, programming time, and processing time were significant cost factors at that time).

Digital Media

As computing power grew and became more wide ranging through networking, and as computer systems became more capable of incorporating visuals, sounds, and moving images, computer-based programs began to be viewed in a new light, as "digital media," discussed in greater detail in chapter 8. The concept of combining all forms of media under the computer umbrella transformed the field of educational technology as well as the entertainment industry.

Creating Hypermedia. The term *hypermedia* emerged in the 1980s as an extension of the term *hypertext* to refer to digital documents in which text, audio, and video are connected with hyperlinks to allow nonlinear navigation among the program elements. This contrasts with *multimedia*, which could combine the same media but in a linear format. Hypermedia required a powerful computer with capacious RAM, a large internal hard drive, and a monitor plus peripheral devices for AV input, such as CD and videodisc players and audio systems. All of these were controlled by a hypermedia program running under an authoring system such as HyperCard™ or Toolbook™. A primary feature of this format was the high degree of interactivity between the learner and the varied information sources.

Hypermedia became possible when computers began to operate with graphical user interfaces (GUIs), incorporating such graphic devices as windows, menus, hyperlinks, and a pointing device (e.g., a mouse). The GUI not only made it easier for novice users to navigate but it allowed teachers and other nonspecialists to create their own materials.

Underlying Research and Theory. In developing digital interactive media or hypermedia, instructional designers began to engage in software design (or in most cases, the somewhat less technical "software authoring"). Just as a shift from print production to film production entails changes in design process, so does a shift from traditional media to digital interactive media (Jonassen & Mandl, 1990).

The design process itself became the subject of research and theory. For example, the concepts of rapid prototyping (Tripp & Bichelmeyer, 1990), user-centered design, and usability methods (Corry, Frick, & Hansen, 1997; Frick & Boling, 2002) became subjects of debate and study. These constructs were borrowed from software design and incorporated into ID to recognize the increased complexity of interactive materials and therefore the increased chance that such materials might be difficult to use, to understand, or to accept. The use of such approaches can be seen as extensions of the traditional emphasis on audience analysis and formative evaluation within ID. The creators of the materials view the eventual users as an audience, and the usability process as a means to ensure that the materials are effective as part of the instructional environment.

The Internet and the World Wide Web

In the 1990s, the rapid growth of the Internet and its most popular protocol, the World Wide Web, fundamentally changed the media environment for instructional designers. Within a decade, more instruction was being prepared for use on the Web than for any other media platform.

Creating Web-Based Learning. In the development and production stages the challenge to the Web producer is to deal with another new set of authoring tools and programming protocols. In the era of CAI, as P. F. Merrill pointed out (2005), authors had to learn and then learn anew a number of different tools over the years: Basic, Pascal, Pilot, TICCIT, and Hypercard™ (p. 4). In the era of the Web, HTML (hypertext markup language) has been the standard authoring application for static text. To add sound, motion, or interactivity, computer-programming code has to be added to the HTML, using a scripting language such as JavaScript or authoring tools such as Flash®, Director, and Authorware (Merrill, P. F., 2005, p. 4). In 2005, it was possible to separate the content from the programming code through the use of XML (extensible markup language). This is viewed as a way to simplify the problem of migrating content and programming from one authoring environment to another one and achieving the goal of handling content as learning objects that can be shared and reused (Merrill, P. F., 2005).

Learning objects. By the 1990s, as the use of Web-based instruction accelerated, designers, particularly in military and corporate training, were seeking a shortcut to creating the thousands of hours of course material needed in the hundreds of Web-based distance learning programs. The key to this problem, many felt, was to create reusable *learning objects*: "small (relative to the size of an entire course) instructional components that can be reused a number of times in different learning contexts" (Wiley, 2002, p. 4). This movement was an extension of the object-oriented programming paradigm that transformed software development beginning in the 1980s.

The technical challenge was to code the learning objects digitally so that they would transfer to and run on each organization's learning management system. In the 1990s, several international efforts began to establish standards for these building blocks. One effort was led by the IMS Global Learning Consortium, Inc., which produced the IMS Metadata Specification ("metadata" are the labels that are put on learning objects, enabling these objects to be stored and retrieved efficiently). The IMS specification, in turn, was incorporated into the Sharable Courseware Object Reference Model (SCORM). By 2000, these specifications were being used in a number of organizations.

The promise of learning objects includes reducing the manpower cost of development and spreading the effort of material creation to the largest possible pool of talent, thus placing well-designed learning materials within the grasp of those who might otherwise not be able to afford them. However, problems both conceptual and technical have slowed the wider adoption of this idea. One conceptual problem is represented in the very name of the concept: bits of content or test items are not *learning* objects if learning is a process that takes place within individuals; they are chunks of *content*. The next question is whether these chunks of content can be removed from their original context, be inserted into a different context, and still have value. It seems to depend on what the "chunk" is and how different the two contexts are. More generic material, such as a worksheet on fractions, might well be usable in a wide range of classes, possibly even across cultures. Smaller or larger granules, presented in more contrasting contexts might be problematic. Those whose beliefs about learning emphasize the importance of contextualization are dubious about the prospect of stripping context out of chunks of instructional material.

On the technical level, critics wonder about the costs and myriad technical hurdles posed by developing a system of cataloging and sharing such media objects that would be both usefully standard and usably flexible. David Wiley (2002; 2006), who helped introduce learning objects into educational technology, was also vocal in supporting both the conceptual and technical criticisms. He continued to support the goal of "increasing access to educational opportunity to people who have been denied that right for any of a variety of reasons," but he suggested that a method more like that of P. F. Merrill (2005), described previously, will ultimately be more useful. So, the concept of reusable digital material will continue to evolve but future directions are unclear.

Underlying Research and Theory. With the ubiquity of the Web and wide diffusion of course management systems (CMSs) and learning management systems (LMSs), it is possible to view Web-based education as a separate genre for design and development. A distinctive feature is that, by its nature, Web-based instruction revolves around *learning*-oriented activities—reading, discussion, construction, expression, reflection, and perhaps inquiry activities—while the face-to-face classroom revolves around *teaching*-oriented activities—lectures, demonstrations, discussions, and tutorial exchanges between teacher and learner. This shifts the focus of research and theory from teaching issues (e.g., presentation variables) to learning issues (e.g., interpersonal communication patterns in collaborative learning).

Mobile Media

The trend in computing hardware is toward miniaturization and wireless operation, leading to a new genre of mobile devices—notebook and tablet PCs; cell phones; digital audio players; handheld game consoles; personal digital assistants (PDAs), which can include the functionality of a computer, a cell phone, a music player, and a camera; and various other combinations of these devices. When such devices can also connect to the Internet, users, in effect, have access to a high-end computer workstation in their hands. They can talk or text message with others and navigate the Web from wherever they happen to be (as long as they are in range of a wireless access point). By 2006, in Europe and Asia, these functionalities were rapidly migrating toward convergence in a cell phone type device, but this movement was emerging more slowly in the United States.

This raises the possibility of a new teaching-learning paradigm—mobile learning, or m-learning. As summarized by Wagner (2005),

> . . . mobile learning represents the next step in a long tradition of technology-mediated learning. It will feature new strategies, practices, tools, applications, and resources to realize the promise of ubiquitous, pervasive, personal, and connected learning. (p. 44)

Some aspects of m-learning are already apparent, based on experiences with earlier technologies, including Web-based distance education. We also have experience in education with some of the mobile technologies, for example, the uses of PDAs as classroom response devices or "clickers." Like

Web-based resources, mobile resources may be used primarily for performance support and to supplement traditional delivery in a new hybrid mode. The functions of laboratories may become more distributed, with many taking place in students' handheld devices (Alexander, 2004). To the extent that they are used to offer stand-alone instruction, they are expected to be used for short programs that can be used during downtime between other work and leisure activities (Wagner, 2005, p. 51).

Applications that are specific to mobile technologies may evolve to fit with their special adaptabilities to interpersonal communications. Alexander (2004) borrowed the concept of "swarming" to speculate about "learning swarms" or ad hoc, temporary learning groups (p. 32). Similar to the groups that form on social networks like Facebook.com, students who develop curiosity about a topic might talk or exchange text messages with others and form a virtual discussion group, which might meet face to face at some times. Or they might simply use the mobile tools to carry out group work assigned in class.

Creating for M-Learning.   At this point we can only speculate about what form m-learning will take and what sorts of creation processes it will demand. We know from our study of other technologies that it is not the technology but the experience that facilitates learning. At the design stage, developers must keep in mind that mobile devices will be able to support certain types of experiences better than others will, so different learning tasks will require different instructional strategies. The constraints of the m-learning setting are

- Computational power of mobile devices is limited.
- The various mobile devices use a wide variety of operating systems, meaning different authoring tools for each device.
- These devices also have very small screens, imposing narrow limits on the size and amount of text and size and resolution of graphic images.
- Likewise, input capabilities are limited (How much text would you like to type with your thumbs?).

Keeping in mind the conditions of use, designers will have to pay special attention to gaining and holding the user's attention in distracting circumstances, suggesting short modules that are highly engaging—such as games, quizzes, or chats.

At the development and production stages, coding is again an issue. Scripting languages and authoring tools (like Flash®) have made it relatively easy to incorporate time-based media and interaction into Web lessons. These extensions to the original protocols have also allowed creators to address mobile technologies by setting up style sheets that display a single content file in multiple ways depending on where that file is to be displayed (different browsers, mobile devices, etc.). At the same time, they have increased the technical expertise required to set up Web-based documents correctly, so the learning curve for developers is once again steep for all but the simplest materials. The payoff is that cross-platform and cross-media development is truly possible.

Blended Learning

Historically, educators have thought of face-to-face instruction and computer-mediated instruction as separate domains. A lesson or course is conducted either face to face or through one of the formats discussed above— CAI, multimedia/hypermedia, Web, or mobile device. The reality is that an increasing proportion of lessons and courses, especially in higher education and corporate and military training, are conducted through a combination of face-to-face and computer-mediated formats, a combination referred to as *blended learning* (Graham, 2006). This trend has been prompted by the ubiquity of the Internet and the Web in the daily lives of students and workers, at least in technologically developed societies. As long as students and workers were already used to communicating through e-mail, instant messaging, and chat rooms and as long as instructors were already used to exchanging files electronically and creating instructional materials with the computer, why not exploit these practices in the classroom?

In higher education, a blended course typically consists of one face-to-face class meeting per week, with students using the Internet and Web to complete group projects and other class assignments (Dziuban, Hartman, Juge, Moskal, & Sorg, 2006, p. 198). In the corporate realm, blending tends more toward a "sandwich" approach: preclass readings and asynchronous discussion, then face-to-face sessions for intensive interaction, followed by online application exercises and mentoring (Lewis & Orton, 2006). In the military, the blending typically involves trainee use of high-fidelity simulations (e.g., firing range and airplane flight) which are integrated with collective field training (Wisher, 2006).

A related concept is *blended learning environment*, creating an immersive total environment that blends aspects of reality, simulation, mixed reality, and virtual reality. This concept is discussed separately later in the chapter.

Creating Blended Learning. The different elements that comprise the blend are each created through the processes appropriate for that format— face-to-face instruction, CAI, simulation, and so forth. The overall blended lesson or course can be designed through the generic ISD process, with special attention to the step of selecting a delivery medium—looking at each objective and deciding if it would be learned best through one of the face-to-face methods or through one of the computer-mediated methods (Hoffman, 2006).

Creating Media: Levels of Sophistication

Creating instructional media can be a very simple or a very complex process. Kemp and Smellie (1994) suggest three levels of sophistication: *mechanical, creative,* and *design.* At the lowest, *mechanical,* level are the simple procedures of, for example, cutting and pasting a picture onto a Web page, photocopying a graph to make an overhead transparency, or video recording a guest speaker for later playback. These are routine actions requiring little planning or creativity.

At the second, *creative,* level the producer has to put thought and planning into the process. A teacher constructing a bulletin board will not only gather or make the materials, but also think about their arrangement, both aesthetically and educationally—to garner attention and make a memorable impact. A trainer may sketch ideas onto index cards and rearrange them for psychological effect before producing a PowerPoint™ presentation. The choice of words and images, their sequence, the visual layout according to good visual design principles—all of these require some level of artistic ability and consideration of the psychological variables that affect audience impact. But production at the creative level does not necessarily entail systematic planning for specific learning outcomes.

The third, *design,* level covers cases in which a designer, or even a design team, plan and assemble materials or a whole learning environment in order to reach a specified learning goal. They will think about the needs of their particular audience and how learners will interact with the material to attain their objectives. The materials themselves might require some technical expertise to produce. For example, an instructional consultant from a campus support service might work with two geography professors to develop an interactive Web exercise for finding, reporting, and interpreting oceanic temperature variations. This would require combining subject-matter expertise, pedagogical methods, visual design knowledge for screen layout, and Web-programming expertise; and since the project could entail multiple people collaborating over a period of time, project management would also come into play.

In the following sections, we will discuss the issues associated with the *creative* level and the *design* level of production. First, at the *creative* level we focus on the search for technical and aesthetic quality and how it is guided by principles from such fields as communication theory, perceptual psychology, and semiotics. Then, at the *design* level, we survey the prominent design methodologies, including the systems approach and some alternatives.

Issues at the *Creative* Level: Technical Quality and Message Design Principles

Media production, even at the mechanical and creative levels, can demand considerable technical expertise, craftsmanship, and artistic ability. There is a long tradition in educational technology of expecting and honoring technical excellence in its products. To go beyond good technical execution alone, the principles that guide creative media production are most often derived from aesthetics and research on message design.

Message Design Theory and Principles

Drawing from communication theory for the concept of instructional messages, Fleming and Levie (1978; 1993) gathered the applicable findings from behavioral science and cognitive science research in search of message design principles. They define a message as "a pattern of signs (words, pictures, gestures) produced for the purpose of modifying the psychomotor, cognitive or affective behavior of one or more persons" (Fleming & Levie, 1993, p. x). The contributors to Fleming and Levie's compendium make a particular effort to translate basic research into usable principles for the creators of instructional media. The perspective of the contributing authors and the editors is that of expecting a message formed according to sound principles to "modify the psychomotor, cognitive or affective *behavior* [italics added]" of those who receive that message, thus combining cognitivist notions under a behaviorist framework.

Houghton and Willows's (1987) two-volume work, *The Psychology of Illustration*, surveys basic research on picture perception and the use of images to enhance learning from text. It offers a model for classifying and discussing pictures according to their own properties and their relationships to instructional text, and demonstration of the application of message design principles. The collection includes semiotic perspectives, particularly concerning the interpretation of images across cultures, and discussion of affective, or emotional, responses to images and their role in promoting engagement in learning. The issue of understanding images is oriented toward perceptual research (how do physical and cognitive processes allow us to recognize pictures) and the conceptual consideration of images themselves (as perceptual data, as symbols in a system, as types of representation). And, in light of the continuing primacy of text as the medium of mediated instruction, the bulk of discussion regarding images in learning is oriented toward the role of images in helping learners to remember, understand, or enjoy text.

Many of the principles gathered by Fleming and Levie (1978; 1993) and Houghton and Willows (1987), along with those developed specifically for the creation of instructional text material (Hartley, 1986, 1996; Jonassen, 1982), remain the primary research-based source of guidelines for creating instructional media despite radical changes in interactive and multimedia technologies.

Message Design for Motion Media. It is assumed that the message design principles discussed above remain viable for moving-image displays in the new interactive media environments, although in the absence of thoroughgoing investigation this is only an assumption. For example, Reeves and Nass's (1996) studies indicate that we respond to the sorts of moving images of people (live or animated) shown on TV monitors as if they were "other people." Application of this understanding to the creation of interactive instructional materials may imply a host of message design principles that modify the perspective from which the original principles were developed, establish ground for the development of new principles, or in some cases, add support to the basis for the original principles.

Semiotic Perspective. The semiotic perspective as it applies to creating instructional materials has been practically articulated for the creators of instructional materials by Sless (1981; 1986), who focuses discussion of creation not on the characteristics of instructional materials themselves, but on the explicit and tacit codes by which people decide what objects (including texts) mean. In document design, a sister field to instructional design, Schriver (1997) speculates that readers of informational texts—verbal, visual, and both—develop and continually refine a hypothesis about the meaning of a text as it relates to themselves as they progress through the material. Her research suggests that readers' past experiences, cultural perspectives, and even their guesses about who created the materials all influence this evolving hypothesis. While these perspectives have gained firmer footholds in document design, technical communications, and visual literacy circles than in the instructional design community, they offer a rich dimension to expand our collective understanding of message design.

Emerging Message Design Principles. Assembly of research-based principles for the explicit guidance of materials creators continues (Clark & Lyons,

2004; Lohr, 2003; Misanchuk, Schwier, & Boling, 2000). These compilations also draw on the application of Gestalt psychology (common in graphic design and fine arts) and on traditional, nonempirical understandings from the professional media design world and they generally offer some process guidance for instructional media designers. However, systematic progress in research on media issues in instructional materials themselves within the field is sparse with the exception of Dwyer's (Moore & Dwyer, 1994) long-standing program of studies comparing learning outcomes with the use of materials that exhibit different formal properties.

Production Standards.   Throughout the evolution of film, video, and AV media the process of converting blueprints into finished presentations has been guided by technical lore built up over time. Wetzel, Radtke, and Stern (1994) refer to these production guidelines as professional tradecraft (p. 113). In film and video, for example, the main issues relate to camera technique, shot composition, editing, and special effects (Mascelli, 1965). Each of these areas has its own cadre of technical specialists who are likely to have learned their tradecraft through years of apprenticeship. Audiences have grown used to a certain level of technical quality and tend to bring these expectations to their viewing of educational media as well.

Trade-Offs on Technical Quality.   As Schiffman (1986) contends, both aesthetics and pedagogy demand that learning materials be clear, attractive, and usable. At the same time, she warns against "a disproportionate emphasis on production standards" (p. 15) when the time and expense of professional level production are out of the proportion to the material's purpose. In certain cases, "quick and dirty" will suffice, reflected in anecdotes about the advertising creative teams that discover that "the storyboard worked better than the finished commercial!"

Issues at the *Design* Level: ISD Models and Alternative Approaches

When dealing with more complex projects, planners are operating at what Kemp and Smellie (1994) termed the *design* level, the level at which some type of serious design thinking is required. The creation of instructional materials and learning environments can be guided by different design mindsets as well as different design procedures. For example, in the genres of educational film, radio, and television the planning process was guided by the paradigm of the *script*, a mindset carried over from entertainment media. The visual arts have a very different mindset for the creative process; engineering has another, and software design has another. Educational technology has borrowed from disciplines such as these and it has evolved its own approaches. The purpose of this section is to survey the broad array of approaches that have been used, beginning with the systems approach, which is usually referred to as the dominant paradigm, and then considering the many alternative possibilities.

Systems Approach to Instructional Design

The essence of the systems approach is to subdivide the instructional planning process into steps, to arrange those steps in logical order, then to use the output of each step as the input of the next. The systems approach traces its origins to concepts that emerged from military research during World War II. An analytical technique that grew out of submarine hunting was called operations research, in which computers were used to make the calculations required. After the war, this approach to analyzing, creating, and managing man-machine operations, now referred to as the systems approach, was applied to the development of training materials and programs.

During the postwar period each of the U.S. military services had developed its own model for training development, all of which were based on the systems approach, a "soft science" version of systems analysis, itself an offshoot of operations research. Alexander Mood (1964), speaking at an early conference on the systems approach in education, explained the distinction:

> *Systems analysis* is often used interchangeably with the term operations analysis and refers to the specific analytical technique which consists of constructing a mathematical model of a phenomenon and optimizing some function of the variables involved in the model. *Systems approach* refers to a much more general and hence less definitive idea. It is simply the idea of viewing a problem or situation in its entirety with all its ramifications, with all its interior interactions, with all its exterior connections and with full cognizance of its place in its context. (p. 1)

The systems approach was viewed in the military as a paradigm for combining the human element with the machine elements in man-machine systems, an antidote to purely mechanistic thinking. From the entry of the systems approach into the field of educational technology, it was recognized by its advocates as a loose set of guidelines which were applicable to the complex problems of human learning only by analogy, and not the sort of completely deterministic and tightly controlled methodology described by some of its detractors. Mood (1964), in the same presentation, cautioned, "One uses it [systems approach] primarily as a guide and as insurance against overlooking an important factor," and later, "This is the most troublesome problem of the systems approach; it is an art—not a science" (p. 14).

Evolution of the Systems Approach in Educational Technology. The concept of systems approach probably was introduced to educational technology at the 1956 Lake Okoboji leadership conference. This annual conference, to which leaders of the field were invited and in which they were expected to produce working papers, often featured a keynote speaker, who provided grist for the following discussions. One of the most influential keynote addresses was the first, "A Systems Approach to Audio-Visual Communication," given by Charles F. Hoban at the 1956 summer conference (Allen, 1960). The conference spotlight coincided with a series of articles by James D. Finn published around the same time (Finn, 1955, 1956a, 1956b). Together, they helped create momentum behind the idea of the systems approach, which eventually became a hallmark of the field.

The vision that drove this new thinking was expressed succinctly by Phillips (1966): "To fashion a coherent assemblage of learning resources, specifically designed *from their inception* to be used with and make possible the implementation of a new curriculum" (p. 373). That is, how much more productive might education be if we could look at the system as a whole—teachers, students, administrators, aides, facilities, hardware, software—and design a total package around a clear goal?

During the 1960s, the systems approach began to appear in procedural models of ID in American higher education. Barson's (1967) Instructional Systems Development project, conducted at Michigan State University and three other universities between 1961 and 1965, produced an influential model and set of heuristic guidelines for developers. During this same period, Leonard Silvern (1965) at the University of Southern California (USC) began offering the first course in applying the systems approach to instruction, "Designing Instructional Systems," which was based on his military and aerospace experience. He also produced a detailed procedural model that influenced later model builders.

The IDI model. These early activities at the consortium that included Syracuse, Michigan State, U.S. International University, and USC (later joined by Indiana University) culminated in a joint project, known as the Instructional Development Institute (IDI). The IDI was a packaged training program on instructional development for teachers, and between 1971 and 1977, it was offered to hundreds of groups of educators. Since it was usually conducted by faculty and graduate students from nearby universities, the IDI became an extremely influential vehicle for disseminating ideas about the ID process among educational technology faculty and students across the United States.

The model divides the creation process into three major phases: (a) the define phase, in which analysis is done to clearly define the problem to be solved and the situational constraints, and a plan of work is organized, (b) the design phase, in which objectives are specified and methods for attaining those objectives are decided up and instantiated in a prototype, leading to (c) the develop stage, in which the prototype is tested and revisions are made based on the prototype tests. The IDI model was quite forward thinking in its emphasis on project management, iterative development, and testing of prototypes.

The military services' ISD model. The Center for Performance Technology at Florida State University was selected in 1973 by the U.S. Department of

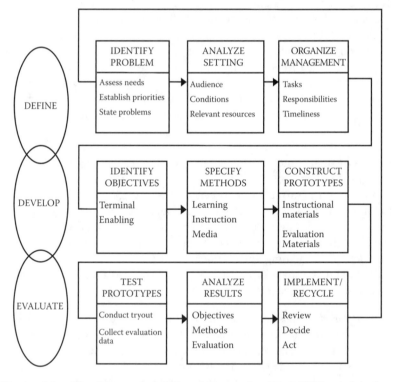

Figure 4.1. The Instructional Development Institute (IDI) model, developed by the National Special Media Institutes (later known as the University Consortium for Instructional Development and Technology), 1971.

Defense to develop procedures to substantially improve Army training. As recounted by Branson (1978), the ID procedures developed for the Army evolved into a model that was adopted by the Army, Navy, Air Force, and Marines, called the "Interservice Procedures for Instructional Systems Development (IPISD)." Shown in Fig. 4.2, IPISD was intended for use in large-scale ID projects. It eventually had enormous influence in military and industrial training because its use was mandated not only in all of the U.S. armed services but also among defense contractors. The seeds of the "ADDIE" acronym can be seen in the top-level elements in Fig. 4.2: analyze, design, develop, implement, and control. As evaluation replaced control, the acronym ADDIE came into being.

The ADDIE Family of Systems Approach Models. The ADDIE stages are sometimes put into the form of a flow chart to show their interrelationships, as shown in Fig. 4.3, giving rise to the misnomer of "ADDIE model" even though there was not and is not an actual, fully developed ADDIE model. However, it can serve as a convenient label for the family of systems-approach models.

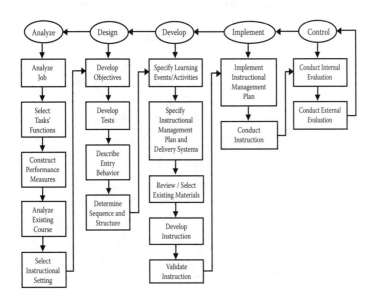

Figure 4.2. The IPISD model. *Note:* From *Interservice procedures for instructional systems development: Executive summary and model* by R. K. Branson. Tallahassee, FL: Center for Educational Technology, Florida State University, 1975. (National Technical Information Service document #AD-A019 486 to AD-A019 490.)

Following the logic of the diagram in Fig. 4.3, the outputs of the *analysis* stage—a description of the learners, the tasks to be learned, and the instructional objectives—serve as input to the *design* stage, where those descriptions and objectives are transformed into specifications for the lesson. Next, the design specifications serve as inputs to the *development* stage, where they are used to guide the selection or production of the materials and activities of the lesson. In the *implementation* stage the instructors, materials, activities, and learners come together to use the products of the development stage. After the instructional program is used, it is *evaluated* to see if the objectives were met and the original problem solved.

In addition to the *summative evaluation* done at the end, along the way decisions made at each stage are evaluated (*formative evaluation*) to determine if that stage was completed successfully and in accord with the original strategic directions of the project. If the results of a step are not satisfactory, for example, if a sample group of trainees are confused by the directions in the prototype of a new simulation exercise, then the *development* step must be repeated, finding ways to clarify the directions. This process of repeating steps until satisfactory results are achieved is referred to as an *iterative* approach.

Gagne, Wager, Golas, and Keller (2005) provided an expansion of basic ADDIE stages into a more detailed procedural guide, shown in Table 4.1.

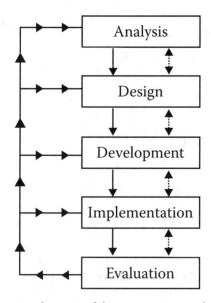

Figure 4.3. The major elements of the systems approach to ID, also known as "the ADDIE model."

Table 4.1. The ADDIE stages expanded

Analysis	a. First determine the needs for which instruction is the solution.
	b. Conduct an instructional analysis to determine the target cognitive, affective, and motor skill goals for the course.
	c. Determine what skills the entering learners are expected to have, and which will impact learning in the course.
	d. Analyze the time available and how much might be accomplished in that period of time. Some authors also recommend an analysis of the context and the resources available.
Design	a. Translate course goals into overall performance outcomes, and major objectives for each unit of the course.
	b. Determine the instructional topics or units to be covered, and how much time will be spent on each.
	c. Sequence the units with regard to the course objectives.
	d. Flesh out the units of instruction, identifying the major objectives to be achieved during each unit.
	e. Define lessons and learning activities for each unit.
	f. Develop specifications for assessment of what students have learned.
Development	a. Make decisions regarding the types of learning activities and materials.
	b. Prepare draft materials and/or activities.
	c. Try out materials and activities with target audience members.
	d. Revise, refine, and produce materials and activities.
	e. Produce instructor training or adjunct materials.
Implement	a. Market materials for adoption by instructors and potential learners.
	b. Provide help or support as needed.
Evaluate	a. Implement plans for learner assessment.
	b. Implement plans for program evaluation.
	c. Implement plans for course maintenance and revision.

Note: Adapted from p. 22 in *Principles of instructional design, 5th ed.* by R.M. Gagne, W.W. Wager, K.C. Golas, and J.M. Keller. Belmont, CA: Thomson/Wadsworth, 2005. Used with permission of Thomson Learning.

Numerous systems approach models have been proposed. They differ in terms of the number of steps, the names of the steps, and the recommended sequence of functions. Gustafson and Branch's (2002) *Survey of Instructional Development Models* includes 18 models. Their list is not intended to be exhaustive, but illustrative of the various ways of implementing a systems approach. Organizations typically use their own homegrown model, often adapting or combining concepts from other models.

The Dick and Carey Model.   One of the best-known systems-approach models is the one developed by Dick, L. Carey, and J. O. Carey (2005), shown in Fig. 4.4.

It is taught in many educational technology programs and it has been adopted or adapted in many organizations as a planning guide. A distinctive feature of the Dick, L. Carey, and J. O. Carey model is that it recommends specifying the assessment instruments prior to developing an instructional strategy. Their concept is that if the developers can be clear enough about what and how they will be testing, they have a much better idea of what sort of instruction will succeed.

Nowadays, there is a general consensus on the main elements of the systems-approach model, according to the authors of *Instructional design competencies: The standards* (Richey, Fields, & Foxon, 2001), representing the International

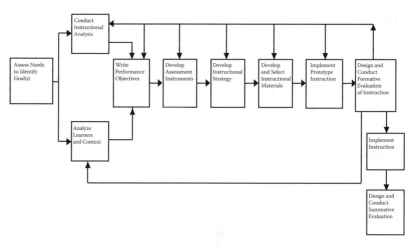

Figure 4.4. The Dick and Carey model of ISD. *Note*: Adapted from illustration on p. 1 in *The systematic design of instruction, 6th ed.* by W. Dick, L. Carey, & J.O. Carey. Boston: Allyn & Bacon, 2005. Used with permission of Pearson Education Ltd.

Board of Standards for Training, Performance, and Instruction (IBSTPI). In specifying the competencies to be expected in professional instructional designers, the IBSTPI standards use the categories of professional foundations (communication skills, research knowledge and skills, self-development, and legal and ethical norms), planning, analysis, design, development, implementation, and management. This list of competencies mirrors quite closely the elements common to most systems-approach models.

Since systems-approach models do not necessarily follow the ADDIE sequence or nomenclature, a more generic name for this family of models is "ISD models." Some authors prefer the term Instructional Systems *Design*, while others prefer Instructional Systems *Development*. We will avoid a discussion of the merits of each term and simply use the acronym ISD.

Stages in the ISD Process

The Analysis Stage. The first priority in analysis is to determine whether instruction is needed at all. A design-development process is undertaken, presumably, because someone has decided that one or more people have a gap in knowledge, skill, or attitude that is important to bridge. The proposed learner could be anyone, from a kindergarten child to an adult organization employee. In the 1970's, Joe Harless, a training designer working in the business sector, realized that many of the people who were successfully "trained," eventually reverted to deficient performance. Harless (1975) found that poor performance was more often caused by lack of incentive or inadequate tools than by lack of knowledge. He developed *front-end analysis*, analytical steps to be carried out at the very front end of the design process to separate the different causes of performance deficiencies, and to make sure that instruction was developed only when instruction was truly needed.

A front-end analysis or needs analysis will gather evidence on the nature and extent of the performance deficiency, determine whether there is a learning need, and determine whether it would be cost beneficial to create some instructional material or system to meet this need. As discussed in chapter 3, other noninstructional interventions can be pursued for the parts of the problem not caused by lack of knowledge or skill.

If the problem is determined to be one of deficiency in knowledge or skill, the next issue in the analysis stage is to determine the types of learning objectives that will need to be pursued—cognitive, affective, interpersonal, or motor skills—and what is the structure of those skills. That is, which are contingent on others? Which should be accomplished first, second, and third? Such an instructional analysis may consist of observations of people at work, behavioral algorithms, focus group discussions, interviews with learners or experts, hierarchical task analyses, or other means. Guides to the many methods of needs and task analysis are found in Zemke and Kramlinger (1982), Rossett (1987), and Jonassen, Tessmer, and Hannum (1999).

Planners will also want to survey the resources they have to work with, including time, money, and people and the constraints bounding their work to determine whether the project is worthwhile. At this stage, planners can also begin to plot out time lines and task assignments for the project.

The Design Stage. In the context of the total creation process, *design* refers to the stage in which content, sequence, strategies, and method s are chosen to meet the specified learning goals. Of all the stages in the ISD process, this is the one that has received the most attention by scholars. Psychological research on human learning and educational research on effective teaching methods have provided a wealth of guidance for these decisions. Design guidance is found in works such as Leshin, Pollock, and Reigeluth (1992) and Foshay, Silber, and Stelnicki (2003).

A major decision at the design stage is to select an overall framework for the lesson or other instructional unit. Many different lesson frameworks have been proposed, often inspired by a particular theory of learning or instruction. Two cognitivist lesson frameworks—Gagne's (Gagne & Medsker, 1996) Events of Instruction and Foshay, Silber, and Stelnicki's (2003) Cognitive Training Model are discussed in chapter 2. Another lesson framework with a more constructivist appearance comes from the work of M. D. Merrill (2002a).

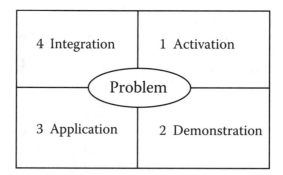

Figure 4.5. Visual model of the major elements of Merrill's "First Principles." *Note*: Adapted from Figure 1 in M.D. Merrill, First principles of instruction, *Educational Technology Research and Development 50*(3), 43–59.

M. D. Merrill (2002a) has developed an eclectic body of instructional principles, which he calls "first principles of instruction" (p. 43). These principles are problem centered and focused on knowledge construction by the learner, as shown in Fig. 4.5. Those particular attributes overlap with some advocated in the constructivist perspective.

M. D. Merrill's (2002a) theory proposes four phases in the instructional process: (1) *activation* of prior experience, (2) *demonstration* of skills, (3) *application* of skills, and (4) *integration* of these skills into real-world activities, with all four phases revolving around (5) a *problem*. Each of these five elements has supporting generalizations or principles, which provide the prescriptions for effective instruction.

M. D. Merrill (2002b) propose a simple framework for applying his "first principles" to learning, called the "pebble-in-the-pond model." The essence of his framework is to begin by imagining the simplest whole version of the task that the learner must be able to perform—the first ripple of the pebble dropped into the pond, then to identify the expanding ripples: "a progression of such problems of increasing difficulty or complexity such that if learners are able to do all of the whole tasks thus identified, they would have mastered the knowledge and skill to be taught" (p. 41). The focus on actual on-the-job problems makes this approach highly suited to application in the workplace.

Many other lesson frameworks are explained fully in Reigeluth (1983; 1999), J. R. Davis and A. B. Davis (1998), and Medsker and Holdsworth (2001).

The Development Stage.   When the term *development* is used as a subset of the larger ISD process, it refers to the stage at which the specifications resulting from the design stage are turned into concrete materials that can be used by instructors and learners. The development stage typically receives little detailed attention in ISD models or their supporting documentation, probably because the authors of ISD models are not themselves expert in the various arts of production and hesitate to spell out these processes in detail.

In the development stage, the design blueprints are first turned into usable prototypes. Pencils, brushes, cameras, microphones, and other creative tools are used to capture or create the words and images needed to carry out the lesson activities. Success is dependent on the artistic and technical skills of specialists in the various media. It is not unusual to find some tension between designers and developers as the production team may struggle to interpret vague or contradictory specifications. Ideally, both understand enough about the other's business to be able to negotiate a mutually satisfactory solution.

Evaluative activities at this stage revolve around tryouts and revisions of prototype materials or processes. Samples of the target population could try out the prototypes one to one or in small groups with observation of the usability of the materials or processes and with assessment of the learning outcomes to determine how nearly the prototype materials met the intended objectives. To enhance the acceptance of the product or process being developed, the goal should be to make it as appealing as possible to the intended users.

After prototype testing and revision, the new materials or processes are ready to be edited and scaled up to their final form. Masters of final products are turned over for mass production, either by an in-house production agency or by external sources. Final versions would be produced in quantity at this point. The output of the production phase is the fully worked out product or program that has been developed, tested, revised, and prepared for large-scale implementation.

Challenges at the interface of design, development, and production. Kerr (1983) noted that ISD novices frequently failed to use the most generic of design procedures common in other fields: generate multiple solutions, accept or reject on some coherent basis, represent design problems effectively to themselves and others, and show a grasp of stopping rules (when is it reasonable to stop). These procedures are not only required for the initial instructional design, they are required throughout the development stage as materials are instantiated either to embody or to support the vision specified in the blueprint.

As the field of educational technology has moved further away from its roots in film and AV production, an identity split has taken place, separating the roles of instructional designer and media developer/producer. With this separation, the processes of instructional design and materials development have tended in many cases to move apart as well (the exceptions often being in school and academic environments where instructional design is happening on a very small scale, and may not be overtly recognized as instructional design at all). The advent of digital tools with shallow learning curves may be bringing the two back together in some contexts, but development and production still tend to be a time-intensive processes, and ones in which the tools learned today have to be relearned tomorrow. Many instructional designers are relieved not to be seen as responsible for, or capable of, direct participation in this stage of the process. In addition, media development when it is carried out well tends to require multiple specialized sets of skills, making the division of labor between pedagogical design and media production all the more inevitable.

Doing digital production yourself. Educational technologists who undertake digital production themselves must master a number of technical skills and processes, specifically, the mechanics of using digital tools of the sorts discussed earlier in this chapter. The challenges are not insubstantial, so it is no wonder that these steps in the overall instructional development process often dominate the designer's thinking even though they are not sufficient to ensure that great, or even usable, materials will result. At this stage, the fit between tools and desired outcomes is vitally important so that efficiency is not sacrificed by struggling against the tools. Production paths, the identification of steps required to create components of materials and ultimately the materials themselves, must be identified and tested and then executed correctly. Frequently a team will discover that a new feature in a tool (the addition of the layers feature in Adobe Photoshop® was a notable example) completely alters a production path and may even eliminate significant numbers of steps that were previously essential.

Supports for doing it yourself. An industry has arisen to fill the gap between the desire to create multimedia products and the technical skills needed to do so. Do-it-yourselfers can take advantage of commercially available support such as clip art, canned color schemes, PowerPoint™ templates, and the like. Unfortunately, these tools have little or no ability to guide the user's choices in terms of pedagogical or artistic sensibility. One can produce a slide show that looks slick but is a disaster in terms of visual appeal, psychological impact, or instructional value.

Outsourcing production. The alternative to doing it yourself is to retain an external contractor or an internal production specialist. Instructional designers hired into very small operations may have to carry out the actual development and production of their materials, but otherwise expect only to have to manage those processes or hire someone else to manage them. In a setting where complex or large-scale development is carried out, the instructional designer is liable to serve as liaison with multiple teams of specialists.

Although it is considered to be vitally important, message design is often an area of disconnect between instructional designers and external or internal media development specialists. Instructional design plans may not include message design direction, in which case these decisions are left up to the specialists who may have had little briefing on the instructional issues involved in the product and therefore no basis on which to make trade-offs or even basic decisions about media forms.

In other cases, instructional design plans may be highly specified in terms of message and media design, but unrealistic from a technical perspective or naïve from a production perspective. Appelman (2005) has proposed a method of analyzing prospective complex learning environments in terms of their affordances that could help bridge the communication gap between instructional design and materials development. Appelman and Boling (2005) have adopted the form-and-function relationship standard in related fields of design to provide a framework within which instructional designers and media specialists can communicate about the functional purposes for decisions of form.

The Implementation Stage. After the prototype material, learning environment, or instructional system has been tested and revised, it is ready for its ultimate purpose—to be used by learners. The programmed instruction movement demonstrated that achievement could be enhanced by thinking of individuals, not groups, as the end users. It was possible to individualize instruction by allowing learners to progress at their own rate and to receive remediation at points in the lesson at which they had difficulty. This led to the idea of *learning for mastery* (Bloom, 1968), that is, expecting that all students can achieve the lesson objectives (as opposed to the bell-shaped achievement curve assumed in most A, B, C, D, F grading systems). The concept of mastery suggests that each learner's outcomes should be compared with some prespecified criteria (not with other learners) and that they should have the opportunity to continue to strive and get help to meet those criteria until they succeed. Only after showing mastery of lower level objectives should they be allowed to attempt more advanced work. This is to prevent ignorance from snowballing and to reduce the failure rate.

Although implementation may entail conducting a whole-class activity, it is more likely to involve students' or trainees' use of self-paced materials or an immersive learning environment. The philosophies and practices surrounding such use are described in detail in chapter 5.

The Evaluation Stage

Origins of evaluation practices. The practice of evaluating the products of design processes before putting them into full-scale use originated in educational radio at station WBOE in the 1930s. There the Cleveland radio producers had developed quite a sophisticated process that was quite comparable to the ISD model. Notably, it featured preparation of a rough draft of the script, which was reviewed by a school principal, then presented to a regular class of students via public address system. The design team members observed the students' reactions, then held a conference to decide on revisions. The revised script was then tested with another audience and revised again before being broadcast throughout the school system (Cambre, 1981).

The World War II military film development process did not include this sort of formative evaluation. Because of time urgencies and the expense of producing films, it was considered impractical to make trial versions for testing. Instead, finished prototypes were reviewed by clients and approved or sent back for editing changes. There was usually some formal summative evaluation, however, consisting of user reports and informal spot surveys. Important films received a more thorough evaluation, including audience testing (Cambre, 1981).

In the postwar period of educational film and television production, procedures for testing audience reactions and learning were progressively refined. Instrumentation for measuring cognitive and affective outcomes was the subject of a great deal of R&D work, for example, in the Navy, Air Force, and Penn State studies discussed earlier in this chapter. However, these procedures and instruments tended to be used in the process of formal research on "learning from television" rather than in the day-to-day development of educational materials.

Formative evaluation received a massive boost in the era of programmed instruction, largely because of the prescription that learners should practice mostly correct responses, so that these desired behaviors could be reinforced. How could one be sure that a sequence of frames would elicit correct responses? Testing was the only answer. In fact, U.S. Department of Defense contracts for programmed training materials required that the producer submit evidence that the material had been tested and that 90% of the learners made 90% correct responses. So testing and revision became part of the culture of programmed instruction. Since programmed instruction design processes morphed into ISD processes, testing and revision are prominent in ISD models. This way of thinking was strongly supported by the systems approach's emphasis on feedback and quality control, thus providing another rationale for evaluation in ISD.

Although Scriven (1967) provided the names for *formative* evaluation—data gathered to improve the product during development—and *summative* evaluation—data gathered to validate the success of the intervention after implementation—, the ideas were well established in educational technology prior to that time.

Formative evaluation.   A commitment to continuous evaluation is one of the hallmarks of ISD. Each phase of the ISD process involves making decisions or creating artifacts, which can be tested through empirical means. Given that the essence of *technology* is "the systematic application of scientific or other organized knowledge" (Galbraith, 1967, p. 12), it is the formative evaluation function that most contributes to making ISD a *technological* process.

ISD models use various graphic devices to illustrate continuous formative evaluation. Morrison, Ross, and Kemp's (2004) model shows this principle by depicting the formative evaluation function as an ellipse, surrounding and interacting with all the other functions. The strategic impact model (Molenda & Pershing, 2004), discussed at greater length in chapter 3, arranges the analysis, design, development, production, and implementation functions around evaluation and revision, depicted as the heart of the ISD process.

Summative evaluation. Summative evaluation intends to determine the ultimate effectiveness of the interventions, often referred to as verification or validation. It is conducted after the artifacts or systems have been implemented with users. A major issue is what should be measured to determine success. A widely accepted framework is that of Kirkpatrick's (1998) four levels (1998), which proposed that one could evaluate program success by any of four criteria: (1) the reaction or satisfaction of learners, (2) the attainment of learning objectives, (3) the transfer of learning to real life tasks, or (4) organizational results, that is, the overall impact of the intervention on the organization's goals. The selection of any of these targets could be justified, depending on circumstances.

Project Management. Instructional development projects that require more than one person or consume more than a few hours of work may demand some attention to their organization and control. The formality of project management usually increases as the scale of the project rises. Projects with large budgets, especially publicly funded ones, or with serious consequences for failure will require tight monitoring and control. For small projects, it is often better to tolerate some false starts and scheduling slippage rather than stifling creativity with officious management. One of the major findings of McCombs's (1986) review of research on ISD was that when military supervisors required training designers to document each step of their ISD work they either developed boring "paint by numbers" lessons or they forged ahead with creative designs and filled out the paperwork afterward. As Maguire (1994) put it, "the surest way to mismanage a project and jeopardize the product is to put so much emphasis on the schedule that it demoralizes the team and drives them to make stupid decisions . . ." (p. 105).

Several generic management issues arise in instructional development projects of larger scope (Foster, 1993). The first set of issues revolve around advance planning: determining the overall project objectives, scheduling for each phase, preparing operating procedures for the project, preparing a budget, and securing funding. For such projects it is particularly important to anticipate project milestones and be clear about what the "deliverables" will be, when the client will receive them, and how quickly reactions and approvals must be received (Morrison et al., 2004).

Instructional design is a social process as much as or more than it is a technology process (Schwen, Leitzman, Misanchuk, Foshay, & Heitland, 1984), so interpersonal issues are the second concern. The process is shaped in important ways by the social relationships among the members of the design team, between the design team, the client, and other stakeholders, and between the design activity and the institutional social setting (Durzo, Diamond, & Doughty, 1979). In practical terms, someone must exercise leadership in establishing a working relationship with the client or sponsor, championing the goal of the project, selecting and motivating team members, and arranging for healthy communication among stakeholders.

The third set of issues has to do with organizing: creating an organizational structure, allocating tasks, delegating responsibility, and maintaining a productive work environment. The fourth large issue is day-to-day monitoring and controlling as the development is actually carried out: deciding on evaluation criteria, conducting formative and summative evaluation, taking corrective actions, and holding to the schedule. The final concern is terminating the project gracefully and preparing a final report, which might include an operations analysis; this analysis of what worked and what went wrong can help the team learn how to do better in the future.

In the complex arena of developing immersive learning environments and similar interactive systems, the thorniest management problems revolve around the actual production process—how to juggle the technical demands of computer programming, animation, graphics, and other specialties while keeping the focus on the learning goals. This issue is discussed later regarding blended learning environments.

Project management software. It is now a routine practice to use project management software to guide ID projects of larger scope. Generic programs such as Microsoft Project™ provide templates for quickly organizing the planning, scheduling, monitoring, and budgeting activities of ID projects. Software for carrying out the actual design and development steps is discussed later in relation to blended learning environments and the automation of ID.

Design Approaches Other Than ISD

Although models based on the systems approach are the most widely discussed and taught, and possibly, practiced, there are numerous alternative ways to think about developing instruction. Dills and Romiszowski (1997) provide extended descriptions of several dozen approaches, including cybernetic, behavioral analysis, situated cognition, semiotic, Direct Instruction, constructivist, existentialist, structural communications, rapid collaborative prototyping, simulations, and intelligent tutoring, among others. Many of these are not intended as fully formed guides to the whole process of planning instruction. Some deal only with parts of the development process, especially offering different options for instructional strategies and tactics at the design stage. Others are larger than instructional development, offering a different philosophical slant on learning or instruction as a whole. Other models address particular types of learning goals. For example, van Merriënboer's (1997) four-component ID model aims to guide the design of learning environments for attaining complex technical skills. It traces the steps of determining and then practicing the cognitive operations needed for mastery of such complex skills. The point is that despite the appearance of an orthodoxy regarding the design-development process, there is actually a broad diversity of viewpoints to draw upon even within the domain of process models.

Alternative Design Traditions. One alternative view of the larger design process is that process models cannot describe fully or direct effectively successful design efforts for any but the simplest situations. In this view, design is seen as a space in which creators of artifacts (e.g., materials, experiences) grapple with multiple tensions and desires from multiple sources. Their efforts at problem solving within this space are based on rich experiential knowledge and training in habits of thought and performance that guide them (Goel, 1995; Rowe, 1987). The distinction between this view and the prevalent model-centric view in educational technology is illustrated by Rowe (1987) as he writes about the proliferation of process models following the systems thinking revolution of the 1950s in architecture. He describes the failure of "phase" or "staged process" models for that field. In his description, these models are similar to those used in instructional design, being "characterized by dominant forms of activity, such as analysis, synthesis, evaluation and so on" (p. 46). Rowe observes, "What seemed necessary [at the time of their development] was a clear and logical procedure for producing designs and plans that could be understood and participated in by all those involved" (p. 111). He acknowledges the conceptual understanding that was gained through the effort, but

> in spite of the very real contributions that were made, at least to our understanding of these processes, in almost all cases the step beyond description

to a normative realm in which process became pursued as an end in itself resulted in abject failure. Attempts to devise the process became exercises in inanity when compared to the great subtlety and profundity of observed problem-solving behavior. (Rowe, 1987, p. 111)

Similar struggles over prescriptive "waterfall" process models have taken place in software engineering. They, too, have applied the ADDIE mentality and found that it can grow into highly prescriptive routines that require large manuals to describe. As in instructional design, people can belabor this approach into one that is slow and cumbersome. Douglas (2006) describes alternative approaches being explored in software engineering, including "agile design"; these alternative approaches are more adaptive to the situation and more people oriented. Seen from these alternative perspectives, the centrality of highly prescriptive step-by-step ISD models within this field might be questioned.

In *The Design Way*, Nelson and Stolterman (2003) detail a philosophy of design as a tradition and culture; that is, a way of life with multiple facets including the internal (development of judgment, a sense of responsibility for the effect of one's designs, etc.) as well as external (data gathering, systematic analysis, etc.). In this view, the designer does not follow a model of design process, or inhabit a "design space" as a skilled actor, but inhabits the world at large as a member of the design tradition. In this view, design is not primarily a matter of solving problems (which, even if they have no clear-cut solution, are by definition solvable), but a matter of shaping the world toward desired, and perpetually unknown, states. This shaping is carried out from a posture of service to that world, which implies equal-status participation from the world—that is, the designer is not the knowledgeable solution provider, but the legitimate collaborator with those on whose behalf design is being carried out. In this view, the character of designers, not just their behavior or skills or knowledge, is fundamental and is the source of a flexible process.

User design. A limitation of traditional ISD is that it involves end users— teachers and learners—very little in the design process. On the one hand, this deprives users of the power to control and learn from their own work of knowledge construction. On the other hand, it handicaps designers in that they miss out on insights that could be offered by users, and their products often face neglect or resistance by users. The concept of user design attempts to rectify this power imbalance. Burkman (1987) was an early advocate of improving the efficacy of instructional design products by involving end users in the design process, on the basis that people are more likely to accept and use solutions that they helped design.

Carr-Chellman and Savoy (2004) portray a range of design approaches from user-based, to user-centered, to truly user-controlled or emancipatory design, which can be transformational for learners and the institutions in which they operate. They also discuss the difficulties of such user-controlled approaches, in terms of time expenditure and tension in the power dynamics among the participants. This is an area in which research has yet to reveal an optimal solution for the benefit of all stakeholders in learning.

Design research. As envisioned by Laurel (2003), coming from a software development environment, the process of design should involve a full spectrum of research tools based on the goals of the particular design enterprise. In the case of instructional design, Carr-Chellman and Savoy (2004) describe a range of learner involvement, from responding to surveys and questionnaires to participating as full partners in action research (p. 712).

Rapid prototyping. The concept of rapid prototyping refers to the early development of a small-scale prototype in order to test key features of the design (Wilson, Jonassen, & Cole, 1993). This idea is not entirely new to educational technology, being foreshadowed in the 1950s in the educational filmmaking practice of preparing "treatments" for review prior to production. And in an early ISD model, Diamond (1975) advocated visualizing an ideal solution and discussing it with the client as an early step in the ISD process. However, Tripp and Bichelmeyer (1990) demonstrated how this notion could be adapted from software engineering to instructional design to address the problem of escalating expense of ID, especially in the corporate realm. They recommend a four-level process including the stages of performing a needs analysis, constructing a prototype, utilizing the prototype to perform research, and installing the final system.

Learning Environments

Using the term loosely, a *learning environment* could be anything from a classroom, to a school, to a state of mind. Within the context of educational technology, it means a physical or virtual space that has been designed to provide optimal conditions for learning, including access to rich resources, possibly focused on a problem and possibly supporting exploratory learning. A computer-based simulation such as SimCity™ could be considered a learning environment. The Math Emporium (described in chapter 3), a physical space with self-instructional, computer-mediated learning resources and live tutors, is another example of a learning environment—a self-contained system that is highly supportive of focused learning.

The creation of sensorially rich, empowering learning environments has a long tradition in educational technology. In the 1940s, Edgar Dale (1946) based his audiovisual pedagogy on "rich experiences . . . flavored with direct sense-experience [having] a quality of newness, freshness, creativeness, and adventure, and . . . marked by emotion" (p. 23).

George Leonard (1968) envisioned an elementary school of the future as a free, open, learner-centered environment. Although Leonard's utopian vision has not materialized as such, he was prescient regarding the realization of many of the elements of the future school. On this future campus, each child has an individual educational plan (mandated by law for special-needs learners in 1975) and pursues a curriculum that includes experiences in the interpersonal, intrapersonal, kinesthetic, and many other domains (à la Howard Gardner's 1983 theory of multiple intelligences, proposed and implemented in some experimental schools). They learn basic skills through interaction with brilliant projected displays (e.g., the plasma screen, developed in Bitzer's PLATO CAI lab, patented in 1971) by means of computer-controlled input devices (also as in the PLATO lab). The subject matter comes from a cross-matrix data bank, which allows random retrieval from a "general cultural data bank" (Leonard, 1968, p. 145; akin to the World Wide Web, operational in 1992). Students share images on their displays with other students (as with the DyKnow Vision™ tablet PC, in use in 2000). Leonard's was a rich environment involving both face to face and mediated activities that cover an array of intellectual, athletic, artistic, spiritual, and moral experiences. More recent developments in technology and pedagogy have fueled new visions of ideal learning environments.

Constructivist, Rich Environments for Active Learning (REAL)

Rich environments for active learning (REALs) are comprehensive instructional systems that incorporate the features considered desirable according to the constructivist perspective, namely, to promote study and exploration within authentic contexts; to encourage individual learner responsibility and initiative; to cultivate collaboration among students and teachers; to support dynamic, generative learning activities; and to use authentic assessment to determine learner achievement (Grabinger, 1996). Cognitive flexibility theory, anchored instruction, and PBL are all theoretical constructs that have inspired the creation of REALs.

Problem-Based Learning (PBL)

Medical education is historically the most prominent venue for PBL, but it is currently being adapted to the school and college settings. In a PBL learning environment, small discussion groups of students accompanied by a facilitator are confronted with a constructed, but realistic, problem. They then engage in inquiry to understand and solve the problem. The learners discuss the issues, derive learning goals, and organize further work (e.g., literature and database searching). Learners present and discuss their findings in subsequent sessions. They then apply the results of their self-directed learning to solve the problem. A PBL cycle concludes with reflections on learning, problem solving, and collaboration (Savery & Duffy, 1996).

Blended Learning Environments: Real, Simulated, Virtual, and Mixed

One of the cutting edge areas for technology-based learning is the creation of immersive environments that blend elements of real life, computer simulation, video games, and virtual reality in various hybrid combinations (Kirkley, S. E., & Kirkley, J. R., 2005). For example, on a field trip to a wetland students who are investigating the effects of pollution might wear headgear showing heads-up displays of information about water quality and wildlife in the area. Or military trainees might practice conducting a search using Arabic language in a Middle Eastern village by way of a notebook PC displaying a 3-D simulation of the village and its inhabitants; the virtual villagers respond to the trainees' questions, which are analyzed with voice-recognition software. By adding a scoring mechanism, this simulation could include a gaming element.

These immersive environments usually include some level of simulation, open-ended representations of "evolving situations with many interacting variables" (Gredler, 2004). The pedagogical value of simulations is that they allow users to play roles, deal with problems, and experience the consequences, thus learning by doing (Gredler, 2004, p. 571).

A common variation on the digital simulation is a microworld—a computer-based exploratory environment that "feels like" a miniature self-contained world in which a participant can explore alternatives, test hypotheses, and discover facts about that world. It differs from a simulation in that the participant is encouraged to view it as a real world in its own right, and not simply as a simulation of some slice of reality. Educational microworlds have been built for the study of physics (ThinkerTools), mathematics (Sim Calc), and genetics (GenScope), among other subjects (Rieber, 2004).

Many other combinations of these immersive elements are possible, for which there are not yet even agreed upon names. What they have in common is the goal of creating environments in which learners experience realistic problems in lifelike settings. Such environments allow learners to manipulate variables that are interconnected as in the real world, allowing them to find patterns and see how different actions affect outcomes, allowing learning to occur inductively. S. E. Kirkley and J. R. Kirkley (2005) see great potential for mixed reality environments, especially when they include game-type activities, but they also acknowledge that such complex immersive simulations could pose challenges for learners who are novices in the subject matter or the technology. They also pose challenges for designers.

Creating immersive environments.   The first challenge of complex immersive environments is that the design process requires multidisciplinary teams, which might include "not only instructional designers and subject matter experts but game and interaction designers as well as graphic designer/modelers, programmers and perhaps even script writers and actors" (Kirkley, S. E., & Kirkley, J. R., 2005, p. 49). Each of these specialties may have its own design processes and technical issues, all of which have to be coordinated. S. E. Kirkley and J. R. Kirkley (2005) have developed an authoring tool, IIPI Create™, which can guide teams through the phases of analyzing the learning needs, translating those needs into objectives and evaluation criteria, and developing activities and environments to achieve those objectives (p. 50). Appelman (2005) recommends focusing on the learner's experience, or experiential mode, and elaborating the details of the learning environment as one would build a concept map (p. 72). Because of the difficulties of creating such programs, they tend to be concentrated in areas where traditional training is too expensive or dangerous or both.

Trends and Issues Related to Creating

Analog-Digital Dilemma

A dominant trend in educational technology since the 1990s is the uneasy coexistence of the whole panoply of analog media (e.g., slides, audiocassettes, videocassettes, films, overheads, etc.) alongside the expanding array of digital (computer-based) media. Both classes of media are characterized by a multitude of incompatible formats and conflicting standards.

Instructors—in schools, colleges and universities, and in corporate settings—still frequently rely on the older, more familiar media formats, such as video, slides, and overhead projection. In schools and universities, VHS videocassettes still constitute the backbone of media collections, and they are still widely used for showing motion images. In corporate training, video recordings are still used in over half the companies responding to the *Training* magazine annual survey (Dolezalek, 2004). Slides in the traditional two-by-two format are still preferred for subjects in which high-resolution visual images are critical. Overhead projectors remain a convenient format for spontaneous creation of verbal or graphic images (Molenda & Bichelmeyer, 2005). Instructors understand the value of communal viewing of certain kinds of materials on a large screen with a high-definition image (as with theatrical films). They resist giving up the analog capability until these sorts of experience can be matched by the digital media.

From the administrative point of view, a lot of capital and human resources are tied up in acquiring, maintaining, and moving around the hardware needed for this usage. Even more time and effort is expended in development projects to produce new customized software in the analog formats. Typically, the output of such projects is too specialized to be adopted or even adapted by other instructors. Therefore, such projects are both expensive and low impact (South & Monson, 2001).

At the same time, educational administrators are struggling to meet the demand for more and more computer-based infrastructure. The hardware needs to be constantly upgraded, while software becomes obsolete at a dizzying pace. The capital and human costs of this proliferation of media formats and the attendant complexities of working are daunting.

The way out of this dilemma being chosen by many institutions is to gradually reduce support for analog media and to shift to a policy of acquiring and producing future materials in digital format. The media production head at a major university reports that "the production tools we now use are all digital in nature. . . . We shoot broadcast quality digital videotape, edit video and author DVDs on computers, and output the complete product to a digital format (DVD or Web)" (R. Zuzulo, personal communication, March 3, 2006).

Further, organizations are looking for standardized formats in order to increase compatibility across departments, even to the point of imagining a single database for all the organization's instructional media. Such standardization would move in the direction of reusable learning objects, thus introducing the possibility of actually reducing the cost of providing instructional media as needed by instructors.

Critique of ISD

From time to time since the 1990s, various voices have questioned the continuing viability of the ISD approach to instructional design. The recent criticisms have come primarily from two directions. The first is from corporate training specialists, who say ISD is too expensive in terms of time and manpower in view of the results given. The second is from academics with a commitment to the constructivist view of teaching and learning, who feel that ISD springs from the behaviorist paradigm and therefore inherently leads to inadequate solutions.

The corporate criticism was probably presented most strongly by Gordon and Zemke (2000), who quoted experts charging that the ISD approach is too slow and clumsy for the fast changing digital environment, fails to focus on what is most important, and tends to produce uninspired solutions. A follow-up article (Zemke & Rossett, 2002) examined these questions more closely and concluded that there were valid points on both sides of these questions, but that shortcomings were more often the fault of the people using the process rather than the process itself. They concluded, "ISD is the best thing we have, if we use it correctly" (Zemke & Rossett, 2002, p. 35).

Other critics have focused on Gordon and Zemke's (2000) first criticism—that ISD is ill suited to the digital environment, which typically requires quick turnaround lest the problem change or disappear before the solution is completed. A recent review of alternative models for online distance education (Schoenfeld & Berge, 2004) indicates that many of them are adaptations of the ADDIE outline, with special features in one or more of the major stages. One popular concept that appears in a number of models is rapid prototyping, discussed earlier in this chapter. It suggests the early creation of a rough prototype of proposed solution, then testing and revising increasingly full and finished versions of the solution. The paradigm at the heart of such a process is successive approximation, rather than the linear process implied in the ADDIE approach.

Constructivism may be viewed as a challenge to ISD either at the level of selecting instructional methods or at the broad philosophical level (Dick, 1997). At the methods level, *constructivism* is a label for a learner-centered pedagogy based on widely accepted principles from cognitive psychology. As such, it is possible to use constructivist prescriptions to design more immersive, problem-centered activities. So, the systems approach remains the guiding paradigm at the strategic level, but at the tactical level some constructivist techniques can be employed.

Viewed at the broad philosophical level, constructivism is an alternative paradigm to previous theories of learning and instruction. Some claim, therefore, that it requires a design and development process that is entirely different. Willis and Wright (2000) propose guidelines for *constructivist instructional design*, which entail a participatory team engaging in a spiral process of progressive clarification of the problem space, the learning strategies to be used, and the very objectives of the lesson. This process would involve rapid prototyping and frequent learner input.

Complexity of Instructional Design and the Need for Automation

Advances in technology have not made planning and producing easier, but more difficult (Spector, 2001) and more labor intensive, with each hour of interactive multimedia instruction requiring 300 person hours of development time (Merrill, M. D., & ID2 Research Group, 1998). As described earlier, creating CAI lessons or immersive learning environments entails huge expenditures of manpower just for the computer programming. In addition, the overall project management can be highly complex and time consuming. Development projects may include not only multiple computer-based media but also databases and performance support systems, requiring communication and co-ordination among a team with widely varying specialties:

> An instructional designer may interact with managers, with people performing training tasks, with subject experts, with system specialists, and so on. A designer proposes solutions and defends project plans, manages a project, chooses media, develops storyboards and other products, conducts evaluations, and so on. . . . As project complexity grows, so does the need to collaborate and to coordinate activities. (Spector, 2001, p. 31)

Many attempts have been made or are under way to manage this complexity with software. First, as described earlier, authoring software was developed to reduce the difficulty of computer programming for routine CAI. In the 1980s and through the 1990s, M. D. Merrill and the ID2 Research Group (1998) worked at automating the process of selecting instructional strategies for different learning needs and also the process of creating routine lessons based on the selected instructional strategies. They later developed a prototype product, "ID2 Instructional Simulator," for building exploratory learning environments (Merrill, M. D., & ID2 Research Group, 1998, p. 261). All of these products are based on the creation and reuse of knowledge objects. However, the use of these systems has not spread far beyond the organizations directly involved in the development. Conceptual and technical issues have continued to impede the automation of instructional design (Spector, Polson, & Muraida, 1993).

Conclusion

The processes related to *creating* in educational technology have evolved greatly over time and changing technologies, as have the theories underlying them. The early mass media that were adapted to educational purposes— film, radio, and television—were largely shaped by the paradigms of their commercial counterparts. Script-based programs followed the protocols of historical reenactments, demonstrations, ethnographies, and the other genres found in the commercial world. Experiments, based first on Gestalt and cognitive theories and later on behaviorist theories, provided insights for refining AV presentations that contributed to cognitive, affective, and motor skill learning. Evaluation procedures also contributed to improvement of individual programs.

A more systemic and systematic procedure for planning and producing instructional media evolved after World War II under the influence of the systems approach and behavioral learning management protocols. Used at first to produce programmed instruction lessons, instructional systems development (ISD) models, which took many locally adapted forms, came to be applied generally to the planning and production of all sorts of instructional materials and systems. The common denominator of most ISD models is the logical progression from analysis to design, to development, to implementation, to evaluation in an iterative cycle.

When mechanical teaching machines were replaced with programmable computers the ISD process remained, but the production stage required a whole new set of skills in computer programming or at least in using authoring software. As the Internet grew in popularity in the 1980s and 1990s, education and training programs searched for ways to incorporate computer conferencing into distance education programs. When the World Wide Web emerged as the dominant Internet service, designers were able to combine student-to-student and student-to-instructor interaction with static text or moving images into one lesson package whose components were connected with hyperlinks, allowing users to explore the resources more or less freely.

This new capability gave impetus to discovery learning and PBL systems inspired by constructivist theories.

In addition to the systems approach, design approaches to creating in educational technology have been borrowed and adapted from many other fields, including visual arts, software design, sociotechnical systems design, organizational development, and cognitive psychology, to name a few. One of the future challenges is to decide whether to retain, adapt, or discard systems approach models and to find ways of thinking about design that are productive for the changing media environment of the 21st century.

With the continued miniaturization and convergence of media under the computer umbrella, instructional developers face new technical challenges in terms of ever-changing programming languages and authoring systems. They also confront whole new mindsets about what a learning environment is and how it should be structured, particularly in regard to the sort of guidance that learners should have as they grapple with problem scenarios and open-ended databases of real or simulated information. Complex immersive environments, which may combine elements of reality, simulation, and virtual reality, hold promise for meaningful, PBL. They also bring new design and development challenges, require orchestration of many different specializations, each with different vocabularies and design approaches.

References

Alexander, B. (2004, September/October). Going nomadic: Mobile learning in higher education [Electronic version]. *EDUCAUSE Review, 39*(5), 29–35.

Allen, W. H. (Ed.). (1960). *Audio-visual leadership: A summary of the Lake Okoboji Audio-Visual Leadership Conferences, 1955–1959.* Iowa City, IA: State University of Iowa.

Anglin, G. J., Vaez, H., & Cunningham, K. L. (2004). Visual representations and learning: The role of static and animated graphics. In D. H. Jonassen (Ed.), *Handbook of research on educational communications and technology* (2nd ed., pp. 865–916). Mahwah, NJ: Lawrence Erlbaum Associates.

Appelman, R. (2005). Designing experiential modes: A key focus for immersive learning environments. *TechTrends, 49*(3), 64–74.

Appelman, R., & Boling, E. (2005). *R541 Instructional development I syllabus.* Retrieved March 29, 2006, from http://www.indiana.edu/%7Eistr541/boling/index.html

Barron, A. E. (2004). Auditory instruction. In D. H. Jonassen (Ed.), *Handbook of research on educational communications and technology* (2nd ed., pp. 949–978). Mahwah, NJ: Lawrence Erlbaum Associates.

Barson, J. (1967). *Instructional systems development, a demonstration and evaluation project* (U.S. Office of Education, Title II-B Project OE 3-16-025). East Lansing, MI: Michigan State University.

Berlo, D. K. (1960). *The process of communication: An introduction to theory and practice.* New York: Holt, Rinehart & Winston.

Bichelmeyer, B., & Molenda, M. (2006). Issues and trends in instructional technology: Gradual growth atop tectonic shifts. In M. Orey, V. J. McClendon, & R. M. Branch (Eds.), *Educational media and technology yearbook 2006* (Vol. 31, pp. 3–32). Westport, CT: Libraries Unlimited.

Bloom, B. S. (1968). Learning for mastery. *Evaluation Comment, 1*(2), 1–5.

Branson, R. K. (1978, March). The interservice procedures for instructional systems development. *Educational Technology, 18*(3), 11–14.

Bruner, J. (1960). *The process of education.* Cambridge, MA: Harvard University Press.

Buck, G. H. (2006). The first wave: The beginnings of radio in Canadian distance education. *Journal of Distance Education, 21*(1), 75–88.

Burke, R. L. (1982). *CAI Sourcebook.* Englewood Cliffs, NJ: Prentice-Hall.

Burkman, E. (1987). Factors affecting utilization. In R. M. Gagne (Ed.), *Instructional technology: Foundations* (pp. 429–455). Hillsdale, NJ: Lawrence Erlbaum Associates.

Carr-Chellman, A., & Savoy, M. (2004). User-design research. In D. H. Jonassen (Ed.), *Handbook of research on educational communications and technology* (2nd ed., pp. 710–716). Mahwah, NJ: Lawrence Erlbaum Associates.

Cambre, M. A. (1981). Historical overview of formative evaluation of instructional media products. *Educational Communication and Technology Journal, 29*(1), 3–25.

Chu, G. C., & Schramm, W. (1968). *Learning from television.* Washington, DC: NAEB, the National Society of Professionals in Telecommunications.

Clark, R. C., & Lyons, C. (2004). *Graphics for learning: Proven guidelines for planning, designing, and evaluating visuals in training materials.* San Francisco: Pfeiffer.

Corry, M. D., Frick, T. W., & Hansen, L. (1997). User-centered design and usability testing of a Web site: An illustrative case study. *Educational Technology Research and Development, 45*(4), 65–76.

Dale, E. (1946). *Audio-visual methods in teaching.* New York: The Dryden Press.

Davis, J. R., & Davis, A. B. (1998). *Effective training strategies.* San Francisco: Berrett-Koehler.

Diamond, R. (1975). *Instructional development for individualized learning in higher education.* Englewood Cliffs, NJ: Educational Technology Publications.

Dick, W. (1997, September/October). Better instructional design theory: Process improvement or reengineering? *Educational Technology, 37*(5), 47–50.

Dick, W., Carey, L., & Carey, J. O. (2005). *The systematic design of instruction* (6th ed.). Boston: Allyn & Bacon.

Dills, C. R., & Romiszowski, A. J. (Eds.). (1997). *Instructional development paradigms.* Englewood Cliffs, NJ: Educational Technology Publications.

Dolezalek, H. (2004, October). Industry report 2004. *Training, 41*(10), 20–36.

Douglas, I. (2006). Issues in software engineering of relevance to instructional design. *TechTrends, 50*(5), 28–35.

Durzo, J. J., Diamond, R. M., & Doughty, P. L. (1979). An analysis of research needs in instructional development. *Journal of Instructional Development, 2*(4), 4–11.

Dwyer, F. M. (1972). *A guide for improving visualized instruction.* State College, PA: Learning Services.

Dwyer, F. M. (1978). *Strategies for improving visual learning.* State College, PA: Learning Services.

Dziuban, C., Hartman, J., Juge, F., Moskal, P., & Sorg, S. (2006). Blended learning enters the mainstream. In C. J. Bonk, & C. R. Graham (Eds.), *The handbook of blended learning: Global perspectives, local designs* (pp. 195–206). San Francisco: Pfeiffer.

Facts you should know about filmstrips. (1965). San Fernando, CA: Frank Holmes Laboratories, Inc.

Finn, J. D. (1955). A look at the future of AV communication. *Audio-Visual Communication Review 3*(4), 244–256.

Finn, J. D. (1956a). What is educational efficiency? *Teaching Tools 3*(3), 113–114.

Finn, J. D. (1956b). AV development and the concept of systems. *Teaching Tools 3*(4), 163–164.

Finn, J. D. (1965). Instructional technology. *Audiovisual Instruction 10*(3), 192–194.

Fleming, M., & Levie, W. H. (1978). *Instructional message design: Principles from the behavioral sciences.* Englewood Cliffs, NJ: Educational Technology Publications.

Fleming, M., & Levie, W. H. (1993). *Instructional message design: Principles from the behavioral sciences* (2nd ed.). Englewood Cliffs, NJ: Educational Technology Publications.

Foshay, W. R., Silber, K. H., & Stelnicki, M. B. (2003). *Writing training materials that work*. San Francisco: Jossey-Bass/Pfeiffer.

Foster, G. (1993). Managing course design. *British Journal of Educational Technology, 24*(3), 198–206.

Frick, T., & Boling, E. (2002). *Effective Web instruction: Handbook for an inquiry-based process*. Unpublished manuscript.

Friend, J., Searle, B., & Suppes, P. (1980). *Radio mathematics in Nicaragua*. Stanford, CA: Institute for Mathematical Studies in the Social Sciences, Stanford University.

Gagne, R. M. (1965). *The conditions of learning*. New York: Holt, Rinehart and Winston.

Gagne, R. M. (1977). *The conditions of learning* (3rd ed.). New York: Holt, Rinehart and Winston.

Gagne, R. M., & Medsker, K. L. (1996). *The conditions of learning: Training applications*. Fort Worth, TX: Harcourt Brace College Publishers.

Gagne, R. M., Wager, W. W., Golas, K. C., & Keller, J. M. (2005). *Principles of instructional design* (5th ed.). Belmont, CA: Thomson/Wadsworth.

Galbraith, J. K. (1967). *The new industrial state*. Boston: Houghton Mifflin.

Gardner, H. (1983). *Frames of mind: The theory of multiple intelligences*. New York: Basic Books.

Godfrey, E. P. (1967). *The state of audiovisual technology: 1961–1966*. Washington, DC: Department of Audiovisual Instruction, National Education Association.

Goel, V. (1995). *Sketches of thought*. Cambridge, MA: MIT Press.

Gordon, J., & Zemke, R. (2000, April). The attack on ISD. *Training, 37*(4), 42–53.

Grabinger, R. S. (1996). Rich environments for active learning. In D. H. Jonassen (Ed.), *Handbook of research for educational communications and technology* (pp. 665–692). New York: Macmillan.

Graham, C. R. (2006). Blended learning systems: Definition, current trends, and future directions. In C. J. Bonk, & C. R. Graham (Eds.), *The handbook of blended learning: Global perspectives, local designs* (pp. 3–21). San Francisco: Pfeiffer.

Gredler, M. E. (2004). Games and simulations and their relationships to learning. In D. H. Jonassen (Ed.), *Handbook of research on educational communications and technology* (2nd ed., pp. 571–581). Mahwah, NJ: Lawrence Erlbaum Associates.

Gustafson, K. L., & Branch, R. M. (2002). *Survey of instructional development models* (4th ed.). Syracuse, NY: ERIC Clearinghouse on Information & Technology.

Harless, J. (1973). An analysis of front-end analysis. *Improving Human Performance, 2,* 229–244.

Hartley, J. (1986). *Designing instructional text* (2nd ed.). London: Kogan Page.

Hartley, J. (1996). Designing instructional text and informational text. In D. H. Jonassen (Ed.), *Handbook of research for educational communications and technology* (pp. 917–947). New York: Macmillan.

Heinich, R., Molenda, M., & Russell, J. D. (1993). *Instructional media and the new technologies of instruction* (4th ed.). New York: Macmillan.

Hoban, C. F., & Van Ormer, E. B. (1970). *Instructional film research 1918–1950.* New York: Arno Press.

Hoffman, J. (2006). Why blended learning hasn't (yet) fulfilled its promises: Answers to those questions that keep you up at night. In C. J. Bonk, & C. R. Graham (Eds.), *The handbook of blended learning: Global perspectives, local designs* (pp. 27–40). San Francisco: Pfeiffer.

Houghton, H. A., & Willows, D. M. (1987). *The psychology of illustration: Vol. 2: Instructional issues.* New York: Springer-Verlag.

Hovland, C. I., Lumsdaine, A. A., & Sheffield, F. D. (1949). *Experiments on mass communication: Vol. 3: Studies in social psychology in World War II.* Princeton, NJ: Princeton University Press.

Jonassen, D. H. (1982). *The technology of text: Principles for structuring, designing, and displaying text.* Englewood Cliffs, NJ: Educational Technology Publications.

Jonassen, D. H., & Mandl, H. (Eds.). (1990). *Designing hypermedia for learning.* Berlin, Germany: Springer-Verlag.

Jonassen, D. H., Tessmer, M., & Hannum, W. H. (1999). *Task analysis methods for instructional design.* Mahwah, NJ: Lawrence Erlbaum Associates.

Kemp, J., & Smellie, D. C. (1994). *Planning, producing, and using instructional technologies.* New York: HarperCollins.

Kerr, S. (1983). Inside the black box: Making design decisions for instruction. *British Journal of Educational Technology, 14*(1), 45–58.

Kirkley, S. E., & Kirkley, J. R. (2005). Creating next generation blended learning environments using mixed reality, video games and simulations. *TechTrends, 49*(3), 42–53, 89.

Kirkpatrick, D. L. (1998). *Evaluating training programs: the four levels* (2nd ed.). San Francisco: Berrett-Koehler.

Laurel, B. (Ed.). (2003). *Design research: Methods and perspectives*. Cambridge, MA: The MIT Press.

Leshin, C. B., Pollock, J., & Reigeluth, C. M. (1992). *Instructional design strategies and tactics*. Englewood Cliffs, NJ: Educational Technology Publications.

Leonard, G. B. (1968). *Education and ecstasy*. New York: Delacorte Press.

Levenson, W. B., & Stasheff, E. (1952). *Teaching through radio and television* (Rev. ed.). New York: Rinehart & Co.

Lewis, N. J., & Orton, P. Z. (2006). Blended learning for business impact: IBM's case for learning success. In C. J. Bonk, & C. R. Graham (Eds.), *The handbook of blended learning: Global perspectives, local designs* (pp. 61–75). San Francisco: Pfeiffer.

Lohr, L. (2003). *Creating graphics for learning and performance: Lessons in visual literacy*. New York: Prentice Hall.

Lumsdaine, A. A., & Glaser, R. (Eds.). (1960). *Teaching machines and programmed learning: A source book*. Washington, DC: Department of Audiovisual Instruction, National Education Association.

Maguire, S. (1994). *Debugging the development process: Practical strategies for staying focused, hitting ship dates, and building solid teams*. Redmond, WA: Microsoft Press.

Markle, S. M., & Tiemann, P. W. (1967). *Programming is a process* [Sound filmstrip]. Chicago: University of Illinois at Chicago.

Mascelli, J. V. (1965). *The 5 c's of cinematography*. Beverly Hills, CA: Silman-James Press.

McBride, W. (Ed.). (1966). *Inquiry: Implications for televised instruction*. Washington, DC: National Education Association.

McCombs, B. L. (1986). The instructional systems development (ISD) model: A review of those factors critical to its successful implementation. *Educational Communication and Technology Journal, 31*(4), 187–199.

Medsker, K. L., & Holdsworth, K. M. (Eds.). (2001). *Models and strategies for training design*. Washington, DC: International Society for Performance Improvement.

Merrill, M. D. (2002a). First principles of instruction. *Educational Technology Research and Development, 50*(3), 43–59.

Merrill, M. D. (2002b). A pebble-in-the-pond model for instructional design. *Performance Improvement, 41*(7), 39–44.

Merrill, M. D., & ID2 Research Group. (1998). ID Expert™: A second generation instructional development system. *Instructional Science, 26*, 243–262.

Merrill, P. F. (2005). Using XML to separate content from the presentation software in eLearning applications. *TechTrends, 49*(4), 34–40.

Misanchuk, E. R., Schwier, R. A., & Boling, E. (2000). *Visual design for instructional multimedia.* Saskatoon, Saskatchewan: M4 Multimedia & Copestone Publishing.

Molenda, M., & Bichelmeyer, B. (2005). Issues and trends in instructional technology: Slow growth as economy recovers. In M. Orey, J. McClendon, & R. M. Branch (Eds.), *Educational media and technology yearbook 2005* (Vol. 30, pp. 3–28). Englewood, CO: Libraries Unlimited.

Molenda, M., & Pershing, J. A. (2004, March/April). The strategic impact model: An integrative approach to performance improvement and instructional systems design. *TechTrends, 48*(2), 26–32.

Mood, A. (1964, April). *Some problems inherent in the development of a systems approach to instruction.* Paper presented at Conference on New Dimensions for Research in Educational Media Implied by the Systems Approach to Education, Syracuse University, Syracuse, New York.

Moore, D. M., Burton, J. K., & Myers, R. J. (2004). Multiple-channel communication: The theoretical and research foundations of multimedia. In D. H. Jonassen (Ed.), *Handbook of research on educational communications and technology* (2nd ed., pp. 979–1005). Mahwah, NJ: Lawrence Erlbaum Associates.

Moore, D. M., & Dwyer, F. M. (Eds.). (1994). *Visual literacy: A spectrum of visual learning.* Englewood Cliffs, NJ: Educational Technology Publications.

Morrison, G. R., Ross, S. M., & Kemp, J. E. (2004). *Designing effective instruction* (4th ed.). New York: John Wiley & Sons.

Nelson, H., & Stolterman, E. (2003). *The design way: Intentional change in an unpredictable world: Foundations and fundamentals of design competence.* Englewood Cliffs, NJ: Educational Technology Publications.

Phillips, M. G. (1966). Learning materials and their implementation. *Review of Educational Research, 36*(3), 373–379.

Reeves, B., & Nass, C. (1996). *The media equation: How people treat computers, television, and new media like real people and places.* New York: Cambridge University Press.

Reigeluth, C. M. (Ed.). (1983). *Instructional-design theories and models.* Hillsdale, NJ: Lawrence Erlbaum Associates.

Reigeluth, C. M. (Ed.). (1999). *Instructional-design theories and models: A new paradigm of instructional theory, Volume II.* Mahwah, NJ: Lawrence Erlbaum Associates.

Richey, R. C., Fields, D. C., & Foxon, M. (2001). *Instructional design competencies: The standards* (3rd ed.). Syracuse, NY: ERIC Clearinghouse on Information & Technology.

Rieber, L. P. (2004). Microworlds. In D. H. Jonassen (Ed.), *Handbook of research on educational communications and technology* (2nd ed., pp. 583–603). Mahwah, NJ: Lawrence Erlbaum Associates.

Rossett, A. (1987). *Training needs assessment.* Englewood Cliffs, NJ: Educational Technology Publications.

Rowe, P. (1987). *Design thinking.* Cambridge, MA: The MIT Press.

Saettler, P. (1990). *The evolution of American educational technology.* Englewood, CO: Libraries Unlimited.

Savery, J. R., & Duffy, T. M. (1996). Problem based learning: An instructional model and its constructivist framework. In B. Wilson (Ed.), *Constructivist learning environments: Case studies in instructional design* (pp. 135–148). Englewood Cliffs, NJ: Educational Technology Publications.

Schiffman, S. S. (1986). Instructional systems design: Five views of the field. *Journal of Instructional Development, 9*(4), 14–21.

Schoenfeld, J., & Berge, Z. L. (2004). Emerging ISD models for distance training programs. *Journal of Educational Technology Systems, 33*(1), 29–37.

Schramm, W. (1962). *Programed instruction: Today and tomorrow.* New York: The Fund for the Advancement of Education.

Schriver, K. (1997). *Dynamics in document design: Creating texts for readers.* New York: John Wiley & Sons.

Schwen, T. M., Leitzman, D. F., Misanchuk, E. R., Foshay, W. R., & Heitland, K. M. (1984). Instructional development: The social implications of technical interventions. In R. K. Bass, & C. R. Dills (Eds.), *Instructional development: The state of the art II.* Dubuque, IA: Kendall Hunt.

Scriven, M. (1967). The methodology of evaluation. In R. E. Stake (Ed.), *AERA monograph series on curriculum evaluation* (No. 1). Chicago: Rand McNally.

Seels, B. B., Fullerton, K., Berry, L., & Horn, L. J. (2004). Research on learning from television. In D. H. Jonassen (Ed.), *Handbook of research on educational communications and technology* (2nd ed., pp. 249–334). Mahwah, NJ: Lawrence Erlbaum Associates.

Seels, B. B., & Richey, R. C. (1994). *Instructional technology: The definition and domains of the field.* Washington, DC: Association for Educational Communications and Technology.

Shannon, C. E. (1949). *The mathematical theory of communication.* Urbana, IL: University of Illinois Press.

Silvern, L. C. (1965). *Basic analysis.* Los Angeles: Education and Training Consultants Co.

Skinner, B. F. (1954). The science of learning and the art of teaching. *Harvard Educational Review, 24,* 86–97.

Skinner, B. F. (1968). *The technology of teaching.* New York: Appleton-Century-Crofts.

Sless, D. (1981). *Learning and visual communication.* London: Croom Helm.

Sless, D. (1986). *In search of semiotics.* Totowa, NJ: Barnes & Noble Books.

South, J. B., & Monson, D. W. (2001). A university-wide system for creating, capturing, and delivering learning objects. In D. A. Wiley II (Ed.), *The instructional use of learning objects: Online version.* Bloomington, IN: Agency for Instructional Technology and Association for Educational Communications and Technology. Retrieved October 4, 2004, from http://reusability.org/read/chapters/south.doc

Spector, J. M. (2001, October). An overview of progress and problems in educational technology. *Interactive Educational Multimedia, 3,* 27–37.

Spector, J. M., Polson, M. C., & Muraida, D. J. (Eds.). *Automating instructional design: Concepts and issues.* Englewood Cliffs, NJ: Educational Technology Publications.

Suchman, J. R. (1966). The pattern of inquiry. In W. McBride (Ed.), *Inquiry: Implications for televised instruction.* Washington, DC: National Education Association.

Television in instruction: What is possible. (1970). Washington, DC: National Association of Educational Broadcasters.

Tripp, S., & Bichelmeyer, B. (1990). Rapid prototyping: An alternative instructional design strategy. *Educational Technology Research & Development, 38*(1), 31–44.

van Merriënboer, J. J. G. (1997). *Training complex skills: A four-component instructional design model for technical training.* Englewood Cliffs, NJ: Educational Technology Publications.

Wagner, E. D. (2005, May/June). Enabling mobile learning [Electronic version]. *EDUCAUSE Review, 40,* 41–52.

Wetzel, C. D., Radtke, P. H., & Stern, H. W. (1994). *Instructional effectiveness of video media.* Hillsdale, NJ: Lawrence Erlbaum Associates.

Wiener, N. (1950). *The human use of human beings: Cybernetics and society.* Boston: Houghton Mifflin.

Wiley, D. A., II. (2006). *RIP-ing on learning objects.* Retrieved April 2, 2006, from http://opencontent.org/blog/archives/230

Wiley, D. A., II. (2002). Connecting learning objects to instructional design theory: A definition, a metaphor, and a taxonomy. In D. A. Wiley II (Ed.), *The instructional use of learning objects: Online version.* Bloomington, IN: Agency for Instructional Technology and Association for Educational Communications and Technology. Retrieved September 11, 2004, from http://reusability.org/read/chapters/wiley.doc

Willis, J., & Wright, K. E. (2000, March/April). A general set of procedures for constructivist instructional design: The new R2D2 model. *Educational Technology,* 5–20.

Wilson, B. G., Jonassen, D. H., & Cole, P. (1993). Cognitive approaches to instructional design. In G. M. Piskurich (Ed.), *The ASTD handbook of instructional technology* (pp. 21.1–21.22). New York: McGraw-Hill.

Wisher, R. A. (2006). Blended learning in military training. In C. J. Bonk, & C. R. Graham (Eds.), *The handbook of blended learning: Global perspectives, local designs* (pp. 519–532). San Francisco: Pfeiffer.

Wood, D. N., & Wylie, D. G. (1977). *Educational telecommunications.* Belmont, CA: Wadsworth.

Zemke, R., & Kramlinger, T. (1982). *Figuring things out: A trainer's guide to needs and task analysis.* Reading, MA: Addison-Wesley.

Zemke, R., & Rossett, A. (2002, February). A hard look at ISD. *Training, 39*(2), 27–35.

5

USING

Michael Molenda
Indiana University

Introduction

> Educational technology is the study and ethical practice of facilitating
> learning and improving performance by creating, *using*, and managing
> appropriate technological processes and resources.

*A*NY DEFINITION OF EDUCATIONAL TECHNOLOGY would be incomplete
without explicit acknowledgement that "using . . . appropriate technological processes and resources" is the end purpose for which the field
exists. The whole point of creating technological resources—instructional
materials and instructional systems—is that they be *used* by learners. This
term also refers to the acceptance and use of technological processes, such as
instructional systems development. It is not enough to study such processes
or to create them, such as when one proposes a new instructional design
model. Educational technology fulfills its mandate when learners actually
use instructional materials and systems and thereby benefit from the analysis and design work that has preceded the use.

This chapter will focus on the concepts and principles related to "using
technological *resources*." The use of technological *processes*, including the

processes associated with creating and managing instructional systems, is addressed in other chapters, particularly chapter 7.

The element of *using* can be understood by examining the theories and practices related to bringing learners into contact with appropriate learning conditions and resources. As such, it is the main arena, where the solution meets the problem. Using begins with the *selection* of appropriate processes and resources—methods and materials, in other words—whether that selection is done by the learner or by an instructor. Wise selection is based on *materials evaluation*, to determine if existing resources are suitable for a particular audience and purpose. If the resources involve new or unfamiliar media or methods, their *usability* may be tested before use. Then the learner's encounter with the learning resources takes place within some environment following some procedures, often under the guidance of an instructor, the planning and conduct of which can fit under the label of *utilization*. When teachers incorporate new resources into their curricular plans in an articulated fashion, this is referred to as *integration*.

In some cases, there is a conscious effort to bring an instructional innovation to the attention of potential users, to market it. Within the context of instructional development projects, this would constitute the *implementation* phase. Viewed in terms of spreading an innovation beyond its original source, to users far and wide, it can be regarded as a *diffusion* process. Thus, the element of *using*" can be viewed as a spectrum of activities ranging from an individual teacher or learner choosing one specific bit of material to a large-scale project shifting an entire organization's training strategy from one format to another, for example, from classroom instruction to online delivery.

Materials Evaluation and Selection

The use of any technology-based resource usually begins with the process of selecting specific materials, either by instructors using "off the shelf" technological materials or by media specialists maintaining collections for others to use. The selection process may begin with a search through reviews of available materials. To aid educators without the time or means to preview audiovisual materials themselves, clearinghouses such as the Educational Film Library Association (later the American Film and Video Association) has systematically collected and published evaluations from respected subject-matter experts. Many other review sources are available for other classes of audiovisual and digital media.

Selection Criteria for Instructional Materials

The decision whether or not to select a particular item depends on many factors. However, there are generic criteria pertinent to instructional materials, regardless of media format:

- Are the objectives of the material aligned with the lesson objectives?
- Does the material match the entry level of the target learners (especially reading and vocabulary level)?
- Is the information accurate and up to date?
- Is the material free from objectionable bias?
- Is the material likely to arouse and maintain learner interest?
- Does the material encourage a high level of mental engagement by the learner?
- Is the technical quality acceptable?
- Is there any evidence of success, such as results of field tests?

Research over the past half-century has examined what attributes of software are most closely connected with effective learning. The net result is an understanding that different criteria must be given priority in different circumstances. For example, a remedial reading teacher might choose a particular vocabulary game because it is likely to spark interest in her students, thus giving them the necessary practice, as a priority over other qualities of the software. On the other hand, an elementary school teacher with a class that is very diverse ethnically might give priority to materials that show special sensitivity to racial and ethnic issues, as a priority over other attributes.

Some selection criteria are specific to certain media formats. For example, video materials raise the issue of pace of presentation, which would not be pertinent to verbal and still-picture formats, such as textbooks or Web pages. On the other hand, a computer-based game or simulation might be judged primarily on how much relevant practice and feedback is offered, which would not be pertinent to teacher-presented media such as a Power-Point™ presentation.

Evaluation checklists evolved in the 1920s and 1930s for teacher appraisal of silent and sound films. Over time, these checklists have been adapted to newer media, to provide more specific guidance for various audiences and distinct subject areas. The practice of using such checklists had evolved to such a level of complexity by the late 1970s that Woodbury (1980) required a three-volume set of books to encompass the subject. In the volume devoted to selection criteria used at the teacher level for instructional materials, she provided criteria and checklists for free materials, federally funded materials, government documents, pictorial media, print materials, nonprint media, games and simulations, toys and manipulatives, television, and film.

Checklists and the selection criteria developed for audiovisual materials have been reinvented for the world of digital media. The *Educational Software Preview Guide*, in its 21st edition in 2004, is published by the International Society for Technology in Education (ISTE, 2004). The criteria listed on the Educational Technology Resource Evaluation Form include familiar considerations:

- Objectives promoted: creativity, collaboration, discovery, higher order thinking, problem solving, memorization
- Grade or ability level . . . Readability level
- Content is current, thorough, age appropriate, reliable, clear
- Content is free of bias
- Motivational qualities
- Technical quality

The checklist does not explicitly ask about evidence of effectiveness. However, it does add questions about embedded learning strategy and about built-in assessment methods:

- Learning strategy incorporated in the design
- Assessment: has pretest/posttest, record keeping by student

Some Realities of Materials Selection

Education theorists propose that teachers should begin lesson planning by focusing on learners and lesson objectives, then proceed from there to selecting the materials and activities that will reach those objectives. Since the 1970s, there have been several major studies of teachers' actual planning processes. The first, by Taylor (1970), found that secondary school teachers first directed their attention to the materials that were already at hand and the time they had in class to use them. Kerr's (1981) later research revealed a similar planning sequence. The thought processes of teachers as they plan become routinized, as Yinger (1979) discovered, in order to save planning time. He found that teachers typically began by gathering the available materials and then thinking of activities based on the materials, not by specifying objectives and conducting a search for materials that would lead toward those objectives. An ethnographic study by McCutcheon (1979) reached a similar conclusion, that elementary school teachers were primarily concerned with immediate, practical issues: Will this help me maintain order? Will this fit in the time allotted? Will these materials be available? Other studies have shown that "available" means immediately accessible, in the classroom or in the building.

So, there is evidence that teachers begin with materials that are immediately accessible, including the old, reliable textbook, and then plan outward to activities and ultimately may possibly link to curricular goals. They do not necessarily select those materials following a systematic selection process.

On the other hand, many materials appraisal and selection decisions are not made by individual teachers but rather by committees. Such committees are a ubiquitous part of the textbook selection process; they are often also used to decide on what nonprint media are purchased at the school or district level. Checklists are critical for committee work for two reasons. First, they provide a more objective way of comparing opinions, providing a framework for discussion. Thus, they ensure that truly pertinent issues will be raised and used as deciding factors. Second, they provide an ex post facto documentation of the committee's decisions, indicating not only the choices made but the rationale for those choices in case those decisions are questioned at a later time.

Usability

Hardware and software that have been created or acquired frequently have qualities that are unfamiliar to users. Users, of course, could be students, teachers, or technology support staff. A new laptop computer being purchased for a high school science club could pose challenges to the technology coordinator in dealing with how to add peripherals and load software. The science club advisor might puzzle over how to navigate through the new version of the physics simulation software. And students might struggle over using a mouse in unfamiliar ways to draw geometric shapes. Each of these could be usability issues.

Usability simply refers to the quality of being easy to use for some purpose. The International Standards Organization defines usability more formally as "the extent to which a product can be used by specified users to achieve specified goals with effectiveness, efficiency and satisfaction in a specified context of user" (Usability Professionals' Association, n.d.). Those who design materials and equipment for use in schools must think about how to make them accessible to teachers with a wide range of technological competency. Students, too, may struggle with computer software that is difficult to navigate, that has flashy graphics that divert attention from the content, or that has an unhelpful help system. If the functions and features are intuitive to use, everyone can focus on the educational value of the material rather than how to make it work.

Usability was an issue long before the computer era. Audiovisual users had to struggle with film projectors that were cumbersome to operate. Synchronized slide-tape players seemed to go out of sync all the time. Opaque projectors could build up heat so high that it caused burns. And it was not just the hardware. Studies of students' reactions to innovative multimedia programs showed that learners often focused more on the novel features of the presentation than on the content. But it was the advent of computers that brought usability problems to the forefront.

Stimulated by the pioneering work of Donald Norman (1988) and Jakob Nielsen (1994), a technology of usability engineering has evolved. The field of usability engineering recognizes many potential sources of usability problems: between user and tool, user and task, user and other users, and user and environment. In terms of software development, concern tends to focus on issues such as

- Consistency, making sure, for example, that specific colors and icons mean the same thing throughout the program and that specific functions are located in the same place
- Simplicity, keeping the layout clear and uncluttered
- Structure, easy to navigate
- Suitability to the needs and abilities of the intended users, including those with vision impairment
- Availability of online help that is actually responsive to problems.

To these issues, Booth (1989) added

- Ease of learning
- Ease of remembering
- Visibility

To ensure that products are as easy to use as possible, designers typically conduct usability testing on prototypes. Ideally, usability testing entails real users working on real tasks in their real environments. Methods such as think-aloud protocols and other observation instruments are used to determine how users react to the prototype so that problems can be detected and resolved before the product is widely distributed (Rubin, 1994). Sometimes surveys and questionnaires are also used to determine users' feelings about the prototype, their satisfaction with it.

Usability *testing* is primarily the province of designers, but judgments about usability are an important part of the job of teachers and technology specialists when making decisions about hardware and software to be acquired or used in a particular context.

Evolution of Research and Theory on Media Use

Post–World War I Period

Utilization may have the longest heritage of any of the elements in the definition, in that the regular *use* of audiovisual materials predates the widespread concern for the systematic design and production of instructional media. During the early years of the 20th century teachers were using theatrical films in the classroom, thus creating a market for films designed specifically for educational purposes. The earliest formal research on educational applications of media was Lashley and Watson's (1922) program of studies on the use of World War I military training films on the prevention of venereal disease with civilian audiences. The focus was on how these films might be used to best effect. And "use" meant instructor use. Indeed, the research during this period and the next half century tended to focus on what the instructor did with media, rather than what students did. Research dealt with media formats such as film, slides, radio, and, later, television and audio recordings. Until convenient self-instructional media formats were developed (e.g., 8mm films and audiocassette recorders) these media were usually experienced as a presentation made to a group, so the "user" was the teacher.

An early large-scale effort to design and produce a set of films specifically for schools was the *Chronicles of America Photoplays*, produced by Yale University in the late 1920s. Knowlton and Tilton (1929) studied the use of these history films in seventh grade classrooms. One of their major conclusions was that the educational value of such films lay not only in the quality of the materials but also in how well teachers *used* them:

> The ability of the pupils to grasp and appreciate these relationships was in no small degree determined by the teacher's own interest in them and the emphasis which she attached to them. However inherently effective the photoplays may be—and the evidence submitted here indicates the potentialities of such material—it will only attain its highest degree of effectiveness when accompanied by good teaching. . . . (Knowlton & Tilton, 1929, p. 91)

This finding, that the instructional value of any media product is determined largely by *how it is used*, would be rediscovered by each succeeding generation with its new media—radio, then television, then programmed instruction, then computer-based instruction.

World War II Period

Later, during the World War II era, the U.S. War Department's Information and Education Division invested tremendous amounts of funds and manpower on the development and use of "audiovisual aids," particularly 16mm films, to support its "rapid mass training" effort. It also invested in research on how to design better films and on how instructors could make better use of the materials provided to them. The findings were used during the war to guide the practice of trainers when using audiovisual aids. The utilization protocols developed by the U.S. Navy, for example, were quite sophisticated and were widely accepted in teacher training programs after the war.

The findings of the social scientists in the Experimental Section of the Research Branch (Hovland, Lumsdaine, & Sheffield, 1949) were reported after the war and were widely discussed in civilian applications as well as being used as the basis for further academic research.

Audiovisual Education Period

The period between World War II and the advent of personal computers in 1982 could be viewed as the audiovisual education period. In this era, educational technology research and practice focused on the design and use of analog media—such as still pictures, slides, overhead transparencies, audio recordings, films, and video recordings—in the teaching-learning process. Johnston (1987) provides a succinct synthesis of research findings in this arena. One of the generalizations he reaches is that it is not the hardware, but the software, that accounts for learning: "The electronic media are vehicles through which programming is passed to a learner. We cannot explore the potential of a medium independent of the programming being carried on it" (p. 3).

The actual rate of use of audiovisual media by K–12 teachers during this era would have to be characterized as moderate. Utilization rates were strongly affected by accessibility. Teachers were very likely to use materials that were stored in their own classrooms, somewhat less likely to use those housed in a center in their building, and even less likely to use items, mainly 16mm films, that had to be delivered from outside the building on a scheduled basis. Surveys in the 1940s and 1950s indicated that about 40% of elementary teachers and 20% of secondary teachers used films "frequently."

Evidence from various sources indicates that the average teacher used about one film per month (Cuban, 1986, pp. 14–18). The reasons for low rate of use of film (and similar media), in addition to accessibility, were lack of training with the technology, unreliability of projection equipment, limited school budgets (for rental of films and purchase of projectors), and difficulty of integrating the material into the curriculum. All of these points have relevance for the technologies that were to come later, especially computers.

Audiovisual education textbooks of this era (e.g., Heinich, Molenda & Russell, 1982) focused on the advantages and limitations of each of the media formats and how instructors could improve the efficiency and effectiveness of their teaching through careful selection of media formats that were suited to learning goals and utilization of materials that engage learners with ideas and learning activities that lead them to the learning goals.

Theoretical Bases for Use of Media in Teaching

Critical Realism

Coherent theories of learning and instruction can be traced back to the classical period in Athens, but the modern history begins with Johan Amos Comenius, a Renaissance era pedagogical theorist (1592–1670) who created an extensive body of work about educational reform. He particularly advocated the use of sensory stimuli to help children achieve meaningful understanding. One of his major books, *Orbis Sensualium Pictus* (The Visible World Pictured; Comenius, 1658/1991), was a richly illustrated textbook meant to be a sort of visual-verbal encyclopedia. The methodology of the book—pairing the descriptions of concepts with pictures of them—exemplified the theory that he embraced: that the primary source of knowledge is experience, which enters through the senses. This philosophical perspective is known today as *critical realism*, which maintains that there exists an objectively knowable reality, independent of human minds, which humans come to know about through sensory data filtered through processes of perception and cognition.

Realists felt that in order to be meaningful and useful to the learner, new knowledge had to be based on the learner's sensory experiences, as opposed to rote memorization, which was the dominant pedagogical paradigm at the time. Comenius followed the inductive method, advocated by Francis Bacon (1561–1626). Since it was not practical to bring all the phenomena of the world into the classroom for children to experience directly, the next best alternative was to provide pictures of those phenomena. In *Orbis Sensualium*

Pictus and *Didactica Magna,* Comenius provided a comprehensive educational philosophy and the most fully elaborated theory of visual media usage until the 20th century.

Early Theories of Cognitive Development

Prior to the 1960s the approaches to media utilization were shaped mainly by early-20th-century psychological theories of cognitive development, particularly Gestalt theory, pioneered by Max Wertheimer (1944) and elaborated by Kurt Koffka and Wolfgang Köhler, which attempted to describe how humans and other primates perceived stimuli and used cognitive processes to understand and solve problems. The Gestaltists insisted that an understanding of human psychology required tools beyond those of scientific observation; they sought a unified study of psychology, rejecting the mind-body dichotomy. The Gestalt perspective, with its emphasis on sensory perception and on how humans construct meaning from bits and pieces of auditory and visual information, had great appeal to audiovisual education advocates.

These Gestalt views were reflected in C. F. Hoban, C. F. Hoban, Jr., and Zisman (1937), who wrote an early influential textbook on audiovisual media applications. They referred to a theory of cognitive development based on the processes of differentiation and integration, and they placed emphasis on the value of concrete experiences in promoting progress in differentiation. Hence, a major rationale for using audiovisual media is to support young learners' mental development through progressive stages, from concrete experiences to abstract generalization.

After World War II, thinking about media utilization was reflected in Edgar Dale's (1946) *Audio-visual methods in teaching,* which continued to influence the field through its third edition in 1969. Dale took a rather eclectic approach, not referring very often to specific theories of learning or instruction. Instead, he emphasized the pursuit of "permanent learning," which was associated with "meaningful learning" coupled with motivation and application (use of the new knowledge). He combined these into the construct of "rich experiences," which formed the basis of his prescriptions for effective teaching: "Rich experiences . . . are often flavored with direct sense-experience. They have a quality of newness, freshness, creativeness, and adventure, and they are marked by emotion" (Dale, 1946, p. 23).

Dale advocated purposeful engagement with ideas in environments that were rich with sensory experience. In this, he foreshadowed the constructivist movement that was to come 40 years later. His construct of the "Cone of Experience" was a way of categorizing teaching methods according to the extent to which they immersed learners in active engagement with concrete, authentic experiences. It reflected a concrete-to-abstract continuum suggested earlier by C. F. Hoban et al. (1937, p. 23).

By the 1960s and afterward, following the so-called Cognitive Revolution, similar ideas were being supported with the cognitive theories of Bruner (1960) and the developmental theories of Piaget.

It is notable that neither before nor after the Cognitive Revolution did textbooks on audiovisual media utilization refer to behaviorist theories—neither to Watson and Thorndike in the early 20th century nor to Skinner at midcentury. Although Thorndike's principles of exercise (repetition), effect (pleasure/pain), and readiness were well known and influential in the literature of education, they were scarcely noted in the literature of educational media. The media advocates were passionate opponents of empty verbalism and rote learning, which were all too often associated with behaviorist approaches.

Behaviorist Influences

Theory base. Behaviorist theories of learning flourished parallel to cognitivist theories through the first decades of the 20th century. Thorndike developed his theory of connectionism in the animal laboratory, but shifted his focus to human learning when he joined the Teachers College faculty in 1899. Thereafter, he developed a comprehensive and influential body of theory in educational psychology. Thorndike was not particularly concerned with audiovisual media, but Saettler (1990) claimed that his "development of a science and technology of instruction unquestionably marked him as the first modern instructional technologist" (p. 56). Thorndike's connectionism was superseded by a more comprehensive theory of behaviorism, represented by Watson, which dominated American psychology in the 1920s and 1930s. Then, in the 1960s, a new interpretation of behaviorism, B. F. Skinner's "radical behaviorism," rose to prominence in American psychology. As discussed in chapter 2, Skinner's theory specified that behaviors were learned when they were followed by reinforcers (and a reinforcer could be anything, whatever worked).

Application of the theory. The primary implication of operant conditioning theory for formal education was that learners needed to be treated individually, so that their responses could be monitored and desired ones reinforced. For Skinner (1954; 1968), this led to the invention of the so-called teaching machine. So, for the practice of utilization, this meant a shift away from students as mass audiences for audiovisual presentations and toward students as individuals working through carefully structured (usually print) materials

Implications for utilization. For about a decade—mid-1960s to mid-1970s—there was a rapid proliferation of materials available in programmed instruction form, whether embedded in some sort of mechanical delivery device or printed in book form. These materials were not widely used in K–12 education outside of experimental settings. To use them as they were intended would have required reorganizing the school into a predominantly independent-study mode, a curriculum model that failed to garner a large or permanent following. However, the Direct Instruction model (based on small groups rather than independent study) enjoyed some success.

At the higher education level, as discussed in chapter 2, the personalized system of instruction (PSI) gained an enthusiastic following, as it provided a practical model for organizing the college classroom around a self-study mode. In corporate training and other nonformal education settings, programmed materials, especially audiovisual modules, gained a strong foothold. These settings were not predesigned to be teacher centered, so it was easier to adapt to a self-study format.

Cognitivist Influences

Theory base. The cognitivist perspective emphasizes the importance of the learners' mental and emotional processes during the course of instruction. From this perspective, learners use their memory and thought processes to generate strategies as well as store and manipulate mental representations and ideas.

All branches of cognitive theory—such as information-processing theory and schema theory—emphasize that learners are active processors of the perceptual information that they encounter in their environment and that the new knowledge must be meaningful to the learner if it is to be retained and used in the future.

Application of the theory. Prescriptions drawn from the cognitivist perspective involve instructional activities that present information to the learner or allow the learner to read or view material and think about it. The concerns revolve around attending to relevant messages, interpreting the new material, relating it to existing mental structures, and remembering it so that it can be retrieved later when needed.

In many cases, it is more efficient to package cognitive instruction for self-study in the form of textbooks, or other text materials, such as Web documents. In any case, as described in chapter 2, cognitivist lesson frameworks are likely to consist of a carefully constructed arrangement of information designed to attract and hold attention and to build the new knowledge onto the learner's previous knowledge. The lesson will likely include opportunities to practice in the form of problems, exercises, or quizzes embedded in the readings, provocative questions asked by the teacher, group discussions, or other types of classroom activities that encourage mental engagement with the material.

Implications for utilization. Teachers who are influenced primarily by cognitive concerns are likely to look closely at the message design of the materials they choose or the documents and presentations they create. They may lean toward the use of novel media formats, such as games and computer-based practice, in order to capture the attention and arouse the interest of learners. However, they are likely to use presentations (e.g., illustrated lectures, videos, and *PowerPoint*™ presentations) and assigned readings (e.g., textbooks, handouts, and Web so-called "tutorials"). They would also employ demonstrations (e.g., how-to-do-it demonstrations and peers or instructors serving as role models), large and small group discussions, and drill and practice exercises.

Constructivist Influences

Theory base. As discussed in chapter 2, constructivist learning theory emphasizes the centrality of learners as constructors of their own idiosyncratic knowledge, especially through negotiation with others in their community.

Application of the theory. A number of design prescriptions can be inferred from constructivist theory. Meixner (as cited in Terhart, 2003, p. 36) recommended a number of design features:

- Place content into a situative context
- Add relevant stimuli that are as authentic as possible
- Make the learner take ownership of the material to be learned
- Use as many motor aspects and different sensory channels as possible
- Place the learning task into a surrounding social field
- Establish Socratic discourse as the form of dialogue in the classroom
- Encourage learners to learn from their own mistakes
- Aim at flexible application of the knowledge
- Generate learning environments which promote knowledge transfer

Implications for utilization. The most obvious implication of the constructivist approach is that the center of control shifts from the teacher to the learner. Instead of teachers using media and technology, learners sit in the driver's seat. Instead of learning *from* media, they are learning *with* media, as proposed by Kozma (1991). In the behaviorist and cognitivist perspectives, the primary user is the teacher; in the constructivist perspective, the primary user is the student. The popularization of digital media has made possible the implementation of all sorts of learner-centered activities that are too labor intensive or too expensive to conduct through traditional face-to-face instruction. Examples include

- Learners producing their own multimedia productions, hypertext documents, and other projects, especially those that are developed collaboratively
- Hands-on participation in business scenarios and social simulations
- Tutorial programs that truly allow variable consequences and multiple branches
- Immersion in *microworlds*, including virtual reality, that allow the learner to visualize and manipulate dynamic interactions, such as experiments in mathematics, biology, chemistry, and physics

Digital technology also makes it possible for reading-type activities to become less passive, more active, and more learner controlled. Examples include

- Web text with links allowing the reader to connect related ideas (hypertext), possibly incorporating sounds and motion images (hypermedia)
- Web-based practice exercises that allow learners to choose different answers in order to experience the consequences of their decisions

Writing-type activities, too, can profit from the digital environment. Examples include

- Creating written documents using word-processing software
- Keeping a journal or blog to provide an outlet for reflections or debriefing after varied sorts of learning activities

It is not an accident that the constructivist view came to popularity around the same time as computer technology began to be widely accessible in schools and universities. The personal computer and the World Wide Web offer many avenues for learner-centered and learner-controlled activities, the sorts of activities promoted by advocates of constructivism.

Eclectic Approaches

Theory base. An eclectic approach (from the Greek *eklektikos*, meaning "selective") simply combines doctrines from different theories without accepting the whole parent theory for each doctrine. Practitioners, no less than philosophers, may adopt an eclectic stance because they find merit in ideas that happen to be promoted by opposing parties. The arbitrary combination of clashing doctrines can produce incoherent theoretical structures in philosophy, but in practical matters, eclecticism often yields useful syntheses.

Application of the theory. In the area of utilization, teachers can easily see that different psychological theories offer guidance for different sorts of learning goals. The theories do not necessarily conflict, but they explain different phenomena better than others. For example, Ertmer and Newby (1993) propose that the behavioral approach is best suited to learners with lower levels of task knowledge and for learning goals requiring lower cognitive processing; the cognitive approach is best suited for middle levels of task knowledge and cognitive processing; and constructivism is best suited for learners with a higher level of task knowledge, working on higher level tasks (pp. 68–69).

Implications for utilization. By the 1980s, textbooks on media use and integration tended to take an eclectic approach in applying theories to support good practice regarding the selection and use of media for instruction. In one typical textbook, the authors counseled, "Instructors and instructional designers need to develop an eclectic attitude toward competing schools of learning psychology" (Heinich, Molenda, & Russell, 1993, p. 15).

A model for teachers' planning for media use, the ASSURE model, recommended these steps:

- Analyze learners
- State objectives
- Select media and materials
- Utilize media and materials
- Require learner participation
- Evaluate and revise (Heinich, Molenda, & Russell, 1993, pp. 34–35)

This model reflects a combination of prescriptions from behaviorism (performance objectives, require learner participation) and the systems approach (analyze learners, evaluate, and revise), while the authors' advice on selecting and using media and materials drew heavily on cognitive and cognitive-constructivist perspectives.

By the late 1990s and early 2000s, traditional media courses in many teacher education institutions were supplanted by courses focusing on the use of computers. By 2000, over 70% of introductory technology courses for teachers had a primary emphasis on use of computers as opposed to the use of traditional audiovisual media (Betrus, 2000). Textbooks aimed at these courses, like the earlier textbooks aimed at audiovisual media, also tended to reflect an eclectic mentality. For example, Lever-Duffy, McDonald, and Mizell (2003), after presenting behaviorist, cognitivist, and constructivist perspectives, advise, "You may choose to use some parts of each theory or accept a learning theory in its entirety. At this point, you should examine all the options and let your own mental model of learning develop" (pp. 16–17).

In another widely used textbook on computer integration, Roblyer (2003) vigorously defends the selective use of what she refers to as directed, constructivist, and combined approaches in integrating technology into curriculum planning. For example, she recommends that "when the absence of prerequisite skills presents a barrier to higher level learning or to passing tests, *directed* instruction usually is the most efficient way of providing them" (p. 73). On the other hand, "Resources such as Logo, problem-solving courseware, and multimedia applications often are considered ideal environments for *constructivist* activities that get students to think about how they think" (p. 73).

In short, there is a widespread consensus that, when instructors are considering ways to facilitate learning with media, an eclectic approach can provide a varied menu of appropriate materials, methods, and activities.

From Utilization to Integration, Implementation, and Adoption

Integration

Media and technology can be viewed as being *integrated* into instruction when they are woven into the fabric of the curriculum in a seamless way, as opposed to simple occasional use, such as using an overhead projector to illustrate a point. In the fullest sense of the term, integration implies a holistic combination of the educational setting, the needs and interests of learners, the curricular content and the objectives related to it, the assessment methods, the abilities of the instructor, the hardware and software resources, and the support system surrounding the operation. The epitome of successful integration would be a learner-centered environment in which instructional resources were selected and used efficiently and effectively to support learning activities aimed at deeper understanding and problem-solving ability.

A concrete example could be imagined using the *Jasper Woodbury* series developed in 1989, an innovative collection of mathematics learning modules stored on laser disc. Each disc immerses the learners in a story entailing mathematical data for problems that learners need to solve to reach a successful ending. The problems address standards of the National Council of Teachers of Mathematics; they apply principles of anchored instruction and active learning and require co-operative work. Evaluation studies showed that students using *Jasper Woodbury* outperformed those using more conventional approaches, they enjoyed mathematics more, and they employed generative methods in solving math problems (Barron et al., 1993).

The path toward such technology integration is shown in such contemporary textbooks as *Integrating educational technology into teaching* (Roblyer, 2006), which suggests a five-phase technology integration planning model. Such an approach to instruction is more likely to be successful when it takes place in a setting that is friendly to a systemic approach.

An example of such a holistic setting for technology integration is the elementary school curriculum known as "Project CHILD," in which three subject-focused specialist teachers form cross-grade clusters (K–2 or 3–5) to facilitate standards-based learning. The teachers and students stay together for three years to enhance continuity. There are three classrooms in a cluster—one each for reading, writing, and mathematics, one of which serves as the student's home base. Each of the three classrooms has at least six learning stations to facilitate diversified learning in three modes—technology, hands on, and paper/pencil. Students rotate through the three cluster classrooms for instruction in each basic subject. Students spend 60–90 minutes in each of the cluster classrooms, returning to their home base for instruction in science and social studies. After a brief whole-group, teacher-directed lesson, students work at the stations to practice and apply the lesson content using a variety of learning modes. The teacher assigns students to their beginning stations, but students move independently as they finish the first assigned task. They set goals and keep track of their station work using a logbook called a "passport" (Butzin, 2004).

The CHILD model, which has won national awards and has been evaluated to verify its effectiveness and cost-benefit, illustrates that successful integration of technology and curriculum is not a utopian ideal, but an everyday reality in some places.

Implementation

One of the largest challenges of educational technology is to ensure that well-developed instructional materials and systems are actually placed into use. There is a long history of exemplary products failing to find acceptance in the marketplace or of being abandoned after being used for a period of time. This problem, discussed in depth by Burkman (1987), can be viewed through various conceptual lenses.

Instructional development lens. First, actual use of an instructional product can be seen as one step in the instructional development process. Looking at a systems approach to instructional development, implementation is the fourth stage of the five-stage ADDIE approach. However, Burkman (1987) and others advise that the probability of successful implementation depends on considerations related to earlier steps. Burkman advises a "user-oriented instructional development process" (p. 439) in which the identity and preferences of the potential adopter are considered from the beginning. With the potential adopter in mind, it is possible to consider the needs and values of the adopter during the design and development stages, with the goal of creating a user friendly product. Later, at the implementation stage it is a matter of making sure that the potential adopters are informed about the innovation and its usefulness to them, and then that they receive support after they adopt the innovation.

Molenda and Pershing's (2004) Strategic Impact model, discussed in chapter 3, suggests a similar approach, but goes one step further by advising that "change management" issues be considered at *every one* of the ADDIE stages, not tacked on at the end. The model suggests that buy-in is most likely to happen if those affected by the change are allowed to participate in planning activities along the way.

Adoption of Innovations

Another view focuses not on prescriptions for increasing user acceptance of instructional materials and systems but on the processes underlying teachers' adoption or rejection of innovations. Holloway (1996) offers an extensive review and critique of research on diffusion and adoption of educational technology.

There are a number of different perspectives on the processes of accepting and using new tools or practices, ranging from an atheoretical view to perspectives based on psychological, sociological, organizational, technological, systems, and ecological theories. Each casts light on different aspects of this complex problem area.

Atheoretical perspective. The early studies of teacher acceptance of audiovisual media in the 1960s and early studies of teacher acceptance of computer-based media in the 1990s and beyond have tended to be atheoretical—that is, factors associated with adoption are sought without reference to an overarching theory of how and why people adopt innovations. Surveys are often used to determine who uses media and what characteristics of the users or their environments seem to explain the pattern of acceptance or rejection. Henry Jay Becker (1991; 1994a; 1994b) and his colleagues at the Center for Research on Information Technology and Organizations (CRITO) conducted a decade of survey-based correlational studies regarding teacher use of computers and the Internet. In their most comprehensive survey, they found that relatively few teachers involved their students in using the Internet in a substantial way. However, the factors most highly correlated with substantial student use were ease of access to classroom connection, computer expertise of the teacher, and belief in "constructivist" pedagogy (Becker, 1999). At that time, about one half of teachers who enjoyed the most favorable working conditions made strong use of the Internet; the usage rates dropped rapidly along with lack of supportive conditions such that very few teachers in the least favorable conditions group used the Internet at all.

Subsequent work by the CRITO group tended to subsume the notion of "favorable conditions" under the umbrella of the extent to which teachers receive support in their efforts to integrate computers into the curriculum. Ronnkvist, Dexter, and Anderson (2000) broke *technology support* into a number of categories: facilities, technical support staff, professional development support staff and activities, and incentives. They studied the correlation between support received and the extent and variety of technology use. Among their findings,

- Both the quality and perceived availability of support are significant predictors of the frequency of teachers' use.
- Teachers in schools with high quality technology support are more likely to engage in a variety of different professional uses of technology.
- Computer skills (expertise) are a strong, positive predictor of variety of uses. (p. 24)

This atheoretical view seems to assume that teachers' adoption of innovations happens naturally under conditions of adequate support.

Systems perspective. Robert Heinich (1967) was among the first to analyze the acceptance and use of media as a problem embedded in the organizational system of schools. He observed that "classroom teachers tend to reduce all media to the status of aids" (p. 19) despite the fact that by then at least two technologies had emerged—instructional television and programmed instruction—that enabled students to learn effectively without the presence of a classroom teacher. He referred to the "craft structure" of teaching and pointed out (Heinich, 1984) that current organizational structures gave teachers the power to decide what media and methods would be used in their classrooms. He further argued that teachers naturally resist the implementation of technologies that would diminish their power by replacing them or placing them in a subservient role. For example, in choosing textbooks, teachers gravitate to those materials that preserve the role of the teacher as the primary deliverer of instruction and avoid alternatives in which the text itself is transformed into instruction, as in programmed instruction. Thus, to maintain their accustomed roles and to preserve their places in the organizational structure, teachers tend to "reduce all media to the status of aids" and to reject applications that require a more systemic rearrangement of power, roles, and structure. That is, teachers would be resistant to technologies or specific applications in which core teaching functions were included in the materials. Textbooks, yes, but programmed texts, no. Supplementary materials on video, yes, but televised lectures, no. Computers for communication and word processing, yes, but complete self-instructional lessons, no.

Heinich (1967) proposed that, if education were viewed as a system the tasks of instruction could be divided more rationally, yielding more effective learning at a lower overall cost. Curricular programs could be developed at a more central level by teams of specialists rather than being reinvented by every teacher in every classroom. Classroom teachers would devote more of their attention to adapting predesigned programs to their students' needs and less to original creation. Of course, such systemic changes would entail changes in power relationships, relationships that are frozen not only into custom but also into law in some cases. Heinich (1967) predicted that such a systemic approach would be resisted by the education profession, whose attitudes "are based on a craft society and the result of a guild approach to production" (p. 16).

The prospect of "unbundling" the functions performed by instructors in order to develop a more rational division of labor is discussed in chapter 3 in the context of distance education at the college level. This is one arena in which the functions of course design, subject-matter expertise, and day-to-day interaction with learners have been relegated to different actors. The motivation for taking a more systemic approach in this case is the demand for cost-effectiveness, since distance education operations tend to be viewed as investments and are expected to meet higher standards of productivity than traditional residential teaching.

In a more recent examination of this problem, a number of theorists have proposed applying "systemic change theory" to education—in which the organization of teachers, learners, and conditions for most effective integration of technology is viewed systemically and with an understanding of change processes (Banathy, 1991; Reigeluth & Garfinkle, 1994; Ellsworth, 1997). The first premise of this theory is that education is a social enterprise and that success depends on maximizing the satisfaction of the people that will affect and be affected by changes (Ellsworth, 1997, p. 2). Ellsworth's second major premise is that "change must be implemented as a package" (p. 3)—that is, that lasting change requires not only actions in the classroom but also in the surrounding system, such as the assessment methods employed, the teacher's reward structure, the technology support system, and possibly the support of parents and administrators. Finally, Ellsworth advises that systemic change requires rethinking of one's assumptions about education; he cites some examples given by Reigeluth (1994):

- Class levels vs. continuous progress
- Covering the content vs. outcomes-based learning
- Norm-referenced vs. individualized testing . . . (Ellsworth, 1997, p. 8)

Thus, the systems perspective involves a change in mind-set about education as well as a different perspective on the process of implementation.

Sociological perspective.   Contrasting with the systems perspective is one that focuses more narrowly on the roles that teachers play within the school (or college). This view tends to take for granted Heinich's (1984) notion of the "craft structure," that is, that the teacher-student relationship is the center of the enterprise and that this is inherently a labor-intensive process. This view is represented in the work of Cuban and his colleagues (Cuban, 1997; Cuban, Kirkpatrick, & Peck, 2001). Cuban (1997) proposes that "The essence of teaching is a knowledgeable, caring adult building a relationship with one or more students to help them learn." He does not expect productivity gains through technology in formal education compared to the gains possible in low-skill manual work consisting of repetitive, routine tasks.

In this view, the important thing about introducing computers into the classroom is that they change social relationships among teachers and students. Teachers' and professors' beliefs about their authority and expectations of control are threatened by hardware and software systems that claim to replace some of the functions of the instructor. If the computer teaches, what is left for the teacher? A symptom of this perception of threat is the teacher's preference for keeping the computer in a separate lab rather than in the classroom. In this view, teachers' and professors' reluctance to embrace new technologies is not simply resistance to the new but "a struggle over core values." This perspective is congruent with the postmodern sensibility, discussed in chapter 11, which views the human-technology encounter with much caution and a good deal of skepticism.

Psychological perspective. A number of models have been developed based on psychological theories of the stages through which potential adopters progress on their way to acceptance and use of an innovation (any idea that is perceived as new to the individual). These *diffusion* models take a psychological perspective, almost a marketing perspective, focusing on the question of why some individuals adopt innovations and others reject innovations, with the decision being seen primarily as a personal, rational choice. The best-known model is that of Everett Rogers (1962) in which the author synthesized findings from 405 studies culled from fields as diverse as education, medicine, public policy, and farming. The synthesis was reported with a model and case histories to substantiate propositions about the stages, process, and variables involved in diffusion, which was defined as the spread, adoption, and maintenance of an innovation. In later editions, Rogers encompassed over 3,000 diffusion studies (1983), then nearly 4,000 studies (1995), and then nearly 5,000 studies (2003).

Rogers (1995) considers the main elements in the diffusion of new ideas to be "(1) an innovation, (2) which is communicated through certain channels, (3) over time, (4) among the members of a social system" (p. 35). He pioneered in analyzing case study data to discern a pattern in the individual's innovation-decision process, finding that an individual passes through the stages of knowledge, persuasion, decision, implementation, and confirmation (p. 36).

Rogers's (1995) diffusion theory deals specifically with technological innovations. However, the situations on which the theory is built are somewhat different from the school or college situation, in that typically Rogers examines rather discrete innovations being adopted by individuals outside the workplace setting for their own benefits. In the school or college setting, the acceptance and use of really consequential innovations tend to require collective decisions involving instructors, midlevel administrators, top administrators, and governing boards. And they entail complex change processes, not just acquisition of equipment or just implementation of a new practice.

A diffusion model that is directed specifically at the school setting is Hall and Hord's (1987) concerns-based adoption model (CBAM). This model views innovation adoption primarily as a psychological process revolving around the teachers' hierarchy of needs. It holds that people considering and experiencing change evolve in the kinds of questions they ask and in their use of whatever the change is. In general, early questions are more self-oriented: What is it? How will it affect me? When these questions are resolved, questions emerge that are more task oriented: How do I do it? How can I use these materials efficiently? How can I organize myself? Why is it taking so much time? Finally, when self and task concerns are largely resolved, the individual can focus on impact. Educators ask: is this change working for students? Is there something that will work even better? The model identifies seven stages of concern: (0) awareness, (1) informational, (2) personal, (3) management, (4) consequence, (5) collaboration, and (6) refocusing. It also goes beyond other models in elaborating a spectrum of levels of use, not just adoption or rejection. Users may fall anywhere on a broad continuum of commitment to and maturity in using an innovation:

- 0. Non-Use: The user has no interest, is taking no action.
- I. Orientation: The user is taking the initiative to learn more about the innovation.
- II. Preparation: The user has definite plans to begin using the innovation.
- III. Mechanical: The user is making changes to better organize use of the innovation.
- IVA. Routine: The user is making few or no changes and has an established pattern of use.
- IVB. Refinement: The user is making changes to increase outcomes.
- V. Integration: The user is making deliberate efforts to coordinate with others in using the innovation.
- VI. Renewal: The user is seeking more effective alternatives to the established use of the innovation. (Hord, Rutherford, Huling-Austin, & Hall, 1987)

In short, these models view acceptance as essentially an individual process of becoming a aware of a new product or practice and gradually accumulating data to make a decision about adoption. The organizational setting is not particularly relevant nor is the impact of the individual's decision on the larger system.

Ecological perspective. An emergent view, presented as "an ecological perspective" (Zhao & Frank, 2003), proposes an umbrella for these divergent visions of how and why instructors accept and use modern information and communication technology (ICT). Zhao and Frank propose the ecological systems of nature as a metaphor for the life cycle through which technology is accepted, adapted, and incorporated into educational institutions. They see a spectrum of qualitatively different uses of ICT, with these different uses finding different niches in the ecological system. Their ecological perspective subsumes earlier views such as rational choice theory: "Teachers use computers in ways that address their most direct needs, bring them maximal benefits, do not demand excessive time to learn, and do not require them to reorganize their current teaching practices. Thus teachers' choices of computer activities minimize costs" (p. 821).

Zhao and Frank (2003) propose that human activities, within their environments, act like other organisms in other environments, seeking niches in which to survive. More precisely, they see different technology uses finding niches suited to them.

Summary on adoption of innovations perspectives. Viewed from these different perspectives, the adoption of ICT can thus be considered several different ways: (a) as a set of resources accepted and used by teachers playing their traditional roles in self-contained classrooms, (b) as a set of tools used by learners, empowering them to take responsibility for their own learning, or (c) as an infrastructure with transformational possibilities, an engine for restructuring the education enterprise. The user, likewise, can be viewed as an independent agent, choosing the best tools for the job, as a player in a larger game of power and authority, or as an interchangeable element in a complex, interconnected system. Each of these visions implies a different approach to implementation or diffusion, different potential adopter, a different client, and a different goal.

Actual Uses of Media and Technology

Media and technology are used differently and at different rates across the various major domains—corporate, higher education, and K–12 education. Since each domain has its own socioeconomic dynamics accounting for these differences, each domain will be addressed separately in the following sections.

Corporate Training

The dynamics of using media and technology in corporate training programs are different from those in formal education. First, the money spent on training is considered a cost of doing business or, at best, an investment that must be recouped through revenue gains later. This leads to a bias toward efficiency that is significantly greater than in formal education. Second, instructors are not necessarily in a position to control the entire instructional process. In larger organizations, the training function is divided among various specialties, including design, production, evaluation, and subject-matter expertise, and major instructional decisions are made on a team basis. Third, businesses often have multiple sites, sometimes in multiple countries, so there is a premium on standardization and mass production of training events. Even without multiple sites, in some industries governmental regulations stipulate the type and frequency of training activities. Fourth, the delivery system for training is often determined by the ICT infrastructure of the organization. If a company builds a videoconferencing system for management communications, there is a bias toward using the excess capacity for other communications, including training.

Given the biases just mentioned, it may be surprising to note that face-to-face classroom instruction incorporating traditional media formats is still the dominant mode in corporate training, according to surveys reported in *Training* magazine over the past decade (Industry Reports, 1996, 1998, 1999, 2000; Galvin, 2001, 2002, 2003; Dolezelak, 2004).[1]

Over this period, the percentage of organizations that report using face-to-face classroom instruction "always" or "often" has remained stable at about 90%. The percentage using manuals and print materials has also been stable at about 80%, and over 50% use video materials "always" or "often."

In addition, some 5–10% of companies were using broadcast or satellite television delivery "always" or "often" during the period of 2001 to 2003.

Computer-based delivery systems played a gradually expanding role in training since the early 1990s. Earlier, this referred to modules delivered via floppy disk or local area network (LAN). Since then computer-based material is more likely encountered by means of CD-ROM or DVD. More recently, it occurs by connecting to the Internet or organizational intranet. In the 2003 *Training* survey, 45% of companies reported using instruction in digital storage media "often" or "always." However, 63% reported that they used Internet or intranet delivery, a major increase over the previous year (Galvin, 2003).

It is interesting that the reported *proportion of time spent* in computer-based training has changed little over the years, reaching 16% in 2003 (Galvin, 2003). The true extent of use of ICT may be obscured by the method of reporting. It appears that "computer only" courses are not replacing "classroom only" courses to any great extent. Rather, hybrid combinations ("blended learning") are becoming more and more common—that is, face-to-face courses may be preceded by readings posted on the Web and followed by a discussion forum conducted through the Web. Corporate training also includes more

---

[1] These surveys are, of course, self-reports by a sample of individuals in various organizations and thus suffer the usual limitations in terms of validity and reliability (for a description of the survey methodology, see Galvin, 2003).

"just in time" instruction, short "help" sessions delivered through LAN or intranet networks to the worker's computer at the time it is needed.

Unlike the corporate realm, in higher education, there is no consistent source of annual data on national trends in the use of media and technology, although there are occasional and partial reports that cast some light on trends in information technology usage. Regarding traditional analog media there is only silence. However, this by no means suggests that college faculty have abandoned audiovisual media. Based on anecdotal reports from university media centers, it seems that audiovisual media are alive and reasonably well. Overhead projectors are still ubiquitous in classrooms. Photographic slides continue to occupy a significant niche. Circulation of projectors is declining, but projectors tend to be built into classrooms and laboratories in departments that make heavy use of slides, such as biology, veterinary medicine, optometry, fine arts, classics, and drama. Demand for video recordings in VHS format remained steady through the 1990s, with thousands of bookings annually at universities with large central collections. As VHS video recordings have become less expensive, many individuals and departments own their own copies; showings of these do not appear on campus circulation records (B. Teach, personal communication, June 21, 2004).

Discussion of technology use in higher education is almost totally focused on computer-based media. During the period of 1997 to 2002, as colleges and universities were expanding their information technology services at a rapid pace, there were national surveys of faculty use. According to annual surveys between 1997 and 2000 (Campus Computing Project, 2000), faculty adoption of certain computer-based teaching applications—such as course Web pages and use of Internet resources—grew each year during that period. However, the percentage increase was *smaller* each succeeding year, indicating a plateau of the adoption rate. Unfortunately, the Campus Computing Project (2000) did not continue to measure these indices. This lack of attention may be an indication of reduced interest in classroom media within the academic computing community. In fact, in the 2004 EDUCAUSE survey, e-learning, distributed learning, and course management systems slipped from near the top to near the bottom of the list of concerns of information technology professionals (Spicer, DeBlois, & EDUCAUSE Current Issues Committee, 2004).

In any event, there are no current national data comparable to those of the Campus Computing Project (2000). Based on a sampling of internal university reports, Molenda and Bichelmeyer (2005) speculated on how faculty members use information technology in teaching, noting that the patterns appear to be quite similar across the cases. Generalizing from these selected cases, they project that nearly 90% of all instructors exchange e-mail with students; some 60% use class listservs to communicate with students; about one half assign students to use Web resources; 40% show digital presentations; about 20% ask students to participate in online discussion forums; and 10–20% provide online simulations or lab experiments. These figures tend to support the earlier theory that, although usage continues to grow, the rate of increase has slowed since the late 1990s.

These findings also support the notion discussed earlier that that faculty incorporation of computer media in their teaching can be viewed as a wide spectrum of adoption decisions, not a single yes/no decision. Applications that require a greater investment of time and energy or that entail fundamental changes in teaching practices are accepted more slowly. Using e-mail is relatively easy to learn and makes work more efficient, but at the other end of the spectrum, the use of online simulations and lab experiments, requires considerable investment of time and special expertise, hence attracting a much lower rate of adoption. As might be expected, professors do not seek out applications that substitute the computer for functions that faculty consider to be core functions, such as lecturing.

A factor that is promoting faculty use of information technology is the nearly ubiquitous adoption of course management systems (CMS). The existence of a CMS motivates faculty to create content to make use of this delivery system. Since the system is there and the university may apply pressure to at least post a syllabus online, many instructors explore other functions of the CMS, typically on an incremental basis, adding applications year by year, leading to the sorts of uses described earlier.

K–12 Education

Traditional audiovisual media. As with higher education, there are no ongoing annual surveys of a national scope to provide a clear picture of how teachers are actually using technology. And, as in higher education, there is virtually no recent research or published literature about rates of use of traditional media. Textbooks are still a mainstay of classroom instruction. They increasingly come with digital ancillary materials, but print still rules. Surveys have revealed that many teachers still use overhead projectors, audiocassette players, and VHS videocassettes. Elementary schools still keep and use record players (Misanchuk, Pyke, & Tuzun, 1999, p. 3).

School district and regional media centers continue to circulate audiovisual materials. Analog media formats, particularly videocassettes, are still widely used. District and regional media center collections include (from greatest to least number of titles): videocassettes, multimedia, curriculum materials, professional books, and digital media. However, purchases of digital media, including internet resources, DVDs, and multimedia, now outpace purchases of most types of analog media (NAMTC, 2003). Data on specific audiovisual utilization patterns are difficult to find, but anecdotal evidence paints a picture similar to the one in higher education: overhead projection and VHS video nearly ubiquitous, and slides used in specific subjects with high visual elements.

Computer-based media. There have been occasional surveys of a national scope regarding teachers' use of computer technologies. However, more recent utilization patterns are perhaps best portrayed by an intensive survey of 19 elementary schools in a single state (Zhao & Frank, 2003). They found that usage could be characterized under the headings of "teacher use" and "student use." The most common *teacher* uses reported (proportion of teachers using weekly or daily) were preparation for instruction (58% used weekly or daily), communication with parents (54%), teacher-student communications (37%), and record keeping (29%).

The most common *student* uses reported were developing basic computer skills, such as keyboarding (53%); core curriculum skills, such as math drill and practice (41%); classroom management, including computer access as a reward (38%); remedial lessons (30%); and student inquiry (14%).

In American schools, access to information technology is ubiquitous and use of that resource is growing incrementally, to the point that it is now the norm for teachers to employ some computer technology at work (U.S. Department of Education Policy and Program Studies Service, 2003). However, as in higher education, the operative principle seems to be gravitation toward applications that pay maximum benefits for the user for minimum investment of time and energy. As Heinich (1967) predicted almost 40 years earlier, applications that entail core teaching functions tend to be less popular than applications that provide labor-saving measures for instructors.

Conclusion

The end purpose of educational technology is *using*, putting learners into contact with appropriate technological resources under conditions conducive to learning. Before using can take place, the resources must be selected and evaluated by an instructor and a plan must be made for utilization. There is a considerable body of theory and research to guide utilization, with current practice favoring an eclectic approach, using behaviorist, cognitivist, and constructivist techniques as dictated by the learning goals and needs of learners. There are numerous lenses through which to view the processes by which instructors become aware of and decide to make use of technological resources. These lenses variously focus on the psychological processes of the user, the sociology of the educational environment, or the total system of the participants, the learning environment, and the surrounding social and political systems.

The extent to which technological resources are actually used depends, first of all, on the setting. The corporate, higher education, and K–12 settings each have different social and economic forces operating on participants. Together with the working theories and technological competencies of instructors, the social and economic forces interact to influence what technologies are used and to what extent.

References

Banathy, B. H. (1991). *Educational systems design: A journey to create the future.* Englewood Cliffs, NJ: Educational Technology Publications.

Barron, L., & Bransford, J. (1993, April). The Jasper experiment: using video to furnish real-world problem-solving contexts. *Arithmetic Teacher, 40*(8), 474–478.

Becker, H. J. (1991). How computers are used in United States schools: basic data from the 1989 I.E.A. Computers in Education Survey. *Journal of Educational Computing Research, 7,* 385–406.

Becker, H. J. (1994a). *Analysis and trends of school use of new information technologies.* Office of Technology Assessment. Washington, DC: U.S. Government Printing Office.

Becker, H. J. (1994b). How exemplary computer-using teachers differ from other teachers: Implications for realizing the potential of computers in schools. *Journal of Research on Computing in Education, 26*(3), 291–321.

Becker, H. J. (1999, February). Internet use by teachers: Conditions of professional use and teacher-directed student use. *Teaching, learning, & computing: 1998 national survey, report #1.* Irvine, CA: Center for Research on Information Technology and Organizations, University of California, Irvine, and University of Minnesota.

Betrus, A. K. (2000). *The content and emphasis of the introductory technology course for undergraduate pre-service teachers.* Unpublished doctoral dissertation, Indiana University, Bloomington, Indiana.

Booth, P. A. (1989). *An introduction to human-computer interaction.* Hillsdale, NJ: Lawrence Erlbaum Associates.

Bruner, J. S. (1960). *The process of education.* Cambridge, MA: Harvard University Press.

Burkman, E. (1987). Factors affecting utilization. In R. M. Gagne (Ed.), *Instructional technology: Foundations* (pp. 429–456). Hillsdale, NJ: Lawrence Erlbaum Associates.

Butzin, S. M. (2004). Project CHILD: A proven model for the integration of computer and curriculum in the elementary classroom. *Asia-Pacific Cybereducation Journal, I*(1), 29–34.

Campus Computing Project. (2000). *The 2000 national survey of information technology in higher education: Struggling with IT staffing.* Retrieved April 13, 2005, from http://www.campuscomputing.net

Comenius (Komenský), J. A. (1991). *Orbis sensualium pictus* [The Visible World Pictured; Electronic version]. Praha, Czech Republic: Trizonia. (Original work published 1658)

Cuban, L. (1986). *Teachers and machines: The classroom use of technology since 1920.* New York: Teachers College Press.

Cuban, L. (1997, May 21). High-tech schools and low-tech teaching. *Education Week on the Web.* Retrieved April 5, 2005, from http://www.edweek.org/ew/vol-16/34cuban.h16

Cuban, L., Kirkpatrick, H., & Peck, C. (2001). High access and low use of technologies in high school classrooms: Explaining an apparent paradox. *American Educational Research Journal, 38*(4), 813–834.

Dale, E. (1946). *Audio-visual methods in teaching.* New York: The Dryden Press.

Dolezalek, H. (2004, October). Industry report 2004. *Training, 41*(10), 20–36.

Ellsworth, J. (1997). Technology and change for the information age. *ERIC Digest.* Retrieved March 26, 2005, from http://www.eric.ed.gov/

Ertmer, P. A., & Newby, T. J. (1993). Behaviorism, cognitivism, constructivism: Comparing critical features from an instructional design perspective. *Performance Improvement Quarterly, 6*(4), 50–72.

Galvin, T. (2001, October). Industry report 2001. *Training,* 40–75.

Galvin, T. (2002, October). 2002 industry report. *Training,* 24–73.

Galvin, T. (2003, October). 2003 industry report. *Training,* 21–45.

Hall, G. E., & Hord, S. M. (1987). *Change in schools: Facilitating the process.* Albany, NY: SUNY.

Heinich, R. (1967). *Keynote address.* Paper presented at Summary Report of the Thirteenth Lake Okoboji Educational Media Leadership Conference, Lake Okoboji, Milford, Iowa.

Heinich, R. (1984). The proper study of instructional technology. *Educational Communication and Technology Journal, 32*(2), 67–87.

Heinich, R., Molenda, M., & Russell, J. D. (1982). *Instructional media and the new technologies of instruction.* New York: John Wiley & Sons.

Heinich, R., Molenda, M., & Russell, J. D. (1993). *Instructional media and the new technologies of instruction* (4th ed.). New York: Macmillan.

Hoban, C. F., Hoban, C. F. Jr., & Zisman, S. B. (1937). *Visualizing the curriculum.* New York: The Cordon Co.

Holloway, R. E. (1996). Diffusion and adoption of educational technology: A critique of research design. In D. H. Jonassen (Ed.), *Handbook of research for educational communications and technology* (pp. 1107–1133). New York: Macmillan.

Hord, S. M., Rutherford, W. L., Huling-Austin, L., & Hall, G. E. (1987). *Taking charge of change.* Washington, DC: Association for Supervision and Curriculum Development.

Hovland, C. I., Lumsdaine, A. A., & Sheffield, F. D. (1949). *Experiments on mass communication: Vol. 3: Studies in social psychology in World War II.* Princeton, NJ: Princeton University Press.

Industry report 1996. (1996, October). *Training,* 37–79.

Industry report 1998. (1998, October). *Training,* 43–76.

Industry report 1999. (1999, October). *Training,* 37–40, 53–54, 73–80.

Industry report 2000: The methods. (2000, October). *Training,* 57–63.

International Society for Technology in Education. (2004). Educational technology resource evaluation form. *Educational software preview guide.* Washington, DC: International Society for Technology in Education.

Johnston, J. (1987). *Electronic learning: From audiotape to videodisc.* Hillsdale, NJ: Lawrence Erlbaum Associates.

Kerr, S. (1981). How teachers design their materials: Implications for instructional design. *Instructional Science, 10,* 363–378.

Knowlton, D. C., & Tilton, J. W. (1929). *Motion pictures in history teaching.* New Haven, CT: Yale University Press.

Kozma, R. B. (1991). Learning with media. *Review of Educational Research, 61*(2), 179–211.

Lashley, K. S., & Watson, J. B. (1922). *A psychological study of motion pictures in relation to venereal disease campaigns.* Washington, DC: U.S. Interdepartmental Social Hygiene Board.

Lever-Duffy, J., McDonald, J. B., & Mizell, A. P. (2003). *Teaching and learning with technology.* Boston: Pearson Education.

McCutcheon, G. (1979, April). *How elementary school teachers plan their curriculum: Findings and research issues.* Paper presented at annual meeting of American Educational Research Association, San Francisco, California.

Misanchuk, M., Pyke, J. G., & Tuzun, H. (1999). Trends and issues in educational media and technology in K–12 public schools in the United States. *Instructional Media,* Number *24,* 3–5.

Molenda, M., & Bichelmeyer, B. (2005). Issues and trends in instructional technology: Slow growth as economy recovers. In M. Orey, J. McClendon, & R. M. Branch (Eds.), *Educational media and technology yearbook 2005* (Vol. 30, pp. 3–28). Englewood, CO: Libraries Unlimited.

Molenda, M., & Pershing, J. A. (2004, March/April). The strategic impact model: An integrative approach to performance improvement and instructional systems design. *TechTrends, 48*(2), 26–32.

NAMTC. (2003). *Bi-annual membership survey results.* Retrieved March 16, 2005, from http://www.namtc.org/pages/member_survey_start.html

Nielsen, J. (1994). *Usability engineering.* Boston: AP Professional.

Norman, D. A. (1988). *The psychology of everyday things.* New York: Basic Books

Reigeluth, C. M. (1994). Introduction: The imperative for systemic change. In C. Reigeluth, & M. Garfinkle (Eds.), *Systemic change in education* (pp. 3–11). Englewood Cliffs, NJ: Educational Technology Publications.

Reigeluth, C. M., & Garfinkle, R. J. (1994). *Systemic change in education.* Englewood Cliffs, NJ: Educational Technology Publications.

Roblyer, M. D. (2003). *Integrating educational technology into teaching* (3rd ed.). Columbus, OH: Merrill, Prentice Hall.

Roblyer, M. D. (2006). *Integrating educational technology into teaching* (4th ed.). Columbus, OH: Merrill, Prentice Hall.

Rogers, E. M. (1962). *Diffusion of innovations.* New York: The Free Press of Glencoe.

Rogers, E. M. (1983). *Diffusion of innovations* (3rd ed.). New York: The Free Press.

Rogers, E. M. (1995). *Diffusion of innovations* (4th ed.). New York: The Free Press.

Rogers, E. M. (2003). *Diffusion of innovations* (5th ed.). New York: The Free Press.

Ronnkvist, A., Dexter, S. L., & Anderson, R. E. (2000, June). Technology support: Its depth, breadth, and impact in America's schools. *Teaching, learning, and computing: 1998 national survey, report #5.* Irvine, CA: Center for Research on Information Technology and Organizations, University of California, Irvine and University of Minnesota.

Rubin, J. (1994). *Handbook of usability testing: How to plan, design and conduct effective tests*. New York: John Wiley & Sons.

Saettler, P. (1990). *The evolution of American educational technology*. Englewood, CO: Libraries Unlimited.

Skinner, B. F. (1954). The science of learning and the art of teaching. *Harvard Educational Review, 24*(1) 86–97.

Skinner, B. F. (1968). Why teachers fail. *The technology of teaching*. New York: Appleton-Century-Crofts.

Spicer, D. Z., DeBlois, P. B., & EDUCAUSE Current Issues Committee. (2004). Fifth annual EDUCAUSE survey identifies current IT issues. *EDUCAUSE Quarterly, 27*(2), 1–23

Taylor, P. H. (1970). *How teachers plan their courses*. London: National Foundation for Educational Research.

Terhart, E. (2003). Constructivism and teaching: a new paradigm in general didactics? *Journal of Curriculum Studies, 35*(1), 25–44.

U.S. Department of Education, Office of the Under Secretary, Policy and Program Studies Service. (2003). *Federal funding for educational technology and how it is used in the classroom: A summary of findings from the integrated studies of educational technology*. Washington, DC: Author. Retrieved April 18, 2004, from http://www.ed.gov/rschstat/eval/tech/iset/summary2003.pdf

Usability Professionals' Association. (n.d.) *What is usability?* Retrieved April 5, 2006, from http://www.upassoc.org/usability_resources/about_usability/definitions_of_usability.html

Wertheimer, M. (1944). Gestalt theory [English translation of "Über Gestalttheorie" 1924/1925]. *Social Research, 11*, 78–99.

Woodbury, M. (1980). *Selecting materials for instruction: Media and the curriculum*. Littleton, CO: Libraries Unlimited.

Yinger, R. (1979, June). Routines in teacher planning. *Theory into Practice, 18*, 163–169.

Zhao, Y., & Frank, K. A. (2003, Winter). Factors affecting technology uses in schools: An ecological perspective. *American Educational Research Journal, 40*(4), 807–840.

6

MANAGING

J. Ana Donaldson
University of Northern Iowa

Sharon Smaldino
Northern Illinois University

Robert Pearson
President, Provinent Corporation, Toronto

Introduction

Educational technology is the study and ethical practice of facilitating learning and improving performance by creating, using, and *managing* appropriate technological processes and resources.

*A*LTHOUGH THE OBJECT OF management has shifted over the years, management has been a critical function in educational technology since the field began in the 1920s. In the first formal definition statement, management was seen as necessary to controlling the products and the processes that were used in the field (Ely, 1963). By the time of the publication of the 1972 definition, the idea of management included the supervision of personnel and the operation of organizations (Ely, 1972). As systems thinking became more pervasive, the systems approach became the dominant paradigm for thinking about management processes in instructional development and technology-based learning systems (Association for Educational

Communications and Technology, 1977) following the theory proposed by Heinich (1970). By the time the 1994 definition was published, management meant planning, coordinating, organizing, and supervising resources, information, and delivery systems in the context of managing instructional design (ID) projects (Seels & Richey, 1994).

This chapter focuses on the concepts and principles related to *managing appropriate technological processes and resources*. Educational technologists, whether acting as instructional designers, school media specialists, or learning consultants, deal with management concerns, much as other professionals do. They work with finite resources to accomplish specified goals in a specified time whether they are managing processes, such as ID, or resources, such as collections of instructional materials. Knowing the most relevant ID models, having the right tools, and having a team of skilled people are crucial to success, but so is a repertoire of management skills. One could say that effective management is an essential ingredient to *getting the job done* regardless of what the job might entail (Kotter, 1999).

The Audiovisual Paradigm of Managing

The management function was central to educational technology in the days of the audiovisual paradigm. From the emergence of the visual instruction field in the 1920s through the expansion of the audiovisual education field in the 1970s, educational technology professionals worked predominantly as directors or coordinators of media service agencies. In elementary/secondary education, they served in such organizations at the school building, school district, regional, or state education agency level. In higher education, media centers served whole institutions or units such as schools and colleges, for example, schools of medicine and colleges of education. Their primary jobs were to acquire audiovisual materials and equipment, to maintain the collection, and to help teachers identify and use materials to enhance their teaching.

Audiovisual Education Directors

The first count of school district visual instruction directors in 1946 showed 164 directors (National Education Association, 1946). That number had grown to nearly 700 by 1954 (National Education Association, 1955). Those numbers continued to grow with the increase in federal government funding of education after 1958. For example, the membership of DAVI, of which nearly half worked as school media directors or coordinators, grew from 3,000 in 1958 to 11,000 in 1970. By 1975, 56% of the membership of AECT (DAVI's successor) held positions that were primarily administrative (Molenda & Cambre, 1977).

The administrative duties carried out during this period were many and varied. They included acquiring, cataloging, storing, retrieving, and distributing audiovisual materials; supervising the production of audiovisual and television resources; acquiring, maintaining, and deploying audiovisual hardware; planning and maintaining facilities for using media in classrooms; promoting appropriate use of media among teachers and organizing in-service professional development programs to that end; managing professional and nonprofessional personnel; budgeting for agency operations; and evaluating the services offered (Erickson, 1968). Overlaid on all of these operations was a clear sense that the larger mission was to bring about change, to modernize and improve teaching and learning. Hence, the other administrative functions were viewed through a lens of change management (Erickson, 1968; Peterson, 1975).

Instructional Design Consulting

Beginning late in the 1970s and continuing through the 1980s, the population of the field shifted from employment in schools toward employment in universities and from administrative duties to ID consulting roles. In most organizations, ID consulting evolved out of the materials production process as educational technologists became more involved with instructional theory. They came to see their role as collaborating with instructors to develop learning environments that were more productive for student learning. As discussed in chapter 2, this role change was stimulated greatly by the programmed instruction movement and continued as cognitivist and constructivist perspectives came to the fore. Concerns with *managing* functions tended to shift toward ID project management rather than managing materials and equipment, but the change management lens remained intact and, in fact, expanded as the ID process came to be seen more and more as a social process as much as a technical one.

Integration of AV and School Libraries

In schools, the 1970s saw a general shift of responsibility for audiovisual materials and services from educational technologists to school librarians, who were already responsible for print-based instructional materials. Audiovisual centers were consolidated with school libraries, and the person with library certification tended to take over as the *school library media specialist*—

the term favored by the American Association of School Librarians (AASL). However, for those who continued to work in school and university media services, the multiple administrative functions persisted, evolving incrementally as computer technology become increasingly pervasive. Schmid (1980) identified the major functions as managing personnel; selecting, acquiring, and deploying equipment; selecting, acquiring, cataloging, and distributing instructional materials; promoting media center services; developing constructive client relationships; and carrying out all these functions with accountability, maintained through obtaining and analyzing data on costs versus services provided.

Recent Diversification

Through the 1990s and early 2000s, the demographics of AECT continued to shift away from media center management. By 2006, only 15% of AECT members worked in media director positions (compared to 56% in 1975). And in 2006, only 11% worked in elementary/secondary education (compared with 39% in 1975), whereas professors and instructional designers constituted 34% and 11% of the membership respectively (Pershing, Ryan, Harlin, Hammond, & AECT, 2006). Instructional media services were still being offered in schools and colleges, but they were likely to be managed by professionals other than AECT members, and they were increasingly overshadowed by services focusing on computer-based delivery.

Contemporary Views of Management, Leadership, and Change

Definitions

Management means effectively orchestrating people, processes, physical infrastructure, and financial resources to achieve predetermined goals. Effective management is important regardless of the setting. Whether a media center that bases the products and services it offers within a school districts' financial limitations, a university-based faculty development program that operates as a cost center, or a privately run e-learning company with daunting profit and return-on-investment (ROI) targets for owners and shareholders, effective management means that goals are achieved and clients are satisfied while budgets are met.

Currently, management is often viewed as being synonymous with project management. Indeed, graduate-level educational technology courses that focus on management often focus on project management. As we will see, project management is an important kind of management activity but is only one of several management functions common to the field.

Seels and Richey (1994) define management as fundamentally a controlling function; control that is exercised as planning, coordinating, organizing, and supervising actions. These actions manifest themselves in terms of four *subdomains* of management theory and practice: project management, resource management, delivery system management, and information management. According to Seels and Richey, managers are leaders who motivate, direct, coach, support, monitor, delegate, and communicate with their colleagues. The view of management as a controlling function has been a part of the concept of educational technology since the publication of its first definition in 1963 (Januszewski, 2001). The view of management taken here expands beyond the view of controlling and into a more comprehensive perspective of the role of manager as leader as well (Woolls, 2004).

Management and Leadership

The management literature draws a distinction between the practice of managing and the practice of leading. As Seels and Richey (1994) point out, managing is fundamentally about control while leading is fundamentally about setting direction for an organization, driving alignment with the stated goals, and motivating every level of the organization to achieve these goals. Both functions are critical and both are intimately intertwined (Woolls, 2004). However, it is hard to imagine great management without great leadership. As Kotter (1999) notes, leadership and management are two distinctive and complementary systems of action. Each has its own function and characteristic activities. "Both are necessary for success in an increasingly complex and volatile business environment" (p. 51). As we have seen, managing is really about dealing with the complex nature of an organization and delivering on its promise to its customers and stakeholders. In the absence of good management, organizations (or projects) descend into chaos. Each system of action tries to accomplish the same thing: deciding what needs to be done, creating networks of people and relationships that accomplish what needs to be done, and ensuring that people actually do the job.

Leadership and change management. Leadership is often about coping with change. According to Kotter (1999), organizations are led through the process of constructive change by first setting a direction—a vision of the future along with the strategies to get there. Leadership accomplishes change by aligning individuals within the organization to achieving the goals as they are set out—by creating committed groups of individuals who understand the vision and are highly committed to seeing it through. Most of all, leadership is about motivating and inspiring people—making sure everyone keeps moving in the right direction even when the going gets tough (Kotter, 1999).

Setting direction is not the same thing as long range planning. When leaders set direction they gather a broad range of inputs—inputs from their fellow employees, inputs from their peers (other leaders inside and outside the organization), inputs from customers, and inputs from the marketplace and market analysts. Leaders carefully assess past performance and look for trends and patterns that may inform the optimal future trajectory. Effective leaders are able to communicate skillfully to all key stakeholders—and in language that they understand—where the organization is and the future state toward which everyone must strive. Kotter (1999) notes that the best organizational strategies are seldom brilliantly innovative; they are grounded in solid data and sound analysis and can be communicated clearly and passionately to all organizational stakeholders.

Strategic planning is a critical task for management within any organization. "Needs assessment typically serves as the first step within strategic planning efforts as it prepares for the process of selecting appropriate solutions to the challenges and opportunities at hand while building shared commitment to the organization's future direction" (Leigh, 2006). Another component of making informed decisions is to apply the essential tools of cost-effectiveness analysis (Levin & McEwan, 2001). Planning is critical, but the analysis of the efficiency and effectiveness of the results on any initiative is also important to the evolutionary cycle as an organization deals with the constant of change.

Achieving sustained change within an organization means getting people all moving in the same direction, aligning their activities with a goal. Aligning is more of a communications challenge than an organizational design challenge. And the communications challenge is not about describing short-term objectives, rather one of communicating a vision for an organization that may be several years in the future. Visionary types of messages are difficult to craft and deliver. Effective leaders must be able to paint a picture of the future that is at the same time sufficiently different from the current state but also seen as achievable by most employees. An effective leader must be perceived as credible and trustworthy by fellow employees, or the vision will fall on deaf ears. It is critical that leaders walk the talk. Effective leaders look for every opportunity to reinforce their message: at every staff meeting, at the water cooler, in the weekly newsletter, and during coaching sessions with individual staff members. Alignment also comes through empowerment. Empowerment is really all about the exercise of leadership throughout the organization—not just in the boardrooms or the corner offices. Effective leaders empower everyone to drive change that is clearly in keeping with the direction even when this change may be momentarily destabilizing.

Finally, since driving change is the function of leadership, a skilled leader must be able to generate a high degree of enthusiasm for the direction of the organization. Hence, motivating staff is all about instilling the right level of commitment for the road ahead that will sustain everyone when things get a little bumpy—and inevitably they will. The management function gets results through systems and control. Good leaders motivate people in different ways. Good leaders always articulate the organization's vision in a way that resonates with the values of the audience they are addressing, making the challenge at hand important and meaningful to others. Leaders also frequently involve peers, subordinates, and superiors in deciding how to achieve the organization's vision. This instills an important feeling of ownership across the organization. Thus, leaders support others through coaching, role modeling, recognizing, and rewarding successful performance (Covey, 1991).

In summary, effective management can be seen as a complementary combination of six management and leadership functions:

Managing	Leading
Planning	Setting Direction
Monitoring	Aligning
Controlling	Motivating

The planning, monitoring, and controlling functions can be viewed as essential components of management. Once planning is completed, the project team is "involved in monitoring the status of the project, communicating progress to stakeholders, and managing variances and risks that emerge during the execution of the project" (Andreadis, 2006, p. 960).

Performance Management

Performance management has had many different meanings over the years. One of those meanings is to monitor and control the quality of the performance of the individuals working in an organization. In the early 1900s, issues related to quality assurance meant inspection, which was the primary method used to ensure a quality product. In the 1940s, quality took on a statistical bent. With the burgeoning of mass production during World War II, it became necessary to apply a more stringent form of quality control:

statistical quality control (SQC). Foundational work for SQC is credited to Walter A. Shewhart (1939) of Bell Labs. Statistical methods were used to control quality within the normal variation of a particular process. Statistical control charts helped operations managers determine if the variability in a product was within the normal range. This approach significantly reduced the need for inspection.

Today, the meaning of performance management has expanded to refer to a perspective that views every facet of an organization in terms of quality standards. Standards of quality are often viewed from a customer perspective: Is the organization's product or service meeting or exceeding customer requirements now and in the future (Schroeder, 2000)? Performance management is often pictured in terms of a quality cycle that requires an iterative sequence of steps:

- Define quality attributes on the basis of customer needs
- Decide how to measure each attribute
- Set quality standards
- Establish appropriate test for each standard
- Find and correct causes of poor quality
- Continue to make improvement (Schroeder, 2000)

The SQC approach is less about finding defects and fixing them and more about ensuring that defects are not made in the first place.

Closely related to SQC is a management philosophy known as "total quality management" (TQM), a strategy developed W. Edwards Deming for improving the quality of products and services by embedding awareness of quality issues among all members of the organization. The Japanese adopted the TQM concept in the early 1950s to aid in the resurrection of their business culture. Corporate America began implementing TQM in the 1980s as a competitive response to the success of Japanese business growth (Mehrotra, n.d.).

Bonstingl (1992) identified the TQM principles which he believed were applicable to schools and which he felt could create "a quality revolution in education." First, TQM requires a focus on customers and suppliers. Bonstingl envisioned teachers and students forming teams that together create the product of the school—each student's continuous growth. The second pillar of the TQM approach is dedication to continuous improvement. As discussed in chapter 3, how progress is *measured* is critical to attaining productivity. Bonstingl advocated that a grading system is destructive to a productive learning environment; he recommended the portfolio approach as an alternative. The third TQM pillar is viewing the organization as a system, thus schools must understand the *processes* that lead to their product before they can advance, a view highly compatible with educational technology's well-established systems perspective. Finally, successful implementation depends on the commitment of top management, meaning that school leaders must show "concerted, visible, and constant dedication" to these principles (p. 7).

Management in Educational Technology

Having reviewed the concepts associated with managing and leading in general, it is well to focus now on management in educational technology *per se*. This may best be done with a framework based on the *objects* of management. Looking across the wide range of organizations in which educational technology is used—from a 10,000-person company, to a five-person learning support office in a community college, to an individual school media center—educational technology program management activities tend to be directed to one of four objects: managing projects, managing resources, managing the performance of people, or managing programs. The goal of *project* management is to ensure that an appropriate solution to a particular performance problem is developed and implemented on time, on budget, and to the specifications established at the outset of the project within a defined time frame (Seels & Richey, 1994). The goal of *resource* management is to ensure that a collection of resources is developed, maintained, and made available as needed through various delivery systems to address the teaching-learning needs of instructors and students. The goal of *personnel* management is to provide the conditions for people to succeed in playing their roles in the work of an organization. All of these management functions—and more—are included in the larger process of *program* management, the overall supervision and control of a set of ongoing related activities within an organization.

Project Management

As noted earlier, project management is a well-understood management function within the field of educational technology. "Good project management saves organizational resources, increases productivity, and increases the likelihood that projects will be successful" (Andreadis, 2006). The projects in question typically entail the design and development of instructional materials and systems; hence, a knowledge base about ID project management grew throughout the 1980s as reflective practice and inquiry advanced in this area (Greer, 1992). ID project management is discussed at greater length in chapter 4.

At the heart of project management is an individual who is ultimately accountable for the successful completion of the project. In large projects, project managers will often need to delegate responsibility for major tasks. While responsibility can be delegated, ultimate accountably cannot. As such, the project manager is the final arbiter of disputes within the project team and acts as the single point of contact with the customer. Project managers are typically not technical experts. As the title suggests, they do not do the work; they manage the work (Kerzner, 2005).

Project management is practiced to ensure that a discrete project, a set of tasks intended to achieve a specific outcome, is completed on time, on budget, and to the client's specifications. A project manager works with a client to establish the project outcomes, budget, and the time line for completing the project. As the project progresses, its scope frequently changes. The project manager must assess the implications of these changes and set new project outcomes, budget, and time lines.

Project managers are responsible for many tasks throughout the life of a project. Project managers typically begin by working with a client to establish a project governance structure. Project governance formalizes the decision-making structure of the project—especially who on the client side has the authority to sign off on completed deliverables as well as making clear the key interface points between the project team and the client. Establishing the ground rules of the project is a critical first step and one that is often neglected.

Project management requires breaking the project down into discrete tasks and subtasks. The project team assigns each task a duration and identifies all important contingencies. Contingencies dictate the sequence in which the project team can complete the tasks. Client sign-offs are among the most important kind of contingencies. Having the client approve a storyboard or an interface design prior to the production of some instructional material, such as an e-learning course, is critical to avoid wasteful project effort downstream. Other types of contingencies include the availability of skilled associates, access to equipment, and the completion of tasks that are themselves key inputs to subsequent tasks.

Project management software, such as Microsoft Project, has become an essential aid for project managers. Project managers use the software to list all project tasks, contingencies, durations, and resources. The visual charts produced by the program (called *Gantt charts*) are invaluable in effectively communicating with the client.

Project managers must also be vigilant to mitigate the risks of project delays. There are two kinds of delays—ones caused by the project team and others caused by the client. In either case, delays force the project manager to adjust the project schedule and changes in the project schedule often result-ing in higher project costs. Delays caused by the project team typically mean that more effort than forecasted has been expended, meaning the project team's cost to complete the project goes up. These costs cannot be passed on to the client. In addition, delays will almost certainly irritate the client. Client-induced delays can buffet a project budget just as much. Client delays typically occur at deliverable sign-off points in the project schedule. A client may be given two days to review a lesson plan, but the client takes a week. This has several implications. The overall project schedule may be thrown off, result-ing in a delay in the project end date. Second, the resources scheduled for the subsequent task will need to be reallocated to other tasks. This is not always possible and idle resources that were not originally accounted for will have an impact on the overall project budget. Like changes, it is the project manager's responsibility to impress upon the client the financial implications of client-imposed delays back on the client. Negotiating an increase in budget due to client delays is among the most difficult tasks for a project manager but client-imposed delays are frequently the biggest contributor to increased costs.

Resource Management

The managing of learning resources adds another dimension to the manager's role. The responsibilities include the multiple tasks related to the information and the delivery infrastructure. In the previous 1994 definition, the subdomains of delivery system management and information manage-ment were addressed in addition to project and resource management (Seels & Richey, 1994). This current focus considers these two responsibilities as an integral component of resource management. Delivery system responsibili-ties include "a combination of medium and method of usage that is employed to present instructional information to a learner" (Ellington & Harris, as cited in Seels & Richey, 1994, p. 51). Information resources look at how the information for learning is stored, transferred, or processed (Kerzner, 2005). The revised interpretation of resource management looks holistically at the *resource systems and services*, the *context* for delivering the resources, and how the *content* is managed for effective learning (Schmidt & Rieck, 2000).

Managing resources also includes the responsibility of overseeing the delivery system and process used to deliver the product. How the required information is stored, transferred, and processed also falls within this man-agement category (Schmidt & Rieck, 2000).

Whether it is product specifications, best practices, or research and devel-opment (R&D), organizations as they operate generate vast amounts of formal and informal knowledge. Knowledge management is practiced to ensure that only the most useful organizational knowledge is captured, documented, and made available to the right people within the organization (Haney, 2006).

Even the simplest modern organization relies on some form of information technology—whether a computer, a simple network, or mobile phones. Information technology management is practiced to ensure that the technology infrastructure of an organization is properly matched to the needs of that organization and is kept up to date for a cost that is reasonable.

Woolls (2004) identified ways in which the school media center and media specialists parallel the business model for resource management. The school media-center facility itself requires a degree of managing. The collection, which is a major component of the facility, requires continual monitoring and updating. The actual physical space must be arranged in both a functional and appealing manner. These resources are continually in need of updating and upgrading, often requiring innovation ways of manipulating a limited budget.

Personnel Management

Projects and programs of any kind require the right people to ensure successful project completion (Schmidt & Rieck, 2000). Personnel management includes ensuring that there are enough people with the right skills to do the work at the right time and for the lowest cost with the necessary resources. Balancing capacity to demand is a major challenge—especially in a professional services firm that must deal with customer imposed delays.

A well-laid-out project schedule is only part of the equation. Having the right project team is just as important. Effective personnel management results in the right people doing the right tasks (Haney, 2006). Personnel management and project management are intimately linked. A skillful breakdown of project tasks will result in the best manpower fit for the task. The costs of poor personnel management can be significant. Assigning a less-qualified project team member to a task will result in the task taking longer than it should and quality may suffer. Assigning a highly qualified project team member to a task may result in higher costs as well if this cost has not been accounted for in the overall project budget. If multiple projects are running concurrently, highly skilled workers often need to be leveraged across several different tasks thus complicating the resource management challenge. Personnel management is also about ensuring that the right people are available just as the task needs to begin. Having a project team member sit idle can add significant cost to the overhead of the organization—costs that are often not factored into project budgets.

In many cases, the school media specialist is responsible for a staff of paraprofessionals and/or multiple media center settings, each with its own staff

(Woolls, 2004). Often, the paraprofessionals serve the facility while the media specialist is involved in other duties or is located in another building. This responsibility generates a need to manage the performance of other personnel (Morris, 2004). As a personnel manager, the media specialist must identify qualified individuals, oversee their performance, assist them in continual professional development, and provide them with motivation to engage in quality service. In addition, the media specialist is often charged with evaluating the staff (Schmidt & Rieck, 2000). Quality is achieved through having staff who are well trained and highly motivated to do the job right the first time (Addison & Haig, 2006).

Program Management

There is sometimes confusion between the terms *project* and *program* management.

> Programs are mission driven, have greater duration, and usually contain multiple projects. Projects are driven by specifications, have finite time limits, and result in a product, package, or service. Project output examples include computer-based education systems, textbooks, and evaluation reports. Projects share such common characteristics as a specified beginning and ending point, a description of the expected outcomes, and specifications for the deliverable products. (Branson, 1996, p. 303)

Project management is considered to involve short-term undertakings without long-term authority (Seels & Richey, 1994), while program management supports a long-term continuing program that is an integral part of the organization's purpose.

Modern approaches to program management are perhaps best epitomized by the work of Crosby (1979). Crosby's notion of *zero defects* suggested that the cost of finding and fixing errors is much greater than the cost of preventing errors from occurring in the first place. Crosby argued that we are conditioned to believe that it is alright to make errors. Hence, modern perspectives on program management have turned from a reactive, inspection-focused activity into a more proactive error avoiding activity. So quality performance is achieved through sound process. A quality performance is achieved through product and service offerings that are well understood and leverage common components as much as possible.

Marketing as a driver of program management. Regarding the principle of continuous improvement, assessing the success of an instructional process has various facets. Measuring results begins with effective communication among the key stakeholders—supervisors and line workers, school administrators and teachers, and so on. The key actors must agree upon who the customer is and how customer needs may be met (Andreadis, 2006). In most businesses, sales and marketing activities are as vital to success as delivering the goods on time, on budget, and at the expected quality level. Fundamentally, marketing and sales is about understanding the needs of customers and being able to align the value of the organization's service or product offering to the customer. Indeed any organizational entity that provides products or services to someone else, whether as a cost center or a profit center, must be concerned with understanding customer needs and shaping an offering that delivers value. Hence, as organizations today are becoming increasingly customer focused, the role of marketing is gaining in prominence. Indeed, marketing is now seen as the key integrative function between the customer and the other major functions of the organization (Kotler, 2003). Organizations that are not in sync with their customer's needs are destined to fail. The stereotypical view of the used car salesman is a long way from the reality of the modern marketing organization. Increasingly the business literature views sales as part of the marketing function within an organization (Brethower, 2006).

The "four Ps" is one of the classic concepts in marketing. *Product, price, place,* and *promotion* define the major preoccupations of the marketing function within management. *Product* is the most basic marketing-mix tool as it represents the organization's tangible offering to the market. Whether the organization's offering is a product or service, it needs to be concerned with shaping the offering in ways that resonate with the client. This means understanding customer needs and translating these needs into a product design, product features, and product quality that satisfy the customer. *Price* is another critical marketing-mix tool that all service providers must be concerned with, regardless of the service providers' positioning as a cost or profit center. Price is the amount of money customers are prepared to pay for a particular service or product. Price must be commensurate with the offer's perceived value. If these are out of alignment, the customer will turn to another source. *Place* includes the various kinds of activities an organization undertakes to make its service or product available to its target customers. For internal service organizations, the issue of place is key, especially as the workplace becomes more and more distributed. In some cases, organizations choose to work through intermediaries (channel partners). Finally, *promotion* speaks directly to positioning value for the customer. Promotion entails a set of activities that communicate the offering to the customer in a way that is clear and compelling.

Contemporary marketing theory suggests that there is value in looking at the four Ps not from the seller's perspective but from the customer's perspective (Lauterborn, 1991). In this view, the four Ps become four Cs:

4Ps	4Cs
Product	Customer Needs and Wants
Price	Cost to the customer
Place	Convenience
Promotion	Communication

Woolls (2004) suggests that the media specialist must also consider ways to market the media center, both within the organization and to the larger community. Often, because of reduced budgets, it is the role of the media specialist to market the media center to ensure its continued success. When funding is difficult, the media specialist must be creative in ways of managing the budget, but also in ways of finding additional resources to add to the media center.

Managing change as a part of program management. Built into quality management and leadership is the need to provide a vehicle for change. Change is an inevitable part of any organization—especially in today's dynamic environment. "Change is highly complex; it is rarely unidimensional or unidirectional and can come from inside the organization as a result of an internally identified need or can be imposed on the organization as a result of external changes" (Malopinsky & Osman, 2006, p. 39). Change management is practiced to ensure an organized and predictable transition occurs from one organizational state to another. Change management often involves organizational, job, and task redesign as well as training, coaching, and performance support.

To be fair, the role of manager needs to be described in terms of *change agent.* Ellsworth (2000) has identified several characteristics of the change agent that can be interchangeable with the descriptions of a manager. Among the characteristics he describes are visionary, guide, planner, and evaluator. As we look at the manager's role in any organization, we see similar types of responsibilities. Clearly, a manager serves as the change agent within the organization.

Program evaluation in program management. Program evaluation is another function within program management. Evaluation ensures that a specific program, service, or product line is delivering the value intended as well as to uncover any important unintended results. In some cases, program evaluation is formative; that is, the purpose of the evaluation is to diagnose problems and recommend changes during the development and testing phases.

Also included in Woolls' (2004) list of responsibilities for the media specialist is that of evaluation of product and process. The media specialist is responsible for continued assessment of the quality of the media center's resources. This evaluation is generally expected to be reported on an annual basis. Further, the process or manner in which the media center is utilized by the clientele is also part of the evaluation cycle. Not only are the circulation records important but also gathering information on the interaction with clients is essential to demonstrate the effectiveness of the media center.

Summary

Even though the role of a manager can cover the four areas described, multiple themes reoccur within each of the areas of responsibility. The role of being a change agent is a critical aspect within each of the spheres. This is the one place where managers need to display their innovative leadership skills in order to move individuals toward innovative approaches. Personal and budget issues are also a constant companion of any type of manager. The emphasis on the client is also a necessary consideration for all managers. The foundations of our profession are based on the belief that all work will be performed in an ethical manner that meets the high standards of the field of educational technology.

As has been discussed, there are many components to effective management. The fundamental thread that is interwoven in all aspects of management is the importance of evaluation. At all stages within a manager's roles is the requirement that the product or service is aligned with the needs of the client. This is true whether the manager is functioning as a project manager, resource manager, personnel manager, or program manager. Evaluation is an ongoing process that must be integrated into every phase of educational technology. With the current emphasis on alignment to standards and legislative guidelines, this process of alignment is even more critical than in the past.

Effective management and leadership are keys to the practice of an educational technologist. Technologists, by definition, work with others to solve real-world problems. This work needs to be planned and delivered and the people doing the work need to be selected, supported, supervised, compensated, and recognized. Educational technologists must effectively interact with their customers and community stakeholders. This means clearly understanding the needs of customers and translating these needs into products and services that resonate in the marketplace. It means positioning the value of the product and services so that customers are willing to pay for them. And because educational technologists are often cast in the role of change agent, they must have the ability to lead effectively—to direct, to align, and to inspire.

References

Addison, R. M., & Haig, C. (2006). The performance architect's essential guide to the performance technology landscape. In J. A. Pershing (Ed.), *Handbook of human performance technology* (pp. 35–53). San Francisco: Pfeiffer.

Andreadis, N. (2006). Managing human performance technology projects. In J. A. Pershing (Ed.), *Handbook of human performance technology* (pp. 943–963). San Francisco: Pfeiffer.

Association for Educational Communications and Technology (AECT). (1977). *The definition of educational technology.* Washington, DC: Author.

Bonstingl, J. J. (1992). The quality revolution in education. *Educational Leadership, 50*(3), 4–9.

Branson, R. K. (1996). Project and program management. In T. Plomp, & D. P. Ely (Eds.), *International encyclopedia of educational technology* (2nd ed., pp. 303–308). Cambridge, UK: Pergamon-Cambridge University Press.

Brethower, D. M. (2006). Systemic issues. In J. A. Pershing (Ed.), *Handbook of human performance technology* (pp. 111–137). San Francisco: Pfeiffer.

Covey, S. (1991). *Principle-centered leadership.* New York: Simon & Schuster.

Crosby, P. B. (1979) *Quality is free: The art of making quality certain.* New York: McGraw Hill.

Ellsworth, J. (2000). *Surviving change: A survey of educational change models.* Syracuse, NY: ERIC Clearinghouse on Information Technology.

Ely, D. P. (1963). The changing role of the audiovisual process: A definition and glossary of related terms. *Audiovisual Communication Review, 11*(1), Supplement 6.

Ely, D. P. (1972). The field of educational technology: A statement of definition. *Audiovisual Instruction, 17,* 36–43.

Erickson, C. W. H. (1968). *Administering instructional media programs.* New York: Macmillan.

Greer, M. (1992). *ID project management: Tools and techniques for instructional designers and developers.* Englewood Cliffs, NJ: Educational Technology Publications.

Haney, D. (2006). Knowledge management, organizational performance, and human performance technology. In J. A. Pershing (Ed.), *Handbook of human performance technology* (pp. 619–639). San Francisco: Pfeiffer.

Heinich, R. (1970). *Technology and the management of instruction.* Washington, DC: Association for Educational Communications and Technology.

Januszewski, A. (2001). *Educational technology: The development of a concept.* Englewood, CO: Libraries Unlimited.

Kerzner, H. (2005). *Project management: A systems approach to planning, scheduling, and controlling* (9th ed.). San Francisco: Wiley.

Kotler, P. (2003). *Marketing management* (11th ed.). Upper Saddle River, NJ: Prentice Hall.

Kotter, J. P. (1999). *What leaders really do.* Boston: Harvard Business School Press.

Lauterborn, R. (1991). From 4Ps to 4Cs. *Advertising Age, 61*(41), 26.

Leigh, D. (2006). SWOT analysis. In J. A. Pershing (Ed.), *Handbook of human performance technology* (pp. 1089–1108). San Francisco: Pfeiffer.

Levin, H. M., & McEwan, P. J. (2001). *Cost-effectiveness analysis: Methods and applications* (2nd ed.). Thousand Oaks, CA: Sage Publications.

Malopinsky, L., & Osman, G. (2006). Dimensions of organizational change. In J. A. Pershing (Ed.), *Handbook of human performance technology* (pp. 35–53). San Francisco: Pfeiffer.

Mehrotra, D. (n.d.). *Applying total quality management in academics.* Retrieved November 5, 2006, from http://www.isixsigma.com/library/content/c020626a.asp

Molenda, M., & Cambre, M. (1977, March). The 1976 member opinion survey. *Audiovisual Instruction*, pp. 65–69.

Morris, B. J. (2004). *Administering the school library media center* (4th ed.). Englewood, CO: Libraries Unlimited.

National Education Association. (1946). Audio-visual education in city school systems. *Research Bulletin, 24*(4).

National Education Association. (1955). Audio-visual education in urban school districts. *Research Bulletin, 33*(3).

Pershing, J. A., Ryan, C. D., Harlin, N. M, Hammond, T. D, & AECT. (2006). 2006 AECT membership salary survey. *TechTrends, 50*(5), 10–19.

Peterson, G. T. (1975). *The learning center.* Hamden, CT: The Shoe String Press.

Schaffer, S. P., & Schmidt, T. M. (2006). Sustainable development and human performance technology. In J. A. Pershing (Ed.), *Handbook of human performance technology* (pp. 1109–1121). San Francisco: Pfeiffer.

Schmid, W. T. (1980). *Media center management: A practical guide.* New York: Hastings House.

Schmidt, W. D., & Rieck, D. A. (2000). *Managing media services: Theory and practice* (2nd ed.). Englewood, CO: Libraries Unlimited.

Schroeder, R. G. (2000). *Operations management: Contemporary concepts and cases.* Boston: McGraw-Hill/Irwin.

Seels, B. B., & Richey, R. C. (1994). *Instructional technology: The definition and domains of the field.* Washington, DC: Association for Educational Communications and Technology.

Shewhart, W. A. (1939). *Statistical method from the viewpoint of quality control.* Washington, DC: The Graduate School, Department of Agriculture.

Woolls, B. (2004). *The school library media manager* (3rd ed.). Englewood, CO: Libraries Unlimited.

7

PROCESSES

Robert Maribe Branch
Christa Harrelson Deissler
The University of Georgia

Introduction

Educational technology is the study and ethical practice of facilitating
learning and improving performance by creating, using, and managing
appropriate technological *processes* and resources.

THE PURPOSE OF THIS chapter is to expand the discussion about technological *processes* in the context of the current definition. While
processes, in general, are common to many professions, educational technologists routinely employ *technological* processes to design, develop, and
implement effective resources for learning. Thus, the processes discussed in
this chapter focus on methods commonly used to facilitate learning goals
and improve performance.

The importance of technological processes within educational contexts
emerges from the need to provide effective communication and co-operation
during the pursuit of mutual goals. The most common portrayal of educational technology processes is the input-process-output paradigm (Fig. 7.1).
The input-process-output paradigm provides a way to think about educational communications during the pursuit of mutual goals.

This chapter extends the work of Seels and Richey (1994) and organizes
the discussion about technological processes into conceptual, theoretical,

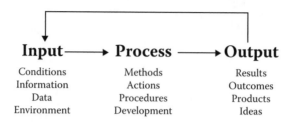

Figure 7.1. The Input – Process – Output paradigm.

and practical frameworks. The conceptual framework for technological processes is based on the notion that a process is series of meaningful activities constructed upon organizing themes. The theoretical framework for technological processes is based on the idea that a process is a series of propositions, based on verifiable evidence, which represent a systematic view of a subject. The practical framework for technological processes is the application of principles or theories to achieve intended results.

Defining Terms

Technology

Technology, in its most generic interpretation, is the application of knowledge for a practical purpose. The definition of technology commonly accepted in the field is taken from Galbraith (1967): "The systematic application of scientific or other organized knowledge to practical tasks" (p. 12). According to Hooper and Rieber (1995), "Technology, by definition, applies current knowledge for some useful purpose. Therefore, technology uses evolving knowledge (whether about a kitchen or a classroom) to adapt and improve the system to which the knowledge applies (such as a kitchen's microwave oven or educational computing)" (p. 156). Most common interpretations of technology focus on the physical products that result from technological research and development, such as computer hardware and software, video recordings, personal digital assistants, and other handheld communications devices, satellites, satellite receivers, and the like. Some people refer to this side of technology as *hard technology*, while reserving the term *soft technology* to refer to the intellectual processes. This chapter focuses on the soft technology side, applying intellectual processes to achieve educational goals.

The venues for educational technology processes typically include teacher planning routines, instructional design operations, curriculum development projects, learning resources administration, and media utilization strategies.

Process

Process is denoted here as a series of actions, procedures, or functions leading to a result. A process typically yields one of two kinds of results: (a) a product or (b) another process. Processes can be naturally occurring, such as the processes involved in the digestion of food and the conversion of that food into energy. The functioning of the human organism depends on many natural, innate processes, including those involved in thinking, learning, communicating, feeling, as well as just staying alive. Humans also *invent* processes in order to achieve their goals more efficiently and effectively. Hunting, cooking and preserving food, and migrating require fabricated procedures that evolve with experience. Nontechnical processes, such as cognitive processes, biological processes, and spiritual processes are extremely important endeavors but exist outside the scope of this chapter. Human-made processes that systematically apply scientific knowledge can be viewed as *technological* processes. This chapter focuses on technological processes applied in the advancement of learning. There are many technological processes that are noneducational, including, for example, those found in mass communication, computer networking, transportation, and energy production. Certainly, an argument could be made that everything is some way educational; however, within the context of this definition, the processes associated with educational technology are interpreted as the methods used to facilitate learning and improve performance.

Assumptions

Technologies should be considered as inventions that extend human capability. Technological inventions are conceivably infinite, and limited only by our creativity. Processes dedicated to creative capacity building are predicated on several assumptions. The Association for Educational Communications and Technology (AECT) assumes that education is a process, technology can facilitate educational processes, and intentional learning environments are complex. The following explanations of these assumptions provide a philosophical orientation to the study and practice of technological processes dedicated to education.

Assumption 1: Education is a process. Education is a series of purposeful actions and operations—a process. The goals of education represent desired learning outcomes; thus, education, in general, can be regarded as a process.

Assumption 2: Technology can facilitate educational processes. Society generally regards technology both as a tool and as a means to an end. Educational processes focus on the systematic application of learning theory to achieving the goals of education. Misperceptions about technology often emerge from the belief that technology is the end. Technology is not a panacea for all of the ills of education, and not a means for replacing people. Technological processes are dedicated means, based on scientific thinking, for communicating ideas and taking action to facilitate teaching and learning. Thus, technology facilitates educational processes.

Assumption 3: Intentional learning environments are complex. Intentional learning environments refer to purposeful educational events that involve learners in multiple, concurrent interactions among people (e.g., teachers and peers), places, content, and media, situated within a context, for a period of time, all seeking a common goal (Fig. 7.2). Complexity is a phenomenon resulting from an increasing amount of information, energy, hierarchy, variability, relationship, and components, which in turn increase possible outcomes and reduce certainty and predictability for a given event. Intentional learning environments produce knowledge and skill. Complexity also emerges from interactions of individual units (Marion, 1999), such as the interactions of the student, teacher, content, and context during educational

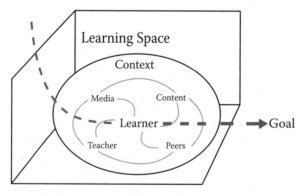

Figure 7.2. The typical components of intentional learning environments.

events. Levy (1992) defined a complex system as one whose component parts interact with sufficient intricacy that they cannot be predicted by standard linear equations. Therefore, processes associated with intentional learning environments should be capable of facilitating, managing, and directing varying amounts of information, variability in relationships, multiple solutions, predictability, unpredictability, certainty, and uncertainty.

Conceptual, Theoretical and Practical Frameworks

The various technological processes dedicated to facilitating learning and improving performance can be viewed as concepts, as theories, or as practices. Table 7.1 presents generic definitions of concept, theory, and practice and offers examples that illustrate the alignment of concepts, theories, and practices.

In the following sections we explore in detail how processes can be considered as concepts, as theories, and as practices and how each of these notions is related to educational technology.

Process Concept

Process as a concept can be defined as a series of activities directed toward a desired result. The results can be tangible or intangible. Process theories are generated to propose relationships among concepts.

Process Theory

A *theory* is a logically self-consistent model or framework for describing the behavior of a certain natural or social phenomenon, thus either originating from or supported by experimental evidence (Wikipedia, 2006). *Process theories* propose that events are the result of particular input states leading to particular output states, following certain processes. Process theories in education deal with activities that facilitate the interpretation, acquisition, construction, and application of knowledge and skills.

Process theories can be descriptive or prescriptive. Descriptive theories are passive, explain phenomena, illustrate relationships, and describe conditional (if-then) statements. Prescriptive theories are active, goal-oriented, rule-based, normative guidelines and strategies, used to construct models, methods, and procedures for practice. A process theory can be a means for achieving an end result, or it can lead to the development of another process theory.

Table 7.1 Definitions and examples of concepts, theories, and practices.

	Concept	Theory	Practice
Definition	A phenomenon that is conceived in the mind. A thought, notion or idea. Characterized as being covert and idiosyncratic and is a socially constructed configuration.	A set of facts and their relation to one another. A body of theorems representing a concise systematic view of a subject. A theory can be descriptive or predictive, and involves induction, deduction and extrapolation. Overt shared frames of reference.	Putting into use; performance of practical work. Application of principles or processes (contextualized). People commonly use concepts and theories to accomplish specific tasks within defined contexts. Implementation of a frame that functions as intended.
Examples			
Ohm's Law	There are dedicated physical relationships among electrons.	Electric current is proportional to voltage and inversely proportional to resistance.	Ohm's Law is practiced when we apply the formula $E = I \times R$ to install a new interior lighting system in a 1,500-square-foot domestic dwelling.
Communication	Meaningful exchanges between people.	Thoughts can be encoded, transferred, and decoded over a variety of distances among individuals and groups of people.	Claude Shannon and Warren Weaver produced a general model of communication. The Shannon–Weaver Model (1947) proposes that all communication must include six elements: Source, Encoder, Message, Channel, Decoder, Receiver.
Visual Literacy	A language of imagery, bound by the explicit juxtaposition of symbols in time and space, can be learned, used and integrated simultaneously into a message.	Visual messages facilitate complex cognitive processing because visuals provide information and opportunity for analysis that is unique.	Visual literacy is often practiced through message design, and characterized by: 1. Typography 2. Layout 3. Elements 4. Color

A popular process theory associated with educational technology is general systems. In the early 1940s, general systems theory emerged within the biological research community as a way to coordinate research and development activities across a variety of disciplines. General systems theory has been addressed in treatises that are beyond the scope of this chapter; however, a brief discussion about the basic theory of a system is necessary to illustrate the relationship between the concept of process, and the practice of process, within the context of educational technology (Table 7.1).

A system is a group of elements inextricably connected and working together for a common purpose. Processes work within systems and can be described in terms of general systems theory as being systematic, systemic, and synergistic. *Systematic* processes follow rules and procedures that apply to all stages or elements of the process. Systematic processes are intended to predictably generate a consistent product or end result. A process is *systemic* when any part of the process has the potential to change any other component of the system, therefore affecting the nature of the entire system. Another attribute of a system being systemic is the system as a whole responding to individual stimuli because of the nonlinear nature of relationships within the system. There is no simple cause and effect relationship, but rather a responsive, systemic relationship. *Synergy* refers to the interaction of two or more elements of a process whose combined efforts yield a force greater than the sum of each individual effort. Consider six individual horses carrying cargo individually to a common location. Consider the same six horses working as a team plus a wheeled cart to transport the same cargo to the same location. The six horses working as a team could carry at least three times as much cargo than if they were working as individuals because of the synergistic nature of the combination of horses and cart. Therefore, the application of general systems theory to specific educational technology processes leads us to the assertion that effective educational technology systems should be systematic, systemic, and synergistic.

Several prominent learning theories inform the study and practice of educational technology, including behaviorism, cognitive information processing, schema, situated cognition, interactional, motivational, and constructivism (see chapter 2).

Some *instructional* theories derived from these learning theories that are frequently applied in the field of educational technology include situated learning, action learning, case-based learning, inquiry-based learning, and student-centered learning. Each of these theories makes a case for applying different processes within the system of education depending on the end purpose. While the aforementioned theories are all related and represent variations on the theme of facilitating learning and improving performance, applying the appropriate technological process to a given setting for educational practice defines the role of educational technology that is unique in comparison with other disciplines.

Process Practice

Process practice, within the context of educational technology, means applying procedures that reflect the concepts and theories of learning and performance improvement. Educational technology procedures are integrated into strategies that are dedicated to effective communication and to creating appropriate instructional strategies. The purpose of process practice is to increase the maximum potential for success among students after they leave the classroom. Educational technology processes are dedicated to increasing the fidelity between the expectations for the student in the classroom (learning space) and the expectations for the student outside the classroom (performance space). The idea is that intentional learning is effective when educational strategies use processes that move a student through a learning space and approaches congruency with a corresponding performance space. Educational technology should provide opportunities for students to experience ever-increasing fidelity between learning space and performance space as a component of the teaching and learning process. The application of effective technological processes move students from a narrow learning space to the broad performance space (Fig. 7.3), thereby increasing the potential for student success. While there are many educational technology processes that facilitate learning and improve performance, several are noteworthy for this definition.

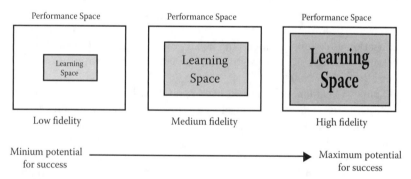

Figure 7.3. Effective technological processes move students from learning space to performance space.

Table 7.2 illustrates the relationship between the conceptual, theoretical, and practical frameworks for technological processes. While every possible process related to each component of the definition is beyond the scope of this chapter, in the following sections, we discuss well-known processes associated with *study, ethical practice, facilitating learning, improving performance, creating, using,* and *managing.*

Study

Technological processes are involved in generating and analyzing data, information, knowledge, and wisdom of all types. Two of the most commonly referenced "types" of data are *quantitative* and *qualitative. Quantitative* refers to data based on quantity, amount, or number, variables that can be manipulated numerically (Vogt, 1993). "This data is commonly utilized in statistical analysis" (Kaufman, Watkins, & Leigh, 2001, p. 164). *Qualitative* refers to variables that are categorical or nominal (Vogt, 1993). "Data is based on quality, kind, or character of information. This data is commonly utilized in interpretive and/or anecdotal analysis" (Kaufman et al., 2001, p. 164). Quantitative and qualitative data may be gathered through a variety of processes employed for the purpose of *study* within the field of educational technology. Specific study processes that educational technologists might choose to employ include formative and summative evaluation, design-based research (Design-Based Research Collaborative, 2003), and case studies (Yin, 1994). All of these different study processes may generate qualitative or quantitative data. The purpose of study, within this context, is to promote iterative processes of design and development toward realizing systemic change. The main message here, related to technological processes, is that the field of educational technology employs a variety of approaches to assess people and evaluate learning materials, such as needs analysis, inquiry, experimental and quasi-experimental designs, formative evaluation, summative evaluation, development research, case studies, and direct observation. While more detailed discussions of inquiry in educational technology are found in chapters 1 and 9, this section emphasizes the idea that educational technologists should be reflective practitioners engaged in systematic inquiry about the effectiveness of the processes that are selected to validate the use of technology for learning.

Ethical Practice

Ethical practice, as it applies to technological processes, refers to the appropriate process for a given situation. Applications of ethics are not

Table 7.2 The concept, theory, and practice alignment for the concept, "Technological Processes."

Conceptual Framework	Theoretical Framework	Practical Framework
Process is conceived as a series of actions directed toward a desired result.	Independent processes and collective processes are dedicated to facilitate learning and improve performance.	An instructional strategy (process) based on situated learning moves the student through a learning space that increases its fidelity to a performance space:

The main concept is the idea of a situated learning strategy.	Processes are united into a set of coherent descriptions and predictions regarding teaching and learning.	**1. Sample Case**
		  a. A well-defined problem
		  b. A simple situation
		  c. Familiar context
		  d. Students already possess all prerequisite knowledge and skills
		  e. Desired outcome is evident
		  f. Conducted within an immediate time period
		  g. Guided by the teacher
		  h. Typically close to one solution is appropriate

	This prescriptive theory posits that learning is achieved most productively when students are guided through a graduated series of cases.	**2. Practice Case**
		  a. A somewhat well-defined problem
		  b. Relatively simple situation
		  c. Somewhat familiar context
		  d. Students possess practically all prerequisite knowledge and skills
		  e. Desired outcome is revealed early in process
		  f. Conducted within a near time period
		  g. A teacher-student collaborative effort
		  h. Relative few solutions are appropriate

		3. Action Case
		  a. An ill-defined problem
		  b. Relatively complex situation
		  c. Context may be unfamiliar
		  d. Students may need to acquire some prerequisite knowledge and skills
		  e. Desired outcome is negotiated
		  f. Conducted within an authentic time period
		  g. Guided by the student
		  h. Typically a variety of solutions are appropriate

limited to censoring inappropriate behavior, but are intended to "provide leadership in identifying new issues that emerge in a rapidly changing techno-logical society that have an influence on ethical professional conduct" (Yea-man, 2004, p. 10). The level to which a process is ethical or not depends upon the context in which it is applied and the result it is intended to achieve. An example might be an intervention that is being considered for a project that is anticipated to take two years to appropriately implement, but political forces demand a result in six months; the deliberate application of a faster process would be unethical. AECT has adopted a code of ethics "intended to aid members individually and collectively in maintaining a high level of professional conduct" (AECT, 2005). The AECT code of ethics is explicitly committed to the individual, the society, and the profession. A regular edi-torial column in *TechTrends* features scenarios and principles dedicated to professional ethics in educational communications and technology to keep professional ethics issues in the minds of AECT members (Yeaman, 2006). While a more detailed discourse about the ethics of educational communi-cations and technology is presented in chapter 11, this section emphasizes the idea that educational technologists should be cognizant of their roles as stewards of processes associated with the ethical practice of facilitating learning and improving performance.

Facilitating Learning

Educational technology proposes that learning can be facilitated through the implementation of certain instructional processes, as discussed in depth in chapter 2. These prescriptive processes are often derived from descriptive theories of how humans learn through interaction with their environment. Cognitive learning theory suggests that certain psychological conditions need to exist in order for various types of learning to occur. Instructional frameworks, such as Gagne's (Gagne, Wager, Golas, & Keller, 2005) Events of Instruction, prescribe a distinct series of instructional events to organize episodes of intentional learning in a way that is consistent with the hypoth-eses posited in theories of learning, as discussed in chapter 2. This is an example of a process derived primarily from cognitive theories of learning. Other instructional process models are derived from other theories.

Behaviorist learning theory posits that learning is determined by the con-sequences that follow people's actions. Behaviorism inspired the process model known as *programmed instruction* with the following steps: specify a learning objective in behavioral terms, show or tell the desired behavior, have the learner practice the desired behavior, and follow the desired perfor-mance with a reinforcer. The design activities needed to create programmed instruction actually evolved into a design process model that merged with the systems theory model to form the instructional systems design (ISD) approach, discussed at greater length in chapter 4.

Constructivist learning theory has inspired further development of a number of instructional process models that had emerged prior to the popularization of the constructivist label, such as anchored instruction, problem based learning (PBL), and collaborative learning. There is not a single "constructivist instructional process" but, rather, a number of strategies that generally follow a common underlying process: Immerse learners in realistic problem spaces and support them through different phases as they struggle to construct their own understanding of the problem. Thus, educational technology calls upon a number of theoretical frameworks to develop instructional processes that help learners reach their goals effectively and efficiently.

Improving Performance

Educational technology claims to improve the performance of students, of teachers and designers, and of organizations, as discussed in chapter 3. Some of the means for performance improvement entail specific processes, such as the competencies promoted by the International Board of Standards for Training, Performance, and Instruction (International Society for Performance Improvement, 2006; Richey, Fields & Foxon, 2001). Educational technology extends successful learning into improved performance for students, first, by focusing learning experiences on authentic goals. Second, technology-enhanced experiences can lead to deeper levels of understanding beyond rote memory. Third, technology-enhanced immersive learning experiences can promote transfer of new skills to genuine problems. Students become doers through these means, with knowledge better connected to performance beyond the classroom setting.

Educational technology can improve the work productivity for teachers and designers by reducing learning time and increasing learning effectiveness. The main process used to achieve more efficient and effective learning is the systems approach to instructional design and development, discussed in chapter 4. Hard technologies have proven capable of effecting economy by delivering instructional materials cheaply over long distances and doing routine operations, such as record keeping less expensively and more reliably than human operators can, particularly for organizations. Soft technologies, though, offer a whole new paradigm for organizing the work of education. This new, technological work paradigm is based on several specific processes: division of labor, specialization of function, and team organization. Corporations and distance education institutions have used such processes to create and offer online courses at affordable prices and at a quality level that is often comparable to the best of face-to-face courses.

Creating

Educational technology proposes that effective instructional materials and systems can be created efficiently through particular development processes. One set of development processes, the systems approach, is distinguished from generic lesson planning in its technological character; that is, it is based on scientific thinking and incorporates empirical data gathering in the process. This approach is often referred to as "instructional system design" or "instructional system development," both abbreviated as ISD. The essence of ISD is to deconstruct the instructional planning process into small steps, to arrange the steps in logical order, and then to use the output of each step as the input of the next. The major stages are analysis, design, development, implementation, and evaluation; hence, the acronym ADDIE is attached to this process, shown in Fig. 7.4.

Each step informs another step, and revision continues throughout the entire process; for example, the output of the analysis stage, a description of the learners, the tasks to be learned, and the objectives to be met serve as input to the design stage, where those descriptions and objectives are transformed into a blueprint for the lesson. Next, the design blueprint serves as input to the development stage for the construction of the materials and activities of

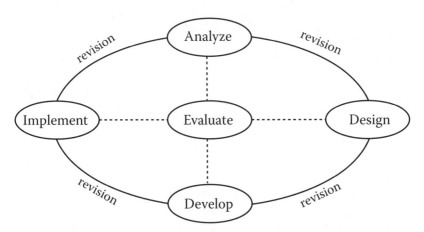

Figure 7.4. The major stages in the ISD process model.

the lesson. There is an opportunity at each major decision point to gather data to test that decision and other prior decisions to verify that the project is moving ahead toward a solution of the originally defined problem.

There are numerous ISD models. They differ in terms of the number of steps, the names of the steps, and the recommended sequence of functions. Gustafson and Branch's (2002) *Survey of Instructional Development Models* includes a number of variations on the ADDIE paradigm that illustrate the variety of ways to implement a systems approach. Regardless of the number of steps shown in a given ISD model, the time devoted to the process can be long or short and the steps can be done rapidly or slowly, depending on the contextual variables relative to specific situations. ISD models provide communication tools for determining appropriate outcomes, collecting data, analyzing data, generating learning strategies, selecting or constructing media, conducting assessment, and implementing and revising the results.

Using

The activities associated with the actual use of media by learners (often referred to as *media utilization*) are also treated as processes and are guided by various process models. The earliest utilization guides evolved during the film era. During World War II, films played an important role in military training in the U.S. armed forces, and considerable research was conducted to determine how films could be used with best effect (see chapter 5). The results were used during the war to guide the practice of trainers when using audiovisual media. For example, the U.S. Navy produced a film called "Film Tactics," which demonstrated how Navy trainers could improve learning by better utilization procedures. The process model given in the film recommended

- Examine the aid to be used
- Prepare the classroom
- Prepare the viewers: tell them the lesson objective; outline the main points
- Present the aid
- Demonstrate the skill
- Review
- Test

This advice is probably as relevant today as it was in 1945.

The pedagogical strategy in the Navy utilization model parallels Gagne's (Gagne et al., 2005) lesson framework, the Events of Instruction, as previously discussed. One major element found in Gagne's prescription is missing from the Navy prescription, that is, learner practice with feedback. The element of practice with feedback became prominent in the years after World War II, influenced greatly by the operant conditioning movement, led by B. F. Skinner. However, behaviorism was not the first or last learning theory to influence teachers' approach to the use of media.

More recent guides to the utilization process attempt to synthesize advice from earlier models and different theories of learning and instruction. One such eclectic guide to using media with learners is the ASSURE model (Heinich, Molenda, Russell, & Smaldino, 2004) as presented in Fig. 7.5. The ASSURE guide and others are discussed in greater depth in chapter 5.

A	S	S	U	R	E
Analyze Learners	State Objectives	Select Methods, Media, and Materials	Utilize Media and Materials	Require Learner Participation	Evaluate and Revise

Figure 7.5. The ASSURE model for media utilization. Note: Based on the text description of the ASSURE model, pp. 34–35 in *Instructional media and the new technologies of instruction* by R. Heinich, M. Molenda, and J.D. Russell. New York: John Wiley & Sons, 1982.

Managing

Educational technology processes promote the proper management of projects. Start, plan, analyze, develop, evaluate, and stop (SPADES) is a project management paradigm based on the fundamental process of initiate, plan, execute, and close out. The primary components are particularly relevant to projects that deal with some form of instructional design or technology. The SPADES components are tasks, events, procedures, products, and processes commonly associated with managing educational technology projects. Selected stakeholders throughout the duration of the project endorse each project component. A comprehensive discussion about managing appropriate technological processes and resources can be found in chapter 6.

Conclusion

The end purpose of technology in education is to promote human learning. Learning itself is a natural biological process, occurring spontaneously in all humans. The use of technology merely for technology's sake is ineffective and often a misuse of resources. Useful processes need to be capable of responding to the emerging trends in educational technology. Some emerging trends in educational technology include the nearly ubiquitous use of computers in society, ever-increasing use of the Internet, the growth of computing technology in the home, advances in distance education, broader advocacy for educational technology, more robust professional development programs than in the past, and new educational delivery systems (Ely, 2002; Molenda & Bichelmeyer, 2006). Processes can be a means for self-efficacy. Processes dedicated to educational communications and technology need to be well-designed means and not an end unto themselves. Many of the important problems of education require political, economic, social, and psychological decisions and actions more than technology. Therefore, various process models should be used to study and practice the operations of creating, using, and managing instructional materials and systems to facilitate learning and improve performance.

References

Association for Educational Communications Technology. (2005). *Code of ethics.* Bloomington, IN: Author. Retrieved March 28, 2006, from http://www.aect.org/About/Ethics.htm

Design-Based Research Collective. (2003). Design-based research: An emerging paradigm for educational inquiry. *Educational Researcher, 32*(1), 5–8.

Ely, D. P. (2002). *Trends in educational technology* (5th ed.). Syracuse, NY: ERIC Clearinghouse on Information and Technology, Syracuse University.

Gagne, R. M., Wager, W. W., Golas, K. C., & Keller, J. M. (2005). *Principles of instructional design* (5th ed.). Belmont, CA: Thomson Wadsworth.

Galbraith, J. K. (1967). *The new industrial state.* Boston: Houghton Mifflin.

Gustafson, K. L., & Branch, R. (2002). *Survey of instructional development models* (4th ed.). Bloomington, IN: Association for Educational Communications and Technology.

Heinich, R., Molenda, M., Russell, J. D., & Smaldino, S. E. (2004). *Instructional media and technologies for learning* (7th ed.). Upper Saddle River, NJ: Prentice-Hall, Inc.

Hooper, S., & Rieber, L. P. (1995). Teaching with technology. In A. C. Ornstein (Ed.), *Teaching: Theory into practice* (pp. 154–170). Needham Heights, MA: Allyn and Bacon.

International Society for Performance Improvement. (2006). *Standards of performance technology and code of ethics.* Silver Spring, MD: Author. Retrieved March 28, 2006, from http://www.certifiedpt.org/index.cfm?section= standards

Kaufman, R., Watkins, R., & Leigh, D. (2001). *Useful educational results: Defining, prioritizing & accomplishing.* Lancaster, PA: Pro>Active Publications.

Levy, S. (1992). *Artificial life.* New York: Random House.

Marion, R. (1999). *The edge of organization: Chaos and complexity theories of formal social systems.* Thousand Oaks, CA: Sage Publications.

Molenda, M., & Bichelmeyer, B. (2006). Issues and trends in instructional technology: Gradual growth atop tectonic shifts. In M. Orey, V. J. McClendon, & R. M. Branch (Eds.), *Educational media and technology yearbook 2006* (Vol. 31, pp. 3–32). Westport, CT: Libraries Unlimited.

Richey, R. C., Fields, D. C., & Foxon, M. (2001). *Instructional design competencies: The standards* (3rd ed.). Washington, DC: International Board of Standards for Training, Performance and Instruction, United States Department of Education.

Seels, B., & Ritchy, R. (1994). *Instructional technology: The definition and domains of the field.* Washington, DC: Association for Educational Communications and Technology.

Vogt, W. P. (1993). *Dictionary of statistics and methodology: A guide for the social sciences.* Newbury Park, CA: Sage Publications.

Wikipedia. (2006). *Theory.* Retrieved March 28, 2006, from http://en.wikipedia.org/wiki/Theory

Yeaman, A. (2004). Professionalism, ethics and social practice. *TechTrends, 48*(4), 7–11.

Yeaman, A. (2006). Professional ethics: Scenarios and principles. *TechTrends, 50*(2), 10–11.

Yin, R. K. (1994). *Case study research: Design and methods* (2nd ed., Vol. 5). Thousand Oaks, CA: Sage Publications.

8

RESOURCES

Anthony Karl Betrus
The State University of New York at Potsdam

Introduction

> Educational technology is the study and ethical practice of facilitating learning and improving performance by creating, using, and managing appropriate technological processes and *resources*.

THE TERM RESOURCES IS understood to include the tools, materials, devices, settings, and people that learners interact with to facilitate learning and improve performance. Both the types of resources (specifically, *technological* resources) and how these resources are used (*appropriately*) serve to differentiate what is done by educational technologists from similar efforts in other fields. The chapter begins with these defining characteristics, and then, it surveys the evolution of the various types of resources and surveys how emerging technologies have affected the field. The second half of the chapter differentiates analog and digital media, examining in greater depth how digital tools have changed the landscape of educational technology. It also discusses how settings and people are used a resources. The chapter concludes with a consideration of ethical issues in the use of resources.

Technological Resources

In the context of the current definition, the term *technological,* as a modifier of *resources* indicates that the resources created and used in educational technology are most often tools, materials, devices, settings, and people. Other resources, such as natural resources or political resources, are not considered to be primarily technological or educational and are therefore not central to the field. While educational technology professionals may indeed understand and account for natural, political, or other types of non-technological resources, including the term *technological* provides a focus on tools, materials, devices, settings, and people as the primary resources that are used to improve learning and performance.

There has recently been significant emphasis in the literature on the use of newer, *digital* resources, almost to the exclusion of historically traditional *analog* resources. Yet in terms of actual practice, analog resources such as textbooks, the overhead projector, and the videocassette recorder (VCR) are still used extensively in both corporate and educational settings. Molenda and Bichelmeyer (2005) illustrate the continuing use of the VCR in schools: "The predominant media formats in the collections of members of the National Association of Media & Technology Centers (NAMTC) continue to be analog, particularly videocassettes. Their collections include (from greatest to least) videocassettes, multimedia, curriculum materials, professional books, digital video disks and CD-ROMs" (p. 27). In corporate training, Dolezalek (2004) reports that manuals and workbooks are used in over three fourths of all training programs and video recordings are used in over one half (p. 34). Analog media are also still in widespread use in higher education. Chalkboards, whiteboards, and overhead projectors serve a valuable function in enabling instructors to generate verbal and visual cues spontaneously to supplement lectures. Photographic slides continue to be used for subjects where high-definition images are critical, such as biology, veterinary medicine, optometry, and the visual arts.

It can be safely concluded that, while current trends do indeed point toward an increased use of digital resources, instructors will continue to use and demand support for a number of the traditional audiovisual (AV) media.

Appropriate Resources

The term *appropriate* is used to modify resources to indicate that the hardware and software used in education should be selected with consideration of their suitability for and compatibility with educational goals. The first criterion of suitability is that they should be selected through a process that meets professional standards. The Association for Educational Communications and Technology (AECT) code of ethics provides many guidelines to professional expectations, as discussed in chapter 11. One of the most fundamental is observance of the relevant law. Section 3 (Association for Educational Communications and Technology, n.d.) requires that members observe copyright laws and other legal protections of intellectual property rights *and* that they inform users of these laws and they encourage compliance. This is not necessarily an easy requirement to meet. Media librarians, technology coordinators, and instructional designers frequently face challenges in these areas; for example

- To have ready access to needed teaching materials, a teacher would like to duplicate a videocassette outside the boundaries of "fair use"
- To enable all the students in his large but poor school district to use current computer applications, a school superintendent would like to use unlicensed software
- To convey important ideas in a distance education course a professor would like to incorporate some visuals downloaded from a website without inquiring about permission

A second professional standard (Section 1.5) is that members follow "sound professional procedures for evaluation and selection of materials and equipment" (Association for Educational Communications and Technology, n.d.). Again, this may seem obvious, but practitioners may find themselves tempted to accept or use resources that have been donated with some expectation of private gain, that may be commercially exploitative, or that are simply easily accessible but would not stand up to close scrutiny.

Other criteria come from political, social, and cultural expectations, such as avoiding content that that promotes gender, ethnic, racial, or religious stereotypes (Section 1.8) and, on the positive side, encouraging the use of "media that emphasize the diversity of our society as a multicultural community." Along the same line, Section 3 calls for providing "opportunities for culturally and intellectually diverse points of view."

These meanings of *appropriate* pertain to educational use in general, for example, the selection of materials for a school media-center collection or a corporate resource-center collection. When the resources are being considered for use as part of an *instructional* lesson or program, other issues come to the fore. The criteria of effectiveness and efficiency now must be included. Effectiveness refers to the suitability and compatibility of a given resource with regard to particular instructional objectives—likelihood of yielding positive results—and sustainability in the local setting. For example, social studies teachers might select a particular social-simulation game if their past experience indicated that it stimulated the sort of topical discussion that they wanted to evoke. It was suitable for their purpose.

Efficiency refers to the wise use of time and resources, including the effort of educational technologists themselves. Since everyone's budget is finite, purchasers have to consider which hardware and software will provide the greatest benefits for the most learners or the greatest benefits to the success of the organization. Especially in the private sector, productivity considerations must be given high priority. For example, a training planner might have to decide whether to choose between face-to-face instruction or Web delivery for a course on copy-machine maintenance. They might first estimate the delivery costs of each option (e.g., travel and per diem costs for the live classroom vs. software and hardware costs for the Web course), then compare those with the benefits in terms of employee performance and meeting company goals. The choice might be a combination of the two modes: two days of Web-based instruction for the general principles and two days of instructor-led workshops for hands-on practice. The point is that efficiency, too, is a critical element of appropriateness.

Resources by Design Versus Resources by Utilization

To explain the types of resources available to help facilitate learning, the authors of the 1972 definition statement (AECT) made a useful distinction between *resources by design* and *resources by utilization*:

> Some resources can be used to facilitate learning because they are specifically designed for learning purposes. These are usually called "instructional materials or resources." Other resources exist as part of the normal, everyday world, but can be discovered, applied, and used for learning purposes. These are sometimes called "real-world resources." Thus, some resources become learning resources by design and others become learning resources by utilization. This distinction is important because it makes clear the position of "noninstructional, real-world" resources as well as designed resources as an area of concern for educational technology. (p. 38)

Without this inclusive definition, those "real-world resources" not necessarily intended for instructional use might not be considered to be resources. This notion was stated clearly in the 1994 definition: "Resources are sources of support for learning, including support systems and instructional materials and environments. . . . Resources can include whatever is available to help individuals learn and perform competently" (Seels & Richey, 1994, p. 12). It is important to include "resources by utilization" in the current definition, especially with the significant increase in the use of this type of resource in information-rich learning environments. Explorations in public television programs such as *Kitchen Chemistry* and *Backyard Geology,* for example, depend on resources not originally intended to be educational, such as baking soda and vinegar.

Whether they are analog or digital, used by design or by utilization, resources play an integral role in facilitating learning and improving performance.

Technological Resources and Early Identity Formation

Prior to the 20th century, education was generally characterized by the organization of teachers, textbooks, chalkboards, and students. This early model of education is typically perceived to have preceded the field of educational technology and is often referred to as *traditional* education. The modern field of educational technology is most often considered to have formed in the early 1900s as a loosely formed group of practitioners with a common interest in using new technological resources as an alternative to traditional education. Throughout the 20th century, new technological resources emerged and faded away, yet even today the differentiation between "traditional education" and technology-supported education has persisted. This differentiation is accepted by Reiser (2007), who defined *instructional media* as "the physical means, other than the teacher, chalkboard, and textbook, via which instruction is presented to learners" (p. 18). While many, even in educational technology, view the textbook and chalkboard as examples of instructional media, it is useful to adopt Reiser's distinction. The emergence of educational resources *after* the chalkboard and textbook—and the corresponding evolution of a field of study concerned with the application of these emerging technologies to instruction—can then be viewed as the catalysts for the establishment of educational technology as a distinct field. In retrospect, it provided a clear identity for the group of people who would evolve into AECT, and who would ultimately attempt to define the concept of *educational technology* that has emerged over the past century.

Early Origins

The use of hand-drawn illustrations in textbooks is among the oldest examples of the use of educational resources and is considered by some to mark the early origins of the field of educational technology. Early, exemplary use of instructional illustrations can be found in the work of Johannes Amos Comenius in the mid-1600s (also discussed in chapter 5). Comenius' *Orbis Sensualium Pictus* (The Visible World Pictured) was a primary school textbook that included illustrations of common concepts to complement the words (Saettler, 1990, pp. 31–32). There is, however, doubt about whether Comenius' work should be regarded as the origin of the field, as he was arguably not consciously working as an educational technologist, and there is little evidence that his work directly influenced the modern founders of the field.

The Visual Instruction Movement

Slides. There is little disagreement that the origins of the modern field of educational technology can be traced to the expanding efforts of practitioners in the late 1800s and early 1900s to use visual images to improve education. The first of the visual media was slide projection, which evolved from 17th century hand-painted slides illuminated by oil lamps. The "magic lantern" provided entertainment for paying audiences throughout the 19th century (Petroski, 2006, pp. 18–19). By the end of the 19th century, lantern slides, which came to be standardized at 3¼ × 4 inches, were in common use in education (p. 19). The slide format was later standardized at the two-by-two-inch frame size, using 35mm film, which was also used for the filmstrip, the most popular format for commercially produced AV materials. After World War II, overhead transparencies became popular for local production of visuals, and overhead projectors were ubiquitous in classrooms at all levels.

Commonly cited as the *visual instruction* movement, this movement was characterized by the emergence and use of still picture and motion picture technology as teaching resources. Initial efforts to make these resources widely available throughout school districts included the collection and organization of visual materials into educational museums. The first in the United States was established in St. Louis in 1905, based on exhibits saved from the World's Fair held in that city in 1904. Circulation of collection materials (e.g., charts, pictures, maps, photographs, stereoscopic pictures, and lantern slides) to schools was carried out by a horse-drawn wagon (Saettler, 1990, p. 129).

Silent Films. The direct ancestors of educational films were the nontheatrical short films that began to emerge around 1910. British and French cinematographers exhibited films showing amazing sights such as microscopic creatures, insects in flight, and underwater seascapes. Films of news events and travel adventures played to rapt audiences. In the United States, Thomas Edison was quick to see the potential of film for classroom instruction. In 1911 and the years after, he released a series of films depicting historical events, natural phenomena, and principles of physics (Saettler, 1990, p. 96). By the 1920s, educators could find many types of films to use—theatrical films edited (often badly) for special purposes, industrial films (often providing biased depictions of their subjects), government films, and a smaller number of films produced specifically for the classroom.

Silent films began to be used in schools as early as 1910 (Saettler, 1990, pp. 98–99). The 1910s and 1920s were times of considerable ferment regarding educational films. Many different individuals, companies (e.g., Ford Motor Company), nonprofit organizations, and government agencies attempted to supplement the existing supply of theatrical films and newsreels. Schools with a progressive image of themselves wanted to be viewed as being current with the new technologies and rushed to establish collections and catalogs of films. Unfortunately, many of the available films were of marginal educational value and teachers often chose to show films for reasons other than their curricular relevance. Nevertheless, interest and usage continued to grow throughout the 1910s and 1920s. By the end of the decade, over half of the state education agencies had units devoted to film or visual education and thick catalogs documented the thousands of films available to educators.

A Professional Association: DVI. Those concerned with using images to improve instruction grouped together into a variety of organizations. In 1923, one of these organizations, the National Education Association's Department of Visual Instruction (DVI), emerged and became the preeminent organization of professionals concerned with the use of still and motion pictures to improve instruction.

Audiovisual Instruction

Sound Recording. The phonograph record, beginning in 1910, was the first widely available format for recorded sound, used almost exclusively for music. Although magnetic tape displaced the phonograph for recording purposes in the 1950s, vinyl records remain in use into the 21st century. As soon as the phonograph was invented, film producers tried various methods of using this new technology to add sound to motion pictures.

Sound Films. By the late 1920s, the technique of adding an optical sound track to the film itself became the winning technology for "talkies." However, there was considerable resistance to sound films in the education community. Some methodologists felt that the practice of having the classroom teacher add narration to silent films added a level of customization and personalization to film showings. Administrators worried about their "installed base," the large investment they had made in silent film projectors. As late as 1936, a survey showed that schools owned ten times more silent film projectors than sound film projectors (Saettler, 1990, p. 234).

Radio. Meanwhile, another audio resource was developing. In the early 1920s, many American universities obtained licenses to operate radio stations, often as technical experiments in electrical engineering. A large proportion of these died out in competition with commercial stations, but some, like the University of Wisconsin's WHA and others in the midwest, put down roots. The operations that prospered were the ones in which radio played an integral part in the university's mission—bringing educational opportunities to the hinterlands in the case of Wisconsin (Wood & Wylie, 1977). "Schools of the air," beaming programs to public schools, were formed in several states and cities in the 1930s. Programs were devoted to health, social studies, home economics, science, music, and many other subjects, including art.

Audiovisual Media. These audio resources were added to the growing base of visual resources. By the 1930s, schools maintained equipment pools that contained (in order of frequency) lantern slide projectors, radio receivers, 16mm silent film projectors, 35mm silent film projectors, filmstrip projectors, opaque projectors, microslide projectors, 16mm sound film projectors, and 35mm sound film projectors (Saettler, p. 234).

DVI to DAVI. The addition of an audio component to existing motion picture technology and the mushrooming of educational radio stations added a major new interest area to DVI, leading in 1947 to a change in the name to the Department of Audiovisual Instruction (DAVI; Saettler, 1990, p. 167). Thus, the first two names of the national organization that is now AECT both have in common a direct link to the emerging technological resources of the time.

Educational Communications and Technology

AV and the Baby Boom. In the post–World War II period, economic prosperity and the aging of "baby boom" children to school age led to a period of rapid expansion of schools and the educational media needed to support modern education. AV equipment and materials were at the heart of the field. In 1961, public schools in the United States owned approximately 7,000,000 filmstrips; 3,000,000 phonograph records; 1,000,000 slides; 700,000 films; and 400,000 tape recordings (Godfrey, 1967). In 1973, the heyday of the AV era, the dominant AV formats in terms of commercial sales were (in order) sound filmstrips, 16mm films, multimedia kits, audiocassettes, and silent filmstrips (Dean, 1975, p. 121).

Television. During the 1950s, dozens of noncommercial television licenses were granted and educational television programs began to be beamed to school audiences. Many of the same parties that had experimented with radio also did so with television, essentially replaying the radio scenario. As had the earlier radio licensees, the television licensees promised that "television will make it possible for the poorest and most isolated public schools to enjoy the benefits of the most modern teaching methods" (Levenson & Stasheff, 1952, p. 155).

Since this was a period of rapid school population growth, there was general shortage of qualified teachers. Television was seen by some as a way to reduce the need for teachers by replacing the presentation function with broadcast lessons. This prospect had special appeal in the southern states, attempting to operate "separate but equal" racially segregated schools. They would be able to offer quality instruction without having to staff two separate schooling systems, each with equally qualified teachers in each classroom. These were the first states to establish statewide educational TV networks. The same rationale was used in later years to justify the use of closed-circuit television in countries outside the United States that operated educational institutions segregated by gender.

In the late 1950s and through the 1960s, there were programs broadcast on a regional basis, such as the Eastern Educational Network and the Midwest Program of Airborne Television Instruction (MPATI), and a national basis, such as *Continental Classroom*. During this period, the Ford Foundation and the federal government were subsidizing the expansion of television in higher education, through grants for closed-circuit TV construction and program production. By the end of the decade of the 1960s, tens of millions of school and college students were receiving televised instruction on a daily basis.

DAVI to AECT. Although utilization of AV and broadcast materials was still the central concern of the profession, other conceptual innovations were making an impact during this period. Theories of communication arising in other academic disciplines were being absorbed; they were seen as providing powerful rationales for using sight and sound to supplement talk and reading as sources of information. The systems approach to instructional design was emerging from the military into the corporate and academic sectors. This approach merged in the 1960s with the behaviorist methods of programmed instruction and behavior management to yield a growing number of systematic models of instructional design, discussed in detail in chapter 4. The field was becoming more process oriented.

By the late 1960s, James D. Finn's promotion of the concept of *instructional technology* merged with B. F. Skinner's followers' popularization of the term *educational technology* to describe reinforcement-centered learning environments to lend momentum to the growing interest in shifting focus from a product-centered view to a process-centered view. This shift was signaled by the change of the name of the national organization in 1971 to the Association for Educational Communications and Technology (AECT), a name that persists today. The new name was adopted to reflect the organization's broader emphasis not only on traditional AV efforts, but also on the application of learning theories (cognitivist as well as behaviorist), communications theory, and systematic approaches to education (Seels & Richey, 1994, p. 13).

Computers. As described in chapters 2 and 4, computers began to be applied to educational uses as early as 1959. The PLATO and TICCIT projects in the 1960s and 1970s explored a broad range of pedagogical methods as well as a broad array of subjects. Computer-assisted instruction (CAI) was available first to college students, but programs for K–12 education emerged almost from the beginning. In the 1970s, one of the early explorers, Seymour Papert at MIT, developed a programming language called *Logo* that was specifically designed to help young children think mathematically. Children used Logo in interesting microworld environments to explore properties of music and physics. Papert later extended this approach to building virtual structures of Lego® blocks, instantiating a "constructionist" theory of learning.

The popularization of stand-alone personal computers in the 1980s released users from being connected to a wired network, creating a market for entertainment and educational products on floppy disks and later CD-ROMs and DVDs. Then the growth of the Internet, with its promise of e-mail and discussion forums, brought users back into the networking mode. In the late 1990s, when the World Wide Web provided a linkage among all the scattered networks and individual users, another level of interactivity was achieved. The Web allowed distance education to take off on a steady climb in popularity, with the promise of anytime, anywhere access. By the end of the 1990s, virtually every university and many school systems were offering menus of distance learning courses via the Web.

Impact on AECT. The changes in modes and delivery systems wrought by the computer revolution have had a deep impact on professional life in the fields related to educational technology. In the case of AECT, it has changed the dialog within the association, with the vast majority of publications and conference sessions being devoted to considerations of instructional computing. And it has brought the challenge of another association, the International Society for Technology in Education (ISTE), which appeals primarily to teachers who play leading roles in computer use in their schools. In many instances, this person, the school technology coordinator, has displaced the AV specialist or the school media specialist as the most visible representative of technology in the building.

It is clear that the early identity and the ongoing identity crises of the field of educational technology are inextricably connected with the emerging technologies of the day. If history is any indicator, educational technology will continue to maintain a central focus on the use of emerging technological resources to improve learning.

Alternative Views of Educational Technology as Resource-Driven

As the field has evolved, some people and organizations—especially those in related fields that are also concerned with using resources for teaching and training—have maintained a narrower focus on resources as hardware driven (rather than theory or processes) as they have outlined and defined their own conception of educational technology. This approach is exemplified by the International Technology Education Association's (ITEA) definition of educational technology: "The use of technological developments, such as computers, audio-visual equipment, and mass media, as tools to enhance and optimize the teaching and learning environment in all school subjects, including technology education" (ITEA, 2003, p. 17). Although the distinction between educational technology as hardware and software and educational technology as AECT has more broadly defined it is important to the AECT community and those it serves, it is equally important to understand other interpretations. In fact, it is somewhat likely that newcomers to the field very recently conceived educational technology to be something similar to the concept outlined by ITEA. The ITEA conception is in fact not only common but also understandable. However, it can be problematic to an organization seeking to promote a broader view of educational technology. The current AECT definition itself can be seen as part of an ongoing effort of the organization to continue to emphasize this broader, more encompassing view.

Emerging Technologies and the "End of Education as We Know It." Maintaining a narrow focus on the technological tools themselves—rather than on the broader and more sophisticated educational systems in which they exist—can lead to a misreading of the potential influence that technological tools can have on education. For example, there have been many educational futurists (often not educators themselves) who have greatly overestimated the potential impact of emerging technologies on education. A good example is Thomas Edison, who in 1922 predicted, "The motion picture is destined to revolutionize our educational system and that in a few years it will supplant largely, if not entirely, the use of textbooks" (Edison, as cited in Wise, 1939, p. 1).

More recently, renowned management consultant Peter Drucker predicted,

> Thirty years from now the big university campuses will be relics . . . Already we are beginning to deliver more lectures and classes off-campus via satellite or two-way video at a fraction of the cost. The college won't survive as a residential institution. (Drucker, as quoted in Lenzer & Johnson, 1997)

Drucker made this prediction before the Web had accelerated the pace of adoption of distance education, so it may be that he would have doubled his bet 10 years later. Or would he? Had residential colleges shown signs of going out of business in that decade?

Predictions of this sort have consistently been made as new technologies have emerged. Understanding the broader systems in which resources are deployed and used (discussed in chapter 3), along with the factors that influence change in these systems (discussed in chapter 5), would help to "soften" these types of overstatements. It is likely that we have not yet, unfortunately, seen the last of the "end of education as we know it" prophecies. When considering these historical overstatements, Reiser (2001) wisely predicted, "In light of the history of media and its impact on instructional practices, I also think it reasonable to expect that . . . changes, both in schools and in other instructional settings, are likely to come about more slowly and be less extensive than most media enthusiasts currently predict" (p. 62). It is indeed possible that there might be some future technology that quickly revolutionizes instruction. Yet when educational technology is viewed through a historical lens, it is clear that we are more likely to continue to see slower, more incremental and evolutionary changes as new technologies emerge.

Tools, Materials, and Devices as Resources

As technologies have emerged and evolved, the terms *tools, materials,* and *devices* have been used in many ways. The specific example of the creation, storage, and use of a training video, which provides one way of conceptualizing the relationship among these terms, is illustrated in Fig. 8.1.

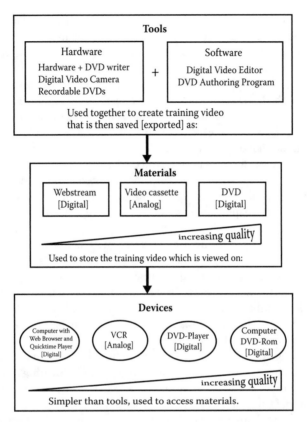

Figure 8.1. The relationship among tools, materials, and devices in creating, storing, and accessing a training video. © A. Betrus, 2006. Used with permission.

The organization of the categories in Fig. 8.1 conveys *typical* conceptions of these terms. That is, *tools* are used to create (and later manipulate if needed) *materials*, which are then accessed by *devices*. Devices are typically somewhat simpler than tools, and they are most often used primarily to access materials—both through viewing and through interacting with them as appropriate.

In the previous example, the hardware and software tools, as well as the devices, are intentionally generic, as there are many ways in which a training video could be created and used. It could be created completely with analog tools instead of using computer hardware and software. The resulting video could be converted into a digital format using yet another tool. Further, some devices needed to access the materials are not listed but are rather implied: The DVD player and VCR, for example, need another device, a television, in order to display the video, and a computer similarly needs a monitor. To complicate things further, a tool typically can double as a playback device. The video could be viewed in the DVD authoring software directly, rather than viewing it with a separate player. In other words, the relationship among these various terms is not always as clear-cut as the diagram would indicate. However, it can be useful as a general way of looking at the relationship among these terms.

In addition to the tools, materials, and devices, the term *media* is also used in the field, although with a number of very different meanings. In popular parlance, "media" refers to mass communication enterprises (e.g., radio, television, or newspaper businesses). In educational technology, the term has historically been used to designate the delivery system through which messages are conveyed to the user—printed matter, still-image projection, television, radio, sound film, and the like. Or it could refer to instructional materials themselves—overhead transparencies, slides, filmstrips, audio-cassettes, and the like. Nowadays the term is most often used to refer to the physical devices that store data (e.g., floppy disks, flash drives, CDs, video-cassettes, or DVD-ROMs). Thus, it tends to be a catchall term that could refer to a tool, a material, or a device.

The training video example is only one of literally thousands of examples of media that have helped to augment the teaching and learning process over the past century. While an attempt is made here to describe some of the tools, materials, and devices that are used at the time that this chapter is being written, perhaps a better place to ascertain more complete and current information can be by examining various surveys and yearbooks (e.g., *Educational Media and Technology Yearbook*) that compile this type of information or by examining the content of introductory technology courses taught to preservice teachers (or the textbooks used in these courses).

Analog Tools, Materials, and Devices

In everyday use, *analog* simply means something that resembles something else. Hence, an analogy is an object or idea that is used as a point of reference to explain some other idea. The term has a more technical meaning in engineering where an analog signal is one that is continuously variable in both time and amplitude—as opposed to digital signals, which are either off or on; they are not continuously variable. Analog sound recordings are made by translating the variations in air pressure striking the microphone into a corresponding variation in voltage amplitude. That is, the voltage fluctuation "looks like" the air pressure fluctuation; it retains all of its dimension. Analog television pictures are similar to film projections in that the entire image is painted on the screen with each frame. So audiocassettes and videocassettes are analog recording media.

By extension, the term *analog* is used to refer to all of the AV media that are not digitized, such as slides, filmstrips, and films, as well as audiocassettes and videocassettes. Some would consider these historical artifacts, as they are largely being replaced by digital equivalents, mainly for greater compression of storage and for easier transmission through computer networks. However, analog media continue to be valued for their high fidelity of reproduction (e.g., slides that project a large image in high definition) and their usability without the intermediation of computers.

An extensive collection of historic analog tools, devices, and materials is currently housed at the University of Northern Illinois in the Blackwell History of Education Museum and can be accessed via the World Wide Web through the AECT Archive Equipment Virtual Tour (AECT, 1999). Along with the use of stand-alone resources, educational technology has a long tradition of combining the functions and features of multiple tools, materials, or devices into one, *integrated technology*. The online AECT Archive Equipment Collection has some historical examples of integrated technologies, including a combined 35mm camera and projector, a combined slide projector and recorder, and a combined phonograph and filmstrip projector, among others. In fact, the affinity of educational technologists for such new, cutting-edge devices (and their application to instruction) largely has traditionally defined our field.

Although current advances in the field are clearly centered on the use of digital resources, the legacy of the field is in the use of analog resources to improve education. Even in the midst of what many see as a digital revolution, analog resources—especially the overhead projector, the VCR, and locally compiled printed materials—remain an integral part of most instructional settings. Critical attributes of analog resources—including high definition, ease of creation, customizability, and lower knowledge barriers for use—will likely ensure that they will continue to be used in a variety of teaching and learning environments well into the future.

Digital Tools, Materials, and Devices

Digital media are those that are stored and transmitted by means of digital codes, usually binary codes—0 or 1, off or on. Unlike analog media, digital representations—a series of zeroes and ones—have no resemblance to the original image or sound, which was probably initially recorded through analog means. The advantage of digital storage is that the data are generally easier to manipulate, more compact to store, and the resulting presentation can be transmitted or reproduced any number of times without loss of quality. For example, think of a photocopy of a photocopy of a photocopy of a newspaper clipping—as opposed to the replication of the same newspaper clipping from a digitally scanned image.

Currently, typical formats of digital media are computer displays, Web pages, compact discs (CD), digital video discs (DVD), video games, and e-books. It is difficult to predict which tools, materials, or devices will be prevalent in five, ten, or twenty years. Current trends and advances in the capabilities of digital tools and devices, along with innovative ways of using them, point to a commercial trend in society to combine functions and features together into one integrated technology. An example of this type of technology is the all-in-one television receiver/VCR/DVD; or the combination of television, telephone, and Internet services through one digital service provider, delivered via one digital cable into the home. Much like their earlier counterparts dealing with analog media, current educational technologists are often working with digital integrated technologies to better facilitate learning and to improve performance in a variety of settings. Currently, the most commonly used integrated technology, from a hardware perspective, is the personal computer.

The Computer. While useful, early versions of analog integrated technologies often had the drawback of being cumbersome and requiring a high level of technical expertise. The computer—especially the home computer—offered the potential of ease of use and convenience that other integrated technologies had lacked. These features, as well as a reduction in size and cost of the computer, led in the 1980s and 1990s to the proliferation of the computer in society at large as well as in education and training. The computer combined the functionality of many previous tools and devices, and delivered instruction in easy and convenient packages to instructors and trainers. Prepackaged instructional software (most often referred to as "computer-assisted instruction," or CAI) largely replaced previous generations of programmed instruction delivered via analog teaching machines. In addition, the computer offered the ability for both instructors and learners to create their own materials.

The computer is currently the primary means by which instructional materials are created. The word processor is perhaps the digital equivalent of the pencil and paper in that it is assumed to be ubiquitously available and a bare necessity for creating instructional materials. The most typical means of storing (saving) digital instructional materials are an internal hard drive, floppy disk, CD, DVD, USB flash drive, or Internet server. A typical scenario might involve instruction being created locally on a computer (the computer is being used as a tool in this case) and saved as Web pages (the materials). Learners then access and interact with the materials from a remote location. In this case, the computer at the remote location is being used as a device to interact with the Web-based materials. Sometimes materials are printed, with the advantage that analog print materials do not require all learners to have access to a computer to use them. Interestingly, there is a tendency among some instructors or learners—even when digital access to materials is available—to print digital materials in order to interact with them.

The societal trend toward the increased use of the computer is very likely to continue into the foreseeable future; and as educational technology is inextricably tied to the technology of the day, it is likely that the educational resources of tomorrow will be increasingly digital and computer based. Currently, the personal computer—and increasingly the Internet—serve as the primary means by which educational technologists store, organize, retrieve, and interact with digital resources.

The Internet and World Wide Web.   Perhaps the most significant added functionality of the computer was access to the Internet in the 1990s. The rapid increase in connections to the Internet in the early 1990s vastly expanded the potential for sharing information at a distance. The advent of graphical user interfaces allowed the World Wide Web to become the most popular Internet protocol. As the World Wide Web became the major service operating over the Internet around 1993, growth of users mushroomed, doubling every year until the end of the decade. Because of its ubiquity, it became the de facto standard platform for sharing resources. Being structured according to hypermedia principles—links and nodes—it largely displaced the earlier concept of hypermedia programs residing in a local computer system. Now the programs resided in the Web and could be tapped from any place in the world that could access the Internet.

In terms of integrated functionality, the Internet-ready computer has generally supplanted the majority of tools and devices that preceded it. While certainly most instructional materials are created with personal computers, and often word processors, current trends point toward an increased emphasis on creating instructional materials for the World Wide Web, with a variety of tools available for this purpose.

Computer-based World Wide Web browsers (e.g., Netscape's Navigator, Mozilla's Firefox, Apple's Safari, and Microsoft's Internet Explorer) are currently the primary means of accessing instructional materials on the Internet. The types of instructional materials available for use with a computer include educational software, educational games, instructional simulations, edutainment software, instructional videos, reference materials, audio recordings, and movies. While much of this material is available via CD or DVD, current trends point toward accessing digital materials directly through the Internet.

Interactive environments. Educators had long appreciated the value of methods that involved learners deeply in realistic problem settings. Simulations and simulation games allow learners to explore complex dynamic situations, such as conflicts among individuals and groups. One of the barriers to development and widespread adoption of such interactive learning environments in the past was that of limited distribution potential. But with the Web as a delivery platform, it is now worthwhile to invest major resources in developing such simulations, simulation games, and other "microworlds" because of the prospect of widespread distribution, dispersing the cost of development. Immersive learning environments are also discussed in chapter 4.

WebQuests. Early on, educators saw the Web as an enormous database that could be used by students to generate their own answers to questions. Dodge (1995) created the WebQuest format for scaffolding information problem-solving experiences. A WebQuest consists of at least four components: an introduction that establishes a context for the student's task, an inquiry task or quest, a set of preselected Web sites with information relevant to the quest, and a suggestion of how to process the information found on the Web, for example, what to look for. A WebQuest is more than a worksheet because it involves scaffolding to guide the learner; scaffolding includes the prior selection of sites that are most relevant and the provision of hints on for what to look. As a generative learning strategy, WebQuests aim for deep learning involving the construction of new knowledge through a critical thinking process (Dodge, 2001).

Web-based distance education. Distance education began in the 19th century using correspondence through the mail. It continued through most of the 20th century with radio, then television, added to the media mix. In the 1980s, as the Internet grew to encompass many home users as well as institutional computer centers it became feasible to offer distance lessons based on computer conferencing for communications among students and instructors.

During the late 1990s and early 2000s, hundreds of universities and businesses adopted the Web platform for their distance education and training, reaching millions of students, and Web-based distance education became the major growth area for educational technology. By 2006, the great majority of all U.S. higher education institutions were offering distance education courses via Web delivery.

To some extent, "Web-based distance education" is a misnomer in that in the typical Web-based course only part of the course activities are based in the Web—perhaps only the information displays, such as the syllabus and list of assignments. Students typically communicate with one another and with instructors through other Internet services, including e-mail, bulletin boards, chats, and file-sharing via file transfer protocol (FTP), and they may use textbooks or online readings for much of the content presentation. However, the Web home page helps students navigate through the course, consistent with the instructor's pedagogical intent and provides ready access to the interactive and communication elements of the course.

Course management systems.   The software application that gave impetus to Web-based instruction was the *course management system* (CMS), developed in the late 1990s and gathering momentum in the early 2000s. The CMS is a suite of applications, tying together all the services previously mentioned, so that students can log in once and have all of their communication services available at a click, without jumping in and out of the Web. For example, once the instructor creates a course, a student can access a course syllabus, assignment materials related Web sites, and posted grades. The instructor and students can communicate with each other in discussion forums.

Blackboard.com introduced its first CMS, CourseInfo, in 1999. By 2006, Blackboard merged with its largest rival, WebCT, and dominated the field of college and university CMSs, although rival open-source software CMS systems were also being developed, such as Moodle and Sakai. On the corporate side, the term *learning management system* (LMS) is preferred, referring to systems that not only provide instruction but also keep records of users' progress, documenting their accomplishments and certifications they may have acquired.

Emerging applications.   There is much educational promise in the emerging functions and features of new Internet and Web applications. Weblogs (blogs), for example, provide a high level of interactivity among users, who could be instructor and learners. Teachers can post online the most current course information for students, and both teachers and students have the ability to communicate in a highly interactive, up to the minute online space where they can place text, pictures, videos, and music to enhance their understanding of given content domain. Similar to Weblogs is the *wiki*, a type of Web site that allows users to freely and collaboratively add, modify, and otherwise manipulate online information towards a collective understanding of an idea or concept.

Mobile media.   Additionally, while the majority of Internet access is achieved through the Internet-ready desktop computer, there has been a trend toward accessing the Internet through smaller, portable devices such as digital phones, watches, laptop computers, compact computers, handheld computers, and personal digital assistants (PDAs). These resources, along with other mobile technology such as game devices and MP3 players, are becoming more and more the norm and may someday supplant the desktop computer as the primary way in which information on the Internet (e.g., e-mail, discussion forums, blogs, wikis, and other applications) is accessed and interacted with.

Electronic Performance Support Systems (EPSS).   Electronic Performance Support Systems (EPSS) might best be described as Web-accessible electronic databases that provide information in a just in time fashion to employees in an organization. An EPSS often takes the form of a "help" system to assist employees solve work-related problems. The EPSS usually stores organizationally specific expert knowledge for employees to reference as needed. Although such cognitive support systems are not designed as educational tools or resources, learning can happen. That is, as users summon the same chunk of information repeatedly they may come to remember and internalize the information. This sort of usage can somewhat blur the distinction between providing information and instructing.

Settings as Resources

Appropriately organizing the context in which problem-solving and performance-improving activities takes place is a critical component of good instructional design. This context includes both internal settings, such as classrooms or learning labs, and external settings, such as museums or zoos. Recently, as constructivist learning theories have become more popular and widespread, there has been an increase in efforts to organize authentic, complex, information-rich learning environments. Whether the setting in which learning and performance take place is internal or external, simple or complex, the ability to organize it appropriately is essential for educational technology professionals.

Community Resources

Community resources such as museums, libraries, zoos, hospitals, police stations, and fire departments can serve to expose learners to authentic examples of concepts and ideas they are studying in a classroom. The means by which students interact with these community resources is typically by traveling to one of these locations, that is, by taking a field trip.

Field Trips. The field trip has maintained itself as a staple in schools. Heinich, Molenda, Russell, and Smaldino (2002) described it as "an excursion outside the classroom to study real processes, people, and objects, [which] often grows out of students' need for firsthand experiences. It makes it possible for students to encounter phenomena that cannot be brought into the classroom for observation and study" (p. 90). The emphasis on unique experiences not available in the classroom (except through lesser, more abstract means) is often the primary criterion used to justify the time, effort, and expense associated with field trips.

Virtual Field Trips. While similar in purpose to conventional field trips, virtual field trips are enabled by technology and offer the possibility of exploring many locations that would otherwise be too expensive or even logistically impossible to visit. The technological resources for a successful virtual field trip include (but are not limited to) one-way video, two-way video teleconferencing, telephones, e-mail, instant messaging, and online discussion forums. Roblyer, Edwards, and Havriluk (1997) explain that virtual field trips "are designed to explore unique locations around the world and, by involving learners at those sites, to share the experience with other learners at remote locations" (p. 230). In practice, while virtual field trips do offer experiences with remote locations (most often via the Internet), they are more likely not to be live and not to involve interaction with peers. Virtual field trips without live interactions could also be classified as reusable learning objects. Regardless of the "flavor" of the virtual field trip, each type has in common the goal of providing students with access to locations that might otherwise be impossible to explore—especially such exotic locations as the ocean floor, outer space, and medieval castles.

Information-Rich Learning Environments. The Cognition and Technology Group at Vanderbilt University has conducted a great deal of research and development around the notion of engaging learners in realistic problematic situations. The group describes four roles for technology in creating and implementing "technology-rich instructional environments that support learning and understanding":

1. Bringing interesting and complex problems into the classroom
2. Providing resources and scaffolds that support learning and problem solving

3. Providing opportunities for feedback, reflection, and revision.
4. Supporting communication and community building (Goldman, Williams, Sherwood, Hasselbring, & Cognition and Technology Group at Vanderbilt, 1999, p. 13).

These sorts of immersive settings are also known as rich environments for active learning (REAL) and are discussed in chapter 4.

People as Resources

Support staff and subject matter experts often serve as resources for instructors and learners. These people have advanced knowledge and experience dealing with specialized learning resources. They are often used when the knowledge and experience of the facilitator and/or learners needs to be augmented with the advanced knowledge and understanding that these specialists can provide. In some cases, these people might be *resources by design,* in other cases, *resources by utilization.*

Educational Technology Specialists

A clear example of a *resource by design,* the educational technology specialist serves to make sense of the sheer quantity of available resource options. Currently, as technology use has become more widespread, and more technical knowledge is required to use technology, the need for educational technology specialists (also known as "technology coordinators" and "instructional support specialists") in K–12 schools, universities, and business and government organizations is even more pronounced than in the past. Additionally, educational technology specialists are often primarily responsible for addressing the ethical considerations that arise when using and managing resources.

In Schools. These technical experts often constitute the primary source of support for the increasingly diverse technological needs of school systems. Most are expected to support the teachers', students', and administrators' use of technology to support learning. They often take on the additional role of maintaining the school technology infrastructure—including Web servers, e-mail servers, and the school data network. While some schools have the resources to support separate network specialists, often funds are limited. In these cases, the educational technology specialist takes on the roles of not only instructional support and technical support but also, increasingly, network support.

Some consider library science a "sister" field to educational technology, and there is a significant overlap in the training that professionals in each field receive. The library media specialist has many of the same skills as the educational technology specialist, as well as highly specialized knowledge of the organization of information and means for gaining access to information databases and other information sources. The library media specialist often receives formal training in instructional design and is charged by the library media profession with being "a primary leader in the school's use of all kinds of technology . . . to enhance learning" (AASL & AECT, 1998, p. 54).

In Colleges and Universities. Colleges and universities typically have some version of a center for the support of faculty using technology. Beginning around 2000, there has been a trend toward creating new centers or reconfiguring old ones into the form of a teaching and learning technology center (TLTC). Hundreds of universities now support a one-stop shopping center where faculty can go to get help preparing technology-based instruction. Typically, according to Long (2001), these TLTCs combine the services of information technology support and faculty development, sometimes adding the library.

Educational technology specialists working in these units tend to specialize in instructional support, leaving the technical support, server maintenance, and data network support to a group of people in a different section or department. The typical model is that the TLTC is available to faculty who are seeking to improve their courses and who look to the expertise of these specialists for advice and help. Recently, activities in these centers have included assisting faculty as they place materials into online course management systems. Other activities include improving PowerPoint™ presentations, setting up and managing listservs, and creating Web pages. In many universities, this center has also assumed the responsibility of assisting instructors as they transform their courses to an online format, although in certain cases separate distance learning agencies take on this role.

In Corporations, Government, and Other Large Organizations. Businesses, corporations, and government organizations often maintain the rough equivalent of educational technology specialists on staff at central locations, satellite locations, or both (depending on the size and mission of the business). These personnel usually work in training departments and typically carry the title of trainer, training specialist, or the like. Like their counterparts in the university, these specialists directly support training and learning within their organizations. They often provide broad technology support, and they generally develop and deliver training to employees at the request of management.

In schools, in colleges and universities, and in commercial and government settings, educational technology specialists offer vital services and expertise. They serve as technological resources for students, teachers, employees, and members of organizations who attempt to enhance learning using technology. While historically they have been seen as important, the increased need for organizations to take full advantage of technological resources has made them essential.

Subject Matter Experts

A good example of a *resource by utilization,* the subject-matter expert (SME) is typically used as a resource for those involved in the design of instruction. SMEs are often used as checks in the creation of instructional processes and resources. Early in the instructional development process, SMEs provide input into the decisions about what content is appropriate to be taught. Later in the instructional development process, the design team will typically present instructional objectives, instructional materials, and assessment tools to SMEs for feedback. The SMEs draw upon their own significant professional and practical experience in the content areas to offer their opinions regarding the quality of the materials.

SMEs are sometimes used *during* instruction as a way to give students direct access to experts. Museum docents, who serve as expert tour guides and are available to answer questions, provide one example of such use. Book authors are another example, as they sometimes agree to communicate directly with students who are studying and analyzing their works. And while teachers in the conventional sense are most often not considered to be resources, expert teachers are often used as resources to deliver guest lectures, speeches, or seminars. In this sense, the expert teacher, serving as a consultant, would be considered an SME and an instructional resource. In each of these cases, the expert knowledge of the SME is used as a resource during instruction.

More recently, SMEs have been used as integral parts of creating and implementing learning environments. Experts in specific subject areas are available within these environments as resources for students to consult as they explore topics and issues. For example, the leader of a local acting troupe could be an SME for middle-school students to consult when putting on a school play. For an amateur violinist, a professional violinist could serve as an invaluable learning resource, at the appropriate time in the amateur's development. Interestingly, the use of SMEs in this manner approximates an apprenticeship model, albeit on a much smaller scale. Historically, apprenticeships have been, especially prior to the development of the printing press, the most pervasive form of passing expert knowledge from generation to generation. While providing learners with access to people with expertise in a given area is often untenable for long periods, offering such access for short periods, or even electronically, can be very valuable in helping learners develop high-level skills in a given area.

Ethical Uses of Resources

Issues such as cost, accessibility, and equitable allocation of resources have become increasingly important in recent years. A central ethical concern for educational technology professionals is the effort to ensure equitable access to resources for all learners. The digital divide and the implementation of the Rehabilitation Act Amendments of 1998 are two areas of ethical concern.

The Digital Divide

As resources become more powerful, and often more expensive, the phenomenon of the *digital divide,* or the disparity between technology haves and have nots, has arisen as a problem for educational technology professionals, as well as for society as a whole. The digital divide is predominantly a socioeconomic divide, and it is most visible in the disparity between wealthy and poor schools and communities. This divide is most obvious when comparing schools in wealthy nations to schools in poorer nations, although even within wealthy nations there remains a pronounced disparity of resources between rich and poor schools. As this is largely an economic problem, the solution is largely economic. By way of addressing this situation, grant money is often used to supplement technology resources in poorer settings, and an educational technology specialist will often serve as the primary grant writer and coordinator, especially for larger grants. Additionally, an educational technologist with proper training in using resources effectively and efficiently can serve as a valuable resource for those schools looking to make the most of the resources they have.

Accessibility and Universal Design

Equal access to information for all U.S. federal employees was addressed when the 1973 Rehabilitation Act was amended in 1998. Included in the amendment was a strengthening of section 508, which now requires that members of the public with disabilities have equal access to technology. This requirement applies not only to federal agencies but also to any school or organization that receives federal funding. Fundamentally, it states that resources, especially electronic and information technology (e.g., Web pages), must be made accessible to people with disabilities. Accessibility can be provided either by proactively designing the resource to be accessible or by retrofitting existing resources using commercially available accessibility tools.

The concept of *universal design*, or the creation of resources that are as broadly useful to as many people as possible, has become increasingly popular as sensitivity to accessibility issues has become more and more widespread. While universal design does not necessarily address all accessibility issues, it does help to ensure that resources are created with the diverse needs of potential users in mind. Universal design of a product alleviates many of the accessibility issues that might otherwise arise.

Conclusion

Resources can take the form of the tools, materials, devices, people, and settings that learners interact with to improve learning and performance. Historically, the identity of the field is intertwined with the use of the emerging technologies of any given era. In the latter half of the 20th century, the field grew beyond the earlier, simpler, resource-driven conception of educational technology. Along with an emphasis on communication and learning theories, a shared focus of the field of educational technology remains on the *appropriate* use of emerging *technological resources* to facilitate learning and improve performance.

References

American Association of School Librarians and Association for Educational Communications and Technology. (1998). *Information power: Building partnerships for learning.* Chicago: ALA Editions.

Association for Educational Communications and Technology. (1972, October). The field of educational technology: A statement of definition. *Audiovisual Instruction, 17*(8), 36–43.

Association for Educational Communications and Technology. (1999). *AECT archive equipment virtual tour.* Blackwell History of Education Museum, University of Northern Illinois. Retrieved September 28, 2004, from http://www.cedu.niu .edu/blackwell/multimedia/high/tour.html

Cuban, L. (1986). *Teachers and machines: The classroom use of technology since 1920.* New York: Teachers College Press.

Dean, J. (1975). Audiovisual media sales trends: EMPC survey. In J. W. Brown (Ed.), *Educational media yearbook 1975–1976* (pp. 119–122). New York: R. R. Bowker.

Dodge, B. (1995). WebQuests: A technique for Internet-based learning. *Distance Educator, 1*(2), 10–13.

Dodge, B. (2001). FOCUS: Five rules for writing a great WebQuest. *Learning and Leading with Technology, 28*(8), 58.

Dolezalek, H. (2004, October). Industry report 2004. *Training, 41*(10), 20–36.

Goldman, S., Williams, S., Sherwood, R., Hasselbring, T., & Cognition and Technology Group at Vanderbilt. (1999). *Technology for teaching and learning with understanding: A primer.* New York: Houghton Mifflin.

Heinich, R., Molenda, M., Russell, J., & Smaldino, S. (2002). *Instructional media and technologies for learning* (7th ed.). Englewood Cliffs, NJ: Prentice Hall.

Information technology: Medical information company announces alliance with school of dentistry. (2005, October 1). *Obesity, Fitness & Wellness Week,* 873.

International Technology Education Association. (2003). *Advancing excellence in technological literacy: Student assessment, professional development, and program standards* (National Science Foundation Grant No. ESI-0000897; National Aeronautics and Space Administration Grant No. NCC5-519). Retrieved April 5, 2006, from http://www.iteaconnect.org/TAA/PDFs/AETL.pdf

Lenzer, R., & Johnson, S. S. (1997, March 10). Seeing things as they really are. *Forbes,* 122–131.

Levenson, W. B., & Stasheff, E. (1952). *Teaching through radio and television* (Rev. ed.). New York: Rinehart & Co.

Long, P. D. (2001, June). Trends: Technology support trio. *Syllabus,* 8.

Molenda, M., & Bichelmeyer, B. (2005). Issues and trends in instructional technology: Slow growth as economy recovers. In M. Orey, J. McClendon, & R. M. Branch (Eds.), *Educational media and technology yearbook 2005* (Vol. 30, pp. 3–28). Englewood, CO: Libraries Unlimited.

Petroski, H. (2006). *Success through failure: The paradox of design.* Princeton, NJ: Princeton University Press.

Reiser, R. A. (2007). A history of instructional design and technology. In R. A. Reiser, & J. V. Dempsey (Eds.), *Trends and issues in instructional design and technology* (2nd ed., pp. 26–53). Upper Saddle River, NJ: Pearson.

Robyler, M., Edwards, J., & Havriluk, M. A. (1997). *Integrating educational technology into teaching.* Upper Saddle River, NJ: Prentice Hall.

Saettler, P. (1990). *The evolution of American educational technology.* Englewood, CO: Libraries Unlimited.

Seels, B., & Richey, R. (1994). *Instructional technology: The definition and domains of the field.* Washington, DC: Association for Educational Communications and Technology.

Wise, H. A. (1939). *Motion pictures as an aid in teaching American history.* New Haven, CT: Yale University Press.

Wood, D. N., & Wylie, D. G. (1977). *Educational telecommunications.* Belmont, CA: Wadsworth.

9

VALUES

Michael Molenda
Indiana University

Rhonda Robinson
Northern Illinois University

Introduction

Educational technology is the study and *ethical* practice of facilitating learning and *improving* performance by creating, using, and managing *appropriate technological* processes and resources.

THE BASIC DEFINITION DISCUSSED in chapter 1 implies that the concept of educational technology entails a number of professional, ethical, and moral values. The centrality of values is even more obvious when viewing educational technology as a field or profession. Codes of ethics, which incorporate the group's core values, are universally considered a critical attribute of a profession. Beyond the moral and ethical dimensions, values statements also have practical utility: organizations with a commitment to specific core values are more effective. Research indicates that they tend to outperform organizations without explicit values (Waterman, 1992; Collins & Porras, 1994).

Earlier definitional works acknowledged certain "common values" but did not discuss them in depth. For example, AECT's most recent prior definition statement held that "instructional technologists, as a community of professionals, tend to value concepts such as: replicability of instruction, individualization, efficiency, generalizability of process across content areas, detailed planning, analysis and specification, the power of visuals, and the benefits of mediated instruction" (Seels & Richey, 1994, p. 87). This chapter is an attempt to make the common values of the contemporary field more explicit.

Educational technology shares many functions, concerns, and values with other fields. For example, cognitive science and educational psychology are also concerned with facilitating learning; performance technology has a central concern for improving performance in the workplace; and the work of teachers surely involves creating, using, and managing many different processes and resources. Educational technology shares not only concerns with other fields but also values. Along with other educators, those in educational technology value the importance of learning and they espouse lifelong learning; they promote equal opportunities for learning for all learners and aim to give learners equitable access to learning resources.

This chapter focuses on those values that are emphasized in educational technology, those that tend to *distinguish* this field from others. Several are explicitly stated in the definition (e.g., "ethical practice," "improving performance," "appropriate," and "technological"); others are implicit. Both the explicit and implicit values are discussed in this chapter. Each of the key terms in the definition will be examined for its values connotation.

Values Related to *Study*

As a field dedicated to the application of organized knowledge to the improvement of learning and performance, research provides the bedrock of practice. Basic research on the variables associated with learning is primarily borrowed from related fields such as psychology, cognitive science, educational psychology, and anthropology. Basic research on instructional message design or learner response to mediated messages falls into the domain of educational technology, as does much of the vast area of visual literacy. Applied research on issues related to the application of technology in education is the most frequent type of inquiry conducted inside the field. Educational technology researchers study ways of analyzing and improving the processes of creating instructional materials and systems (instructional design), creating media and computer-based learning environments, using media and information technology in the classroom (utilization and implementation), and managing all the associated activities (project management, technology service administration).

Inquiry Approaches

The knowledge base can be expanded by many means of inquiry in addition to formal research. Formative and summative evaluation of specific products can inform subsequent design and selection decisions within an organization. Action research related to the implementation of an innovation can provide valuable "lessons learned" for practitioners and other change agents. Case studies of success, or especially of failure, can cast light on technology implementation processes in complex settings. The disciplined study of failed systems is a major method of knowledge development in the related field of engineering (Petroski, 1992).

Programmatic research in educational technology is valued and continues to be needed, with results shared, so that research results are best able to inform practice. Even with the current data-driven, decision-making policy at the federal level, this includes recommendations for research inquiry in all methods of research, focusing on questions regarding both success and failure, and the varied effects of technology on learners and learning. With an emphasis on studying global issues related to educational innovation and appropriate uses of technology for learners, the research on educational technology can continue to support improved practice worldwide.

Communication Through Scholarly Journals

The habits of reflection and of documentation of experiences are distinguishing characteristics of true professionals (Schön, 1995). The sharing of personal professional experiences is facilitated by contemporary communication tools such as e-mail, Web pages, and blogs. The more traditional venue for sharing findings and opinions, the scholarly journal, continues to play an important role in the discourse of the field, although in recent years many journals are distributed online through the World Wide Web rather than through printing and mailing. Molenda and Kang (2004) found several dozen educational technology journals of national or international scope. Of these, 14 were distributed online only and 16 were distributed both in print and online formats (p. 36).

Another study (Holcomb, Bray, & Dorr, 2003) identified a pool of over 100 periodicals with some connection to educational technology and narrowed that list to 30 that were deemed most relevant. Readers rated these 30 according to academic prestige, usefulness in keeping up to date in practice, and usefulness as assigned class reading. The researchers found that different clusters of periodicals emerged for each purpose, indicating that many different publications had value, but for different purposes. Highest in prestige were

- *Educational Technology Research and Development*
- *Human-Computer Interaction*
- *Cognition and Instruction*
- *Memory and Cognition*
- *Journal of Educational Computing Research*

Highest in keeping up to date were

- *Educational Technology Research and Development*
- *Educational Technology*
- *Cognition and Instruction*
- *TechTrends*
- *Webnet Journal*

Highest in classroom use were

- *Educational Technology*
- *Educational Technology Research and Development*
- *TechTrends*
- *Technology and Learning*
- *Computers in the Schools*

A study of the publications of recently tenured educational technology professors (Carr-Chellman, 2006) revealed an extraordinary range of different journals in which they published: the 17 respondents had articles in 120 different periodicals (p. 9). Since much of the research in educational technology is applied to specific subject matter in specific learning settings, it is reported not just in educational technology journals but also in other professional journals, such as

- *Early Childhood Research Quarterly*
- *Elementary School Journal*
- *Reading Research and Instruction*
- *Journal of Research in Science Teaching*
- *Journal of Teacher Education*
- *Studies in Art Education*

However, the journals in which educational technology scholars published most frequently were

- *Educational Technology Research and Development*
- *TechTrends*

- *Journal of Educational Computing Research*
- *Journal of Research on Computers in Education*
- *Computers in Human Behavior*
- *Educational Technology*
- *Journal of the Learning Sciences* (Carr-Chellman, 2006, p. 11)

Thus, because educational technology is such an interdisciplinary field and its artifacts are used in such a wide array of settings, it is not surprising that its literature is distributed over a diverse range of periodicals. Nevertheless, there are several journals—such as *Educational Technology Research and Development*, *TechTrends*, and *Educational Technology*—that serve multiple purposes for large segments of the field, providing some stability and continuity in the ongoing conversation among scholars in the field.

Values Related to *Ethical Practice*

Although no field advocates unethical conduct or omits ethical guidelines, the ethical issues that are of special concern to educational technology are distinguishable from those of other fields. Educational technology's distinctive ethical concerns focus on the processes of creating instructional materials and learning environments and on relations with learners during the use of those materials and environments.

As discussed in chapter 1, critical theory is particularly vocal in reminding researchers and practitioners to think about power relationships—whose welfare is primary, who controls events, and who has a voice in the process. Sensitivity to power relationships extends to those who design learning environments, those who use them, and those who manage and evaluate the overall process. Since learners are the supposed beneficiaries of education, it is incumbent upon professionals to accord them a fair share of power in the teaching-learning process.

Protecting the interests of learners is of high priority in critical theory, but so is it in other perspectives. Behaviorism proclaims "the learner is never wrong," insisting that any failure should be blamed on poor design or use of the instructional system. The application of behaviorist learning theory in the form of programmed instruction and structured tutoring helped break from the group-based instruction model toward an individualized model by regarding each learner as having a different stimulus history, a different reinforcement history, and a different level of mastery of the target skill. Hence, each learner required a customized program of instruction and reinforcement. Further, the techniques of programmed instruction and structured tutoring allowed learning to be individually paced.

The cognitivist perspective on teaching and learning also prescribes careful attention to individual needs since this theory posits that each person develops internal cognitive structures or schemata that are necessarily unique, since each person has a different life experiences.

The constructivist perspective goes one step further than the cognitivist position, positing that even when two people participate in the same event each person constructs a different and unique interpretation of that experience. Thus, the constructivist position places exceptional emphasis on the necessity of viewing each learner individually.

Of course, regard for the interests of individual learners, special needs learners, and learners with cultural or language differences is not limited to adherents of any particular "isms." These are merely examples of the rationales that join to form a solid platform for this value.

One way that learners are empowered through educational technology is through the employment of user-centered design. When this concept originated as "user-oriented development" (Burkman, 1987), the primary user in mind was the teacher, the person who either accepted or rejected the product of the instructional design process. But more recently this idea has come to encompass the learner as well. By giving teachers and students a voice at stages throughout the development process, it is more likely that the final product will be effective and that it will be accepted for use. At some levels, especially adult education, it is possible to have learners actually create the instruction. For example, production supervisors working in small groups could brainstorm a list of ways to handle conflicts in the workplace. They could compare items across groups and agree on the best solutions, which then constitute the content of the lesson. In this view, a user-centered design process is not only a surer path to instruction that is usable in the end but also a way of empowering learners and teachers within their own world and a way of creating content that has high credibility with the audience.

In addition to caring for learners, ethics demands that practitioners carry out all their duties informed by current knowledge of "best practices" in the field. Keeping up to date with research and advances in knowledge is an expectation of all professional fields, but it has special importance in educational technology because educational technology claims to be based on the application of scientific and other organized knowledge to education. Efforts to make professional advances accessible include the Web sites and blogs of many educational technologists and research programs, the theory into practice journals such as *TechTrends* and the many reports on practice given at international conferences, which may be shared in proceedings of those conferences.

Values Related to *Facilitating Learning*

To begin with, educational technology shares the central commitment of education to helping people learn. Further, by promoting "learning how to learn," educators give people the habits and attitudes to enable them to continue to pursue their own educations under their own initiatives. This is critical to forming lifelong learners, one of the goals of education.

Educational technology has the implied mission of helping people learn better than they would through their own devices or through the intervention of others who lack educational technology qualifications. Why ask for recognition as a distinct field unless there is claim to something better than other fields do? Providing better facilitation of learning means creating experiences and providing environments in which learners are more motivated to learn, advance more rapidly, retain more, are able to apply their knowledge better, and experience greater satisfaction—all of this within the constraints of the time, money, and human resources available. Educational technology does this through technologies that provide access to more people and that promote learning more effectively.

Enhancing Access to Learning

Although the concept of access to learning does not explicitly appear in the definition, educational technology has implicit commitment to using information and communications technologies (ICT) to expand the reach of education to those who might not otherwise be served. For example, broadcast radio has been used to extend educational opportunities to rural residents in many less developed countries in Asia, Africa, and Latin America. Television has also been used to bring quality instruction to classrooms in areas—both in developed and less developed countries—with a shortage of qualified teachers. Videoconferencing is used every day, particularly in the corporate setting, to bring training opportunities to learners located far from central training facilities.

Not only is it possible to extend access to learning through ICT, it is a moral imperative to work toward equalizing educational opportunities across ethnic and geographic communities, regardless of distance or economic disadvantage. The equalization of social and economic development contributes to global peace and stability. Educational technologists have a key role to play in the development of equitable learning opportunities in the United States and throughout the globe. The concomitant improvement of economic status for underserved learners is a vital part of the future for which they work.

Values Related to *Improving Performance*

As discussed in chapter 1, for a field to have any claim on public support it must be able to make a credible case for offering some public benefit. It must provide a superior way to accomplish some worthy goal. In this section, the focus is on ways that educational technology contributes to efficiency and effectiveness in pursuing the goals of learning and performance. Performance is discussed in terms of learner performance, teacher/designer performance, and organizational performance. The concepts of *efficiency* and *effectiveness* are not simple or easy to define. As discussed in greater depth in chapter 3, efficiency does not mean simply the fastest or cheapest means. Efficiency (and effectiveness) can only be determined with regard to agreed-upon goals and the means of measuring their attainment. That is, slower or more expensive means may be justified if they lead to the accomplishment of goals worthy of the cost.

Improving Performance of Learners

As elaborated in chapter 3, the goal in facilitating learning is not just short-term recall of information, but long-term ability to apply knowledge, skills, and attitudes in real-world settings. In the past, those who were designing and using instructional materials or learning environments tended to measure success in terms of scores on immediate posttests, tests that typically demanded only short-term recall of verbal information. In more recent years, research in cognitive psychology and neuroscience has expanded our understanding of the dynamics of the learning process. We can recognize a qualitative difference, in terms of physical changes in the brain, between superficial knowledge and knowledge that is ready for active use (Bransford, Brown, & Cocking, 1999). Weigel (2002) contrasted surface learning with deep learning. Surface learning is characterized by mere memorization of facts, carrying out procedures thoughtlessly, seeing little value or meaning in the knowledge, treating material as unrelated bits of information, and studying without conscious purpose or strategy (p. 6). By contrast, in deep learning, learners relate ideas to previous knowledge, look for underlying patterns, examine claims critically, and reflect on their own understandings (p. 6).

A trait associated with deep learning is the ability to transfer the new knowledge to novel situations, especially those outside the learning environment. From research on situated cognition, we now realize that what is learned in the classroom or online context tends to be confined to use in that setting unless instructors consciously provide opportunities to practice the new skill in contexts that resemble the real world. Students find it perfectly natural to abandon their in-school knowledge as they depart the schoolhouse door. They even segregate knowledge learned in one subject from application to other subjects: "Do we have to remember algebra in chemistry class?" In contemporary educational technology, transfer of learning to settings outside the classroom is a conscious concern. Designs and utilization practices should promote transfer. Promoting transfer, therefore, is a value that is emphasized in educational technology.

Improving Performance of Teachers and Designers

Besides improving the performance of learners, educational technology aims to improve the performance of teachers and designers. The tools of instructional design are meant to help planners develop instructional materials and systems more efficiently and effectively. The goal is to help average practitioners attain above average results.

In addition to giving them better tools, educational technology strives to give practitioners better professional preparation. This means, for example, the use of authentic assignments, authentic assessment, and internship experiences as part of the training program. These are ways of contextualizing the training, therefore making it more likely to transfer to real-world practice.

Improving the Performance of Organizations

Finally, besides improving the performance of learners and practitioners, educational technology aims to improve the performance of organizations themselves. Primarily, it does this by increasing the productivity of learning processes, helping people within the organization gain new skills more rapidly and at less expense, thus saving time and money for the organization. But there are ways of improving organizational performance beyond just training. The people in organizations can be helped to be more productive by getting better tools, having better working conditions, being motivated to work harder, and having access to job aids or other sorts of cognitive support on demand. Noninstructional interventions such as these fall within the field of human performance technology (HPT). HPT is an umbrella concept that incorporates educational technology plus all of the other ways of improving human performance in the workplace. This concept is discussed in chapter 3 and in greater depth later in this section.

Promoting Efficiency and Effectiveness. As discussed at length in chapter 3, efficiency in education is a delicate subject because efficiency is often associated with cost cutting without regard to its effect on learners or educational institutions. In the context of educational technology, efficiency in education and training refers to designing, developing, and implementing instruction in ways that make wise use of resources, both human and monetary. Effectiveness has to do with the degree to which learners attain worthy learning goals; that is, the school, college, or training center facilitates the learning of knowledge, skills, and attitudes that are desired by their stakeholders, including the learners themselves.

Educational technology values instruction that is both efficient and effective. The two must go hand in hand. Instruction that is merely cheap is a waste of scarce resources if it misses the goal of producing worthy learning outcomes. Similarly, instruction that produces desired learning results but consumes excessive resources, is not timely, or does not reach learners is also a waste of scarce resources. Regardless of one's preferred teaching-learning perspective there is a common desire to find ways to help people learn better (effectiveness) and to find ways to do that without wasting effort and expense on the part of instructors or learners (efficiency). For example, both behaviorists and constructivists believe that their practices are superior in achieving learning results (effectiveness) and both believe that learners will accomplish worthy goals faster and easier if their methods are used (efficiency).

Measuring inputs and outcomes. Judgments about efficiency and effectiveness depend heavily on how costs and benefits—human and monetary—are calculated. As discussed in chapter 3, cost-benefit equations can be set up to include anything the stakeholders may agree upon regarding what counts as a cost and what counts as a benefit. Is the learner's time part of the cost? Is the learner's social development part of the benefits? People may conscientiously differ on issues such as these. In fact, there will always be debate, in businesses and educational institutions, about what goals are worth pursuing and what indicators should be used to measure progress toward those goals. And educational technologists, like all educators, have a stake in the outcome of those debates.

Human Performance Technology

Some educational technology professionals, especially those involved in corporations and other large organizations, view their work under the larger umbrella of HPT. In HPT, the technological approach is applied to not just instructional activities but all interventions that affect people in the workplace. That is, organizational productivity can be improved through several types of interventions in addition to training: offering incentives, providing job aids, adapting tools to the task, redesigning jobs, and altering the organizational structure. In that sense, HPT incorporates educational technology and goes beyond it.

Since educational technology is so closely linked to HPT, it may be useful to examine the culture of HPT to find out what values are dominant in that field, beyond those discussed in educational technology.

The International Society for Performance Improvement (ISPI; 2002) supports a set of performance technology standards to guide the practice of HPT. These standards give indications of the values that are salient in HPT, most of which could also be considered implicit in the work of educational technology, especially for those who work in the sorts of organizational settings that most HPT practitioners do: businesses and other large organizations, including government, military, and nonprofits. The distinctive HPT values are

- A focus on *results*—measuring the impact of the interventions on the target problem
- Adding *value*—results must be worth the cost, yielding positive cost-benefit solutions
- Working in *partnerships* and collaboration—clients and stakeholders work together recognizing that people accept the changes that they help create.

Values Related to *Creating, Using, and Managing*

Educational technologists believe that decisions made in the creation and use of learning resources can and should be enlightened by empirically derived knowledge. At the same time, they recognize that creation and use of learning resources require leaps of imagination as they are carried out. Instructional designers cannot "cut and paste" previously created material all of the time; they usually have to generate new solutions and new material. Instructors using designed materials have to make on the spot adaptations, since each situation has unique aspects. Thus, educational technology incorporates both art and science in its practice, and it accepts the values of connoisseurship as well as the values of empirical inquiry. The reflective practitioner mentioned earlier is an important aspect of our field; reflection on practice is vital to the active role teachers and designers must play in the creation and use of educational technology materials and strategies.

Values Related to *Appropriate*

As discussed in chapter 1, both processes and resources are meant to be modified by the term *appropriate,* meaning suitability and compatibility with intended purposes and ethical guidelines.

Work Processes

Appropriate work processes are addressed by ethical standards that require the use of sound professional practices. Just as physicians are expected to follow "standards of care," so other professionals are obliged to know and adhere to the current best practices in their fields. A number of these expectations are specified in the AECT code of ethics.

For work processes for instructional design to meet the standard of appropriateness, they should be suited to needs of the organization—such as school, college, or business—and its learners. It would be contrary to the interests of a university for educational technology professionals who offer instructional consulting services to advocate instructional design practices that increased the costs of the university without commensurate benefits or that increased the workload of faculty without comparable payoffs. Further, those instructional design practices would also be expected to improve learning opportunities for students who experience the instruction. In short, design processes should be efficient and effective. The same would pertain to work processes involved in selection and use of instructional systems. Practitioners are expected to know, recommend, and use utilization techniques that are up to current standards. Those techniques ought to be justifiable on the basis of proven results, thus reminding them of the need to access and understand the results of published research inquiries.

Technologies

Different technologies may be evaluated in terms of their appropriateness for a particular age group or for a particular socioeconomic or cultural setting. For example, since computers became widely available, controversy has raged about the appropriateness of computer use by very young children. Montessori schools and Waldorf schools explicitly exclude computers from their early childhood education programs (Kaminstein, n.d.; Association of Waldorf Schools of North America, n.d.). Their rationale is that children need multisensory experience, they need to move, they need discovery and experimentation, they need varied repetition, and they need "the thrill of accomplishment that comes from hard work" (Kaminstein, n.d.). Children may be deprived of these experiences during whatever time they spend with computers. Monke (2005) extended this argument into the area of play, claiming that free, unstructured physical play is a developmental necessity for young children and that the computer can lure them away from such play. Monke's claim that "even relying on books too much or too early inhibits the ability of children to develop direct relationships with the subjects that they are studying" (p. 38) was strikingly consistent with Edgar Dale's (1946) advocacy for direct, purposeful experiences.

Healy (1999) summarized this "deprivation of play time" argument:

> If a child spends an inordinate amount of time on video games (or tele-vision, or even other types of computer use) instead of playing and experimenting with many different types of skills, the foundations for some kinds of abilities may be sacrificed. These losses may not show up until much later, when more complicated kinds of thinking and learning become necessary. (p. 206)

Apologists for computer use by young children begin by criticizing the practice of lumping together all computer uses under one heading; different applications have different effects. They are then able to point to findings of particular studies or of meta-analyses that show, for example, that children can have positive emotional experiences with computers, frequently use them collaboratively, and participate in a lot of peer interaction around the computer (Clements & Sarama, 2003).

It is possible that advocates on both sides have defensible claims. Young children need a range of first-hand, direct, physical experiences for proper development. Assuming that they have adequate time and opportunity for such direct experiences, there may also be occasions in which certain uses of computers may be of great benefit. It comes back to educational technology's commitment to make judgments of appropriateness of technologies on the basis of the needs of specific learners in specific circumstances.

Likewise, critics view with alarm the export of advanced technologies into countries or subcultures that are perceived as not ready for them. New tech-nologies might expose indigenous people to foreign mores or ideas that conflict in some way with traditional ones. New technologies might be unsustainable in terms of the local infrastructure or they might impose harmful financial burdens on the local economy. They might exacerbate political domination or "cultural imperialism."

Educational technology's value position is that technological solutions should be evaluated for their sustainability, their cultural suitability, and their economic impacts. Neither high technology nor low technology is good or bad in itself. Either one—or none at all—might be appropriate in a given situation.

Specific Resources

When applied to specific resources, appropriateness can be judged by numerous criteria. Are the materials suited to the developmental levels of learners? to their reading levels? to their current levels of subject-matter mastery? to the objectives of a particular lesson? At times, racial or ethnic aspects of the material could be important. Sensitivity to learners' interests and cultural and experiential backgrounds is required, and attention to the equal positions of power and authority, equal access, and equalization of opportunity for disadvantaged learners is vital. Deciding on and applying criteria of appropriateness are parts of the professional expectations of educational technologists.

Values Related to *Technological*

As discussed in chapter 1, the term *technological* is meant to apply to both processes and resources. One of the hallmarks of the field is its commitment to approaches that accord with "the systematic application of scientific or other organized knowledge to practical tasks" (Galbraith, 1967, p. 12). This term is the key one in the name *educational technology*. It indicates the unique perspective of this field compared to others. Other fields apply processes to education, but those processes are neither necessarily systematically conducted nor based on scientific grounds. Other practitioners—teachers, professors, and trainers—develop, select, and use resources for instruction, but they do not necessarily focus on technological resources. This field does.

Another label for technological processes and resources is "soft and hard technologies." The former refers to ways of thinking about teaching, learning, and using adroit problem-solving methods. The latter refers to the hardware and software used to actually communicate with learners. It is a given among educational technology professionals that the hard technologies in themselves are not panaceas. Information and communication technologies (ICT), although potentially of tremendous power in terms of increasing access to education as well as lowering cost and reducing time expenditure, are just carriers of educational messages and methods. The robustness of those messages and methods ultimately determines the value of the program.

Further, it is the special responsibility of this field to consider the unintended consequences of pervasive use of ICT. Overuse or inappropriate use of ICT can lead to the isolation and alienation of users, as mentioned in the previous example regarding preschoolers and computers. Syntheses of the plethora of research on the impact of television viewing on children provide ample guidelines for dealing with these issues (Seels, Fullerton, Berry, & Horn, 2004). More recent experiences with learners ubiquitously using digital technologies (e.g., wireless access, cell phones, PDAs, and other increasingly miniaturized and mobile technologies) certainly suggest that the sense of alienation may grow, or may be affected by this increasing ability to be in touch with others electronically although not physically. In the end, the human touch is an indispensable ingredient in any well-rounded educational program.

Summary

Educational technology shares many values in common with related fields, such as education, but there are a number of values that are more distinctive to educational technology and that are prominent in theoretical and practical writings in the field. Each element of the basic definition carries with it one or more distinctive values.

Study

Practice in educational technology is based on inquiry of several types—basic research on learning; applied research on the processes of design, utilization, and management; formative and summative evaluation of specific materials; action research on projects in the field; case studies, especially of failed systems; and personal reflections on experiences with technology.

Ethical Practice

Codes of ethics are themselves statements of values, so many specific instances of value statements can be found in formulations such as AECT's code of ethics. These tend to revolve around the relationships among educational technologists, learners, and the materials and systems with which they are involved. A major ethical requirement for practitioners is simply knowing and observing best practices.

Facilitating Learning

Beyond the goal of simply helping people learn, educational technology strives to help them learn better than they could on their own or through means other that educational technology. Strategies that promote engagement, inquiry, and reflection help learners learn how to learn, better understand themselves as learners, and become equal partners in the learning equation. Additionally, one of the implicit goals of educational technology is to improve access to learning through ICT. Through these technologies, more people can have access to learning regardless of distance, boundaries, or economics, thus contributing to social equality.

Improving Performance

Educational technology strives to help people not only learn more deeply, but also to retain skills longer, and to apply them in settings beyond the classroom. The values of efficiency and effectiveness, although they apply to all the elements of the definition, are especially pertinent to improving the performance of individual learners, of teachers and designers, and of the organization as a whole. Educational technology helps individuals and organizations accomplish their goals while making best use of the time and resources available.

Creating, Using, and Managing

The educational technology approach to the creation of instructional materials and learning environments generally embraces systemic, systematic, and scientific procedures. At the same time, it recognizes and values artistry within this process.

Appropriate Processes and Resources

To be appropriate, work processes must first be state of the art. Currency of knowledge and competence in action are minimal values to be assumed. To be appropriate, work processes should also be suitable for the situation in which they are used—beneficial to the institution and to the learner.

Resources that are selected or created for use with learners may be judged on many different criteria. The value that should drive the development and application of such criteria is sensitivity to the needs and interests of learners.

Technological Processes and Resources

Nothing could be more logically central to the meaning of educational technology than the connotations of the term *technology*. It implies a commitment to solutions that are systematically and scientifically based (the "soft" aspect of technology) and/or that incorporate ICT as means of involving learners in learning activities (the "hard" aspect of technology). Of the latter aspect, educational technology encourages critical analysis of the unintended consequences of hard technology proliferation, demanding that human interests have primacy over technical ones.

Overall

Adding together the values found in relation to each of the elements, taken overall, educational technology values applied as well as basic research, ethical practice, learner empowerment, sensitivity to individual learner needs, deep learning, learning how to learn, transfer of learning, learner access to resources, state-of-the art practice, efficiency with effectiveness, empirically-based decision making, artistry, technological approaches to problem solving, technological resources (being sensitive to balance), and humaneness.

References

Association of Waldorf Schools of North America. (n.d.). *What about computers and Waldorf education?* Retrieved October 5, 2005, from http://www.awsna.org/awsna-faq.html#computers

Bransford, J. D., Brown, A. L., & Cocking, R. R. (Eds.). (1999). *How people learn: Brain, mind, experience, and school.* Washington, DC: National Academy Press.

Burkman, E. (1987). Factors affecting utilization. In R. M. Gagne (Ed.), *Instructional technology: Foundations* (pp. 429–455). Hillsdale, NJ: Lawrence Erlbaum Associates.

Carr-Chellman, A. A. (2006). Where do educational technologists really publish? An examination of successful emerging scholars' publication outlets. *British Journal of Educational Technology, 37*(1), 5–15.

Clements, D. H., & Sarama, J. (2003). Strip mining for gold: Research and policy in educational technology—a response to "Fool's Gold" [Electronic version]. *Educational Technology Review, 11*(1), 7–69.

Collins, J., & Porras, J. (1994). *Built to last: Successful habits of visionary companies.* New York: Harper Business.

Dale, E. (1946). *Audio-visual methods in teaching.* New York: The Dryden Press.

Galbraith, J. K. (1967). *The new industrial state.* Boston: Houghton Mifflin.

Healy, J. M. (1999). *Endangered minds: Why children don't think and what we can do about it.* New York: Simon & Schuster.

Holcomb, T. L., Bray, K. E., & Dorr, D. L. (2003, September/October). Publications in educational/instructional technology: Perceived values of ed tech professionals. *Educational Technology, 43*(5), 53–57.

ISPI. (2002). *ISPI's performance technology standards.* Silver Spring, MD: International Society for Performance Improvement.

Kaminstein, M. (n.d.). *Why our Montessori classrooms are computer-free.* Retrieved October 5, 2005, from http://www.montessori.org/?defaultarticle=&default node=46&layout=29&pagefunct

Molenda, M., & Kang, S. P. (2004). Online journals in educational technology: Status and directions [Electronic version]. *Asia-Pacific Cybereducation Journal, 1*(1), 35–44.

Monke, L. (2005, September/October). Charlotte's Webpage. *Orion.* Retrieved October 5, 2005, from http://www.oriononline.org/pages/om/05-5om/Monke.html

Petroski, H. (1992). *To engineer is human: The role of failure in successful design.* New York: Vintage. (Original work published in 1985 by St. Martin's Press, New York.)

Schön, D. A. (1995). *The reflective practitioner: How professionals think in action.* Aldershot, UK: Arena.

Seels, B. B., Fullerton, K., Berry, L., & Horn, L. J. (2004). Research on learning from television. In D. H. Jonassen (Ed.), *Handbook of research on educational communications and technology* (2nd ed., pp. 249–334). Mahwah, NJ: Lawrence Erlbaum Associates.

Seels, B. B., & Richey, R. C. (1994) *Instructional technology: The definition and domains of the field.* Washington, DC: Association for Educational Communications and Technology.

Waterman, R. (1992). *Adhocracy: The power to change.* New York: Norton.

Weigel, V. B. (2002). *Deep learning for a digital age.* San Francisco: Jossey-Bass.

10

A HISTORY OF THE AECT'S DEFINITIONS
OF EDUCATIONAL TECHNOLOGY

Alan Januszewski
The State University of New York at Potsdam

Kay A. Persichitte
University of Wyoming

Introduction

*T*HE PURPOSE OF THIS chapter is to provide a historical context for the current definition of educational technology. We will do this in several stages. First, we will review the primary purposes and considerations for defining educational technology. Then, we will review each of the four previous definitions, paying particular attention to the primary concepts included in each definition. We will examine the context and rationales for decisions made regarding each of these primary concepts. We will also present some of the historical criticisms of the definitions which provided the impetus for changing the definitions.

The criteria and purposes for producing a definition were discussed at the time of the writing of the first definition in 1963.

> A satisfactory definition of instructional technology will let us find common ground, will propose tomorrow's horizons, and will allow for a variety of patterns that specific individuals may follow in specific institutions . . . Research must be designed in terms of clear understanding of instructional technology. Superintendents of schools are requesting criteria for new personnel

needed in various phases of instructional improvement. Teacher-education institutions need assistance in planning courses for pre-service and in-service instruction that will provide the skills and understanding which will be required in tomorrow's classrooms . . . Let us consider the criteria for useful definitions. They should (a) clarify the description of the field in ordinary language; (b) summarize existing knowledge; (c) mediate applications of knowledge to new situations; and (d) lead to fruitful lines of experimental inquiry. . . . This report aims to provide a working definition for the field of instructional technology which will serve as a framework for future developments and lead to an improvement in instruction. (Ely, 1963, pp. 7–8)

Those involved in the writing of the 1963 definition obviously believed that there were a lot of things to consider when defining educational technology. Or put differently, the existence of such a definition would have far reaching consequences, sometimes with implications that the authors might not intend. Acknowledging this opened the door to criticisms of the definitions and the purposes cited for redefining educational technology. The authors of subsequent definitions all seemed to adhere, at least in part, to the purposes and criteria identified in the 1963 definition.

The 1963 Definition

The leadership of the Association for Educational Communications and Technology (AECT) recognized the 1963 definition of *audiovisual communications* as the first formal definition of educational technology (AECT, 1977). This definition, the first in a series of four officially sanctioned definitions, was developed by the Commission on Definition and Terminology of the Department of Audiovisual Instruction (DAVI) of the National Education Association (NEA) and supported by the Technological Development Project (TDP). In 1963 *audiovisual communications* was the label that was used to describe the field as it was evolving from the audiovisual education movement to educational technology:

Audiovisual communications is that branch of educational theory and practice primarily concerned with the design and use of messages which control the learning process. It undertakes: (a) the study of the unique and relative strengths and weaknesses of both pictorial and nonrepresentational messages which may be employed in the learning process for any purpose; and (b) the structuring and systematizing of messages by men and instruments in an educational environment. These undertakings include the planning, production, selection, management, and utilization of both components and entire instructional systems.

Its practical goal is the efficient utilization of every method and medium of communication which can contribute to the development of the learner's full potential. (Ely, 1963, pp. 18–19)

A footnote that was included as part of this definition read "the audiovisual communications label is used at this time as an expedient. Another designation may evolve, and if it does, it should then be substituted" (p. 18).

Conceptual Shifts Signaled in Definitions

There are three major conceptual shifts that contributed to the formulation of the definitions of educational technology as a theory: (1) the use of a "process" concept rather than a "product" concept; (2) the use of the terms *messages* and *media instrumentation* rather than *materials* and *machines*; and (3) the introduction of certain elements of learning theory and communication theory (Ely, 1963, p. 19). Understanding these three ideas and their impact on each other is essential to understanding the idea of educational technology in 1963.

A technological conception of the audiovisual field called for an emphasis on process, making the traditional product concept of the field of educational technology untenable. The Commission believed, "The traditional product concept in the audiovisual field views the 'things' of the field by identifying machines, use of particular senses, and characteristics of materials by degrees of abstractness and/or concreteness" (Ely, 1963, p. 19). Members of the Commission preferred a process concept of the field which included "the planning, production, selection, management, and utilization of both components and entire instructional systems" (p. 19). This process conception also emphasized "the relationship between events as dynamic and continuous" (p. 19).

The Commission argued that "materials" and "machines" were "things" or products and opted not to use those terms in the definition. Instead, the Commission used the terms *messages* and *instruments*. The Commission further argued that materials and machines were interdependent elements. "A motion picture and projector are inseparable as are all other materials requiring machines for their use" (Ely, 1963, p. 19). One was of little practical use without the other.

The Commission used the concept of *media instrumentation* to explain *instruments*. The Commission said, "Media-instrumentation indicates the transmission systems, the materials and devices available to carry selected messages" (Ely, 1963, p. 20). The concept of media instrumentation also included the people who utilized the instruments in the educational environment as well as the transmission systems. The idea that both people and instruments comprised media instrumentation was based in the broader concept of the man-machine system (Finn, 1957).

In discussions of the relationship and integration of learning theory and communications theory to instructional technology, the Commission stated, "Certain elements of learning theory and communications theory offer potential contributions [to the field of educational technology]; e.g., source, message, channel, receiver, effects, stimulus, organism, response" (Ely, 1963, p. 20). The Commission integrated learning theory and communications theory by identifying and combining the two systems basic to the process view of the field: the learning-communicant system and the educational-communicant system. These two systems use concepts from both learning and communications theories that delineated and specified the roles of the individuals involved in the use of these systems. The learner-communicant system "refers to the student population" and the educational-communicant system "refers to the professional persons in the school" (p. 23). These two systems could be of any size, ranging from a single classroom to large school systems (Ely, 1963). Merging the two communicant systems into a single model of the educational process provided the field of audiovisual communications with a theoretical framework (Ely, 1963) and a model that allowed educational technology to be viewed as a theoretical construct (AECT, 1977).

The fundamental doctrine advanced by the writers of the first definition was that it was a "branch of educational theory and practice." The word *theory* was particularly important in this definition because it had a special place in the history of the audiovisual field, because of the status that it conferred on the field, and because of the expectation for further research to influence the evolution of that theory.

Finn's Characteristics of a Profession

The 1963 definition was heavily influenced by James Finn's (1953) six characteristics of a profession:

> (a) An intellectual technique, (b) an application of that technique to the practical affairs of man, (c) a period of long training necessary before entering into the profession, (d) an association of the members of the profession into a closely knit group with a high quality of communication

between members, (e) a series of standards and a statement of ethics which is enforced, and (f) an organized body of intellectual theory constantly expanded by research. (p. 7)

Of these six characteristics of a profession, Finn (1953) argued that "the most fundamental and most important characteristic of a profession is that the skills involved are founded upon a body of intellectual theory and research" (p. 8). Having established the importance of theory and research for a profession, Finn further explained his position by saying that ". . . this systematic theory is constantly being expanded by research and thinking *within* the profession" (p. 8). Finn was arguing that a profession conducts its own research and theory development to complement the research and theory development that it adapts/adopts from other academic areas. If educational technology was to be a true profession, it would have to conduct its own research and develop and its own theory rather than borrowing from more established disciplines like psychology.

Finn (1953) evaluated the audiovisual field against each of the six characteristics and determined that the audiovisual field did not meet the most fundamental characteristic: an organized body of intellectual theory and research. "When the audiovisual field is measured against this characteristic . . . the conclusion must be reached that professional status has not been attained" (Finn, 1953, p. 13). This argument was largely accepted by, and had a profound effect on, the leadership of the audiovisual field in the late 1950s and early 1960s.

Finn (1953) laid a foundation that the audiovisual field was troubled by a "lack of theoretical direction" (p. 14). He attributed this to a "lack of content" and the absence of "intellectual meat" (p. 14) in the contemporary meetings and professional journals of the field. In his argument promoting the development of a theoretical base for the audiovisual field, Finn warned,

> Without a theory which produces hypotheses for research, there can be no expanding knowledge and technique. And without a constant attempt to assess practice so that the theoretical implications may be teased out, there can be no assurance that we will ever have a theory or that our practice will make sense. (p. 14)

Finn dedicated his career to rectifying this deficiency in the field, and the resulting impact of his work on the 1963 definition is evident.

Advancing an argument that audiovisual communications was a theory was an attempt to address the "lack of content" cited by Finn (1953). The Commission identified "the planning, production, selection, management, and utilization of both components and entire instructional systems" (Ely,

1963, p. 19) as tasks performed by practitioners in the field directly related to Finn's (1953) discussion of the "intellectual technique" of the audiovisual field—Finn's first criterion for a profession.

The first official definition of educational technology can be viewed as an attempt to bring together remnants of theory, technique, other academic research bases, and history contained in the audiovisual literature, into a logical statement closing the gap on the "poverty of thought" (Finn, 1953, p. 13) that characterized the audiovisual education movement. The evolution of audiovisual communications (and later, educational technology) as a theory began to add "intellectual meat" to audiovisual practice. By merging the audiovisual communications concept with the process orientation of the field into a new intellectual technique grounded in theory, the Commission strengthened the professional practice and offered a direction for further growth as a profession.

Emergence of a Process View

Included among the many factors contributing to the development of the process view of educational technology were the two beliefs held by the most influential and prominent individuals involved with the audiovisual field: (1) that technology was primarily a process (Finn, 1960b) and (2) that communication was a process (Berlo, 1960; Gerbner, 1956). The conceptual view of educational technology as a way of thinking and a process was established by the 1963 definition.

The intention of the Commission that produced the first official definition of the field was "to define the broader field of instructional technology which incorporates certain aspects of the established audiovisual field" (Ely, 1963, p. 3). Not unexpectedly, the 1963 definition drew some critique as it was applied to the emerging field of the 1960s and 1970s.

Prominent individuals involved with audiovisual education, such as James Finn (1957; 1960a) and Charles Hoban (1962), had previously used the term *technology* when referring to the activities of the audiovisual field. Donald Ely (1973; 1982) observed that the use of the word *control* in the 1963 definition was problematic for many individuals involved with educational technology. Ely (1982) explained, "The strong behavioral emphasis at the time seemed to call for the word 'control'" (p. 3). He noted that the word *facilitate* was substituted by many professionals "to make the definition more palatable" (Ely, 1973, p. 52). Perhaps equally important was the desire by members of the field to move away from a behaviorally based psychology to a more humanistic psychology (Finn, 1967).

Criticisms of the 1963 Definition

As noted in the introduction, no *one* definition can be *the* definition, and there were criticisms of the 1963 definition. James Knowlton (1964), a faculty member at Indiana University, was a consultant for the 1963 Commission on Definition and Terminology. In an essay that reviewed the 1963 definition, Knowlton stated that the definition itself was "couched in semiotical terms" (p. 4) but that the conceptual structure used in the rationale for the 1963 definition "was couched in learning theory terms [and] this disjunction produced some surprising anomalies" (p. 4). Knowlton's argument was based on a need for conceptual and semantic consistency in the definition. Knowlton argued that failing to pair the language of the definition with the language of the conceptual structure in the rationale resulted in a general lack of clarity about this new concept. This lack of clarity in turn caused confusion in the direction of research and practice in the field.

Less than a decade later, Robert Heinich (1970) saw a need to redefine the field of educational technology for two reasons. First, he was critical of the "communications" based language used in the 1963 definition. Heinich argued that this language was too complicated for school personnel to interpret and apply. Second, Heinich argued that the power to make many of the decisions regarding the use of technology in schools should be transferred from the teacher to the curriculum planners. Heinich's argument for changing the definition was based on both linguistic concerns and evolutionary changes in the functions of practitioners in the field. Heinich promoted an approach to schooling where specialists would decide when and where schools would use technology. This position was different from that which was discussed in the rationale for the 1963 definition. In the rationale for the 1963 definition, teachers were viewed as partners of educational technologists rather than as their subordinates (Januszewski, 2001).

Forces Impelling a New Definition

Other contemporary issues emerged which began to influence the field. The report of the Presidential Commission on Instructional Technology (1970) stated that instructional technology could be defined in two ways:

> In its more familiar sense it means the media born of the communications revolution which can be used for instructional purposes alongside the teacher, textbook and blackboard. In general, the Commission's report follows this usage . . . the commission has had to look at the pieces that

make up instructional technology: television, films, overhead projectors, computers and the other items of "hardware and software." (p. 19)

The second and less familiar definition . . .

> (Instructional technology) . . . is a systematic way of designing, carrying out, and evaluating the total process of learning and teaching in terms of specific objectives, based on research in human learning and communication and employing a combination of human and nonhuman resources to bring about more effective instruction. (Commission on Instructional Technology, 1970, p. 19)

Educational technology professionals responded to this report in a special section of *Audiovisual Communications Review* (1970). The professional reviews of the government report were mixed at best. Ely (Ely et al., 1970) of Syracuse University thought that the Commission's overall effort was commendable given its lofty charge. Earl Funderburk (Ely et al., 1970) of the NEA called the recommendations a balanced program. But David Engler (Ely et al., 1970) of the McGraw-Hill Book Company disapproved of the Commission's effort to relegate the process-based definition of instructional technology to some "future" role. Leslie Briggs (Ely et al., 1970) of Florida State University accused the Presidential Commission of providing a "two-headed image" of instructional technology by stressing both a hardware and a process orientation of the concept.

The contributors to this special section of *Audiovisual Communications Review* (1970) were generally dissatisfied with the "two-headed" orientation primarily because of the confusion it might cause among the potential client groups of educational technology. They viewed the hardware orientation favored by the Presidential Commission as a setback for the profession. It meant the unacceptable return to the "audiovisual aids" and "technology as machine" conceptions of educational technology. This orientation also implied the de-emphasizing of research and theory. Given these professional discussions and developments, professionals in the field believed that a new definition of educational technology was necessary.

The 1972 Definition

By 1972, through evolution and mutual agreement, the DAVI had become the AECT. Along with the organizational change came a change to the definition.

The newly formed AECT defined the term *educational technology* rather than the term *audiovisual communications* as

> Educational technology is a field involved in the facilitation of human learning through the systematic identification, development, organization and utilization of a full range of learning resources and through the management of these processes. (Ely, 1972, p. 36)

As a member of the group that wrote several of the early drafts of the 1972 definition, Kenneth Silber (1972) was successful in including changes in many of the roles and functions of the practitioners of the field as part of that definition. Silber introduced the term *learning system* which combined ideas of the open classroom movement with some of the concepts of educational technology. Like Heinich's (1970) perspective, Silber's (1972) "learning system" (p. 19) suggested changes in the roles of the teacher and the educational technologist. Unlike Heinich, Silber supported the idea that learners should make many decisions regarding the use of educational technology themselves. Educational technologists would produce a variety of programs and designs that learners would use or adapt to meet their own "long-range learning destination" (p. 21). Silber's position was that the teacher should be more a "facilitator of learning" and less a "teller of information."

A Definition Based on Three Concepts

There are three concepts central to the 1972 definition characterizing educational technology as a field: a broad range of learning resources, individualized and personalized learning, and the use of the systems approach. "It is these three concepts, when synthesized into a total approach to facilitate learning, that create the uniqueness of, and thus the rationale for, the field" (Ely, 1972, p. 37). Examining these three concepts along with the idea of educational technology as a "field" is crucial to understanding the AECT's (1972) definition of educational technology.

It is particularly important to recognize that different interpretations of these three concepts would result in differing conceptions of the field through the next three decades. The different interpretations and relative emphases of these concepts were due in large part to differences in educational philosophy and educational goals. Differing interpretations of these concepts would also have the more visible effect of substantially different products and processes developed in the field.

The writers of the 1972 definition seemed to be aware that the major concepts could be interpreted differently, and they seemed to be interested in including individuals with different philosophical and academic backgrounds in the field. The writers of the 1963 definition and its supporting rationale seemed less concerned with accommodating divergent educational philosophies. Perhaps this was due to the fact that the 1963 definition was the first formal attempt to define educational technology. Such an undertaking was formidable enough. Perhaps it was because the writers of the 1972 definition paid more attention to the discussions of educational philosophy in the literature from the rest of the field of education. Perhaps it was because the 1963 definition viewed educational technology as an educational theory and, potentially, as an educational philosophy itself. Regardless, there is no doubt that by 1972, the authors of the definition of educational technology chose to consider educational technology a field of study and not as a specific theory (Januszewski, 1995, 2001).

Educational Technology as a Field

The decision to refer to *educational technology* as a field of study rather than a theory or a branch of theory had at least four results: (1) we acknowledged that there was more than one theory of educational technology, more than one way to think about the role(s) of educational technology; (2) the definition prompted significant philosophical discussions by members of the profession; (3) the use of the word *field* encompassed both the "hardware" and "process" orientations of instructional technology described by the Presidential Commission (1970); and (4) this definition was based on the "tangible elements" (Ely, 1972) that people could observe. The 1972 definition essentially defined educational technology by role and function rather than as an abstract concept, as was the case for the 1963 definition, where educational technology was viewed as a theory.

The concept of "field" has been a thorny one for educational technologists. Like many areas of study within education, it is very difficult to discuss educational technology without using the word *field* as a descriptor. Certainly audiovisual professionals used the term to describe the "audiovisual field" before the terms *instructional technology* or *educational technology* were ever used. The 1963 definition statement frequently used *field* (Ely, 1963) to move the discussion along, even though it was argued that educational technology was a theory or branch of theory. On the surface, the use of *field* seems a rather inescapable semantic problem when speaking of educational technology. But it is significant that the writers of the 1972 definition chose to use *field* rather than *theory* in the definition because the use of the word *field* established a territory. It also provided certain legitimacy to efforts to advance both products and processes. The consequences of this decision were anticipated by Finn (1965), who proclaimed

> Properly constructed, the concept of instructional or educational technology is totally integrative. It provides a common ground for all professionals, no matter in what aspect of the *field* they are working: it permits the rational development and integration of new devices, materials, and methods as they come along. The concept is so completely viable that *it will not only provide new status for our group, but will, for the first time, threaten the status of others* [italics added]. (p. 193)

Criticism of the 1972 Definition

The 1972 definition was not the object of numerous criticisms as was the 1963 definition, probably because it was considered only an interim definition (Ely, 1994). Only one such article appeared in the literature of the field of educational technology—a critique was written by Dennis Myers, then a graduate student at Syracuse University, and Lida Cochran, a faculty member at the University of Iowa (Myers & Cochran, 1973).

The brief analysis by Myers and Cochran (1973) articulated at least five different criticisms. First, they proposed including a statement in the rationale for the definition stating that students have a right of access to technological delivery systems as part of their regular instruction. Including such a statement follows from Hoban's (1968) discussion on the appropriateness of technology for instruction in a technological society. Second, Myers and Cochran argued that the 1972 definition statement was weakened by neglecting to include a theoretical rationale for the definition. This criticism, which correctly pointed out that the definition is lacking a unified theoretical direction, supported Heinich's (1970) assertions in his philosophical view of the field.

In a third point, Myers and Cochran (1973) criticized the limited role that the educational technologist was provided in the description of the systems approach provided in the definition. In a fourth point, they discussed the shortcomings of the terminology used to discuss the domains and roles in educational technology.

Perhaps the most interesting point made in this analysis concerned the relationship of educational technology to the rest of the field of education. In noting the problem of defining the field by the functions performed, Myers and Cochran (1973) pointed to the importance of considering the purpose of education.

> What is important is that certain functions get done in education. That generalization is important because it conveys an attitude that transcends narrow professional interests and strikes a note of community and cooperativeness, qualities which are essential to the solution of problems facing education and society. (p. 13)

Here, Myers and Cochran (1973) seemed to be chastising the writers of the 1972 definition for being overly concerned with intellectual territory and the roles performed in the field of educational technology. This particular criticism lost only a little of its sharpness when it was viewed in light of earlier comments made about the inappropriateness of the limited role assigned to educational technologists in the definition (Januszewski, 2001).

In summary, by 1972, the name of the concept had changed from audio-visual communications to educational technology. The organizational home for professionals in the field had changed name: from DAVI to AECT. There had been substantial changes in our schools, hardware, and other technological innovations during the nine years since the writing of the first definition. Educational technology was now identified as a field of study, open to interpretation by those who practiced within it. The 1972 definition reflected these interpretations but was intended to be only a temporary measure. Almost as soon as it was published, work began on the next definition.

The 1977 Definition

In 1977, the AECT revised its definition of educational technology with its third version:

> *Educational technology* is a complex, integrated process, involving people, procedures, ideas, devices and organization, for analyzing problems and devising, implementing, evaluating and managing solutions to those problems, involved in all aspects of human learning. In educational technology, the solution to problems takes the form of all the *Learning Resources* that are designed and/or selected and/or utilized to bring about learning; these resources are identified as Messages, People, Materials, Devices, Techniques, and Settings. The processes for analyzing problems, and devising, implementing and evaluating solutions are identified by the *Educational Development Functions* of Research Theory, Design, Production, Evaluation Selection, Logistics, Utilization, and Utilization Dissemination. The processes of directing or coordinating one or more of these functions are identified by the *Educational Management Functions* of Organizational Management and Personnel Management. (AECT, 1977, p. 1)

The *Definition of Educational Technology* (AECT, 1977) was a 169-page book intended to accomplish two things: (a) systematically analyze the complex ideas and concepts that were used in the field of educational technology, and (b) show how these concepts and ideas related to one another (Wallington, 1977). This publication included the definition of educational technology (which comprises 16 pages of the text), a history of the field, a rationale for the definition, a theoretical framework for the definition, a discussion of the practical application of the intellectual technique of the field, the code of ethics of the professional organization, and a glossary of terms related to the definition.

Educational Versus Instructional Technology

The conceptual difference between the terms *educational technology* and *instructional technology* constituted a large portion of the analysis of this book. Understanding how the authors of the 1977 definition viewed the relationship of instructional technology to educational technology is essential to understanding the 1977 definition and its theoretical framework. The basic premise of this distinction was that instructional technology was to educational technology as instruction was to education. The reasoning was that since instruction was considered a subset of education then instructional technology was a subset of educational technology (AECT, 1977). For example, the concept of educational technology was involved in the solution of problems in "all aspects of human learning" (p. 1). The concept of instructional technology was involved in the solution of problems where "learning is purposive and controlled" (p. 3).

Educational Technology as a Process

Two other complex conceptual developments were also undertaken by the authors of the 1977 definition, which were interrelated. First, the 1977 definition of educational technology was called a "process" (AECT, 1977, p. 1). The authors intended the term *process* to connote the idea that educational technology could be viewed as a theory, a field, or a profession. Second, the systems concept was infused throughout the entire definition statement and in all the major supporting concepts for the definition in both its descriptive and prescriptive senses. The authors of the 1977 definition connected these two conceptual developments by saying that the use of the systems concept was a process (AECT, 1977).

As one of the three major supporting concepts for the 1972 definition of educational technology, the systems approach had become the basis for the definition itself by 1977. Through their efforts to reinforce the process conception of educational technology, the leadership of the field now assumed that all of the major supporting concepts of the definition were tied to, or should be viewed in light of, the systems approach.

The three major supporting concepts of the 1977 definition were learning resources, management, and development. Learning resources were any resources utilized in educational systems; a descriptive use of the systems concept the writers of the 1977 definition called "resources by utilization." Authors called the resources specifically designed for instructional purposes, a prescriptive use of the systems approach, "resources by design" or "instructional system components" (AECT, 1977).

Like the concept of learning resources, management could be used in a descriptive fashion to describe administrative systems or in a prescriptive way to prescribe action. The concept of management was often used as a metaphor for the systems approach in education (Heinich, 1970). The term *instructional development* was frequently used to mean the "systems approach to instructional development" or "instructional systems development" (Twelker et al., 1972). The fact that the management view of the systems approach to instruction often included an instructional development process and the fact that instructional development models frequently included management as a task to be completed in the systems approach to instructional development further intertwined the systems concept with the process view of educational technology. These descriptive and prescriptive interpretations of the 1977 definition would influence future definitions.

As previously noted, the predilection that educational technology was a process was not new when the 1977 definition was written. *Process* was one of the three major supporting concepts incorporated into the rationale of the 1963 definition (Ely, 1963). Believing that educational technology was a process provided one of the major reasons that the leadership of the profession tended to reject the report of the Presidential Commission on Instructional Technology (1970), which focused heavily on the hardware of the field in its first definition of instructional technology.

The authors of the 1977 definition, who purposefully used the term *process* to develop a systematic and congruent scheme for the concept of educational technology, said,

> The definition presented here defines the theory, the field, and profession as congruent. This occurs because the definition of the field of educational technology is directly derived from, and includes, the theory of educational technology, and the profession of educational technology is directly

derived from, and includes, the field of educational technology. (AECT, 1977, p. 135)

In the end, the effort to demonstrate the congruence of the major concepts involved with educational technology created as many issues for the field as it resolved. Five immediate advantages for describing educational technology as a process were (1) the use of the term *process* reinforced the primacy of the process view of educational technology over the product view of educational technology. The process view had been outlined in the 1963 definition statement, but the report of the Presidential Commission on Instructional Technology (1970) appeared to reverse this emphasis. (2) The term *process* would ground the definition of educational technology in the activities of its practitioners, activities that could be directly observed and verified. (3) The term *process* could be used to describe educational technology as a theory, a field, or a profession. (4) The term *process* allowed the further evolution of thought and research around the concept of systems. Finally, (5) an organized process implies the use of research and theory, which would reinforce the idea that educational technology was a profession.

Educational Technology as Field, Theory, or Profession

The authors of the 1977 definition argued that educational technology could be thought of "in three different ways—as a theoretical construct, as a field, and as a profession" (AECT, 1977, p. 17). They continued, "None of the foregoing perspectives is more correct or better than the others. Each is a different way of thinking about the same thing" (p. 18). The writers of the 1977 definition argued that the theoretical construct, the field, and the profession were all process based. The term *process* described and connected all three of these perspectives of educational technology with a single word.

Educational technology had been called a theory in the 1963 definition (Ely, 1963), and it had been called a field in the 1972 definition (Ely, 1972). New to the 1977 definition was the argument that educational technology was also a profession. Prior to the publication of the 1977 definition, the term *profession* was used in passing as it related to educational technology. Since Finn (1953) had argued that the field had not yet reached professional status, members of the field (e.g., Silber, 1970) had made few attempts to analyze educational technology systematically as a profession. Using Finn's criteria, the writers of the 1977 definition argued that educational technology was now a profession.

Depending upon the interpretation and application of the systems concept, educational technology could be explained as a theory, a field, or a profession in the 1977 definition. The impact of using the term *process* to describe educational technology as a theory, a field, or a profession hinged on these differing interpretations of the systems approach, once again prompting discussions and philosophical debates among prominent educational technologists. The period of the 1980s was not so focused on criticism of the 1977 definition as much as characterized by broad academic wrangling over the interpretation and application of the definition (Januszewski, 1995, 2001).

The three major supporting concepts of the 1977 definition—learning resources, management, and development—could also be interpreted differently based on divergent conceptions of the systems approach. The different interpretations of learning resources, management, and development also provided the writers of the 1977 definition with a rationale to distinguish between educational technology and instructional technology.

The 1994 Definition

By 1994, the definition of educational technology had nearly come full circle. The definition that was produced in 1994 read, "Instructional technology is the theory and practice of design, development, utilization, management, and evaluation of processes and resources for learning" (Seels & Richey, 1994, p. 1).

There are no new concepts included in the 1994 definition. What was new was the identification of multiple theoretical and conceptual issues in the explanation of the definition. The 1994 definition was intended to be much less complex than the 1977 definition. The extent to which the writers were successful can be judged in part by reviewing the criticisms of the 1977 definition.

The attempt by the writers of the 1977 definition to show the congruence of educational technology and instructional technology revealed a conceptual problem for the field. The definition of educational technology, which was concerned with "all aspects of human learning" (AECT, 1977, p. 1), had become so broad that some individuals in the field of education pointed out that there was no difference between educational technology and curriculum, school administration, or teaching methods (Ely, 1982). Saettler (1990) wryly pointed out that the definition had become everything to everybody, and he dubbed the 1977 definition the "omnibus definition."

Logical Problems

There were also serious flaws in the reasoning and the conceptual interpretations used in the theoretical framework and rationale for the 1977 definition of educational technology. Establishing the difference between education and instruction, the authors argued, "Education, then, includes two classes of processes not included in instruction: those processes related to the administration of instruction . . . and those processes related to situations in which learning occurs when it is not deliberately managed" (AECT, 1977, p. 56). An example of learning not deliberately managed given in the discussion was "incidental learning" (p. 56). It was reasonable for the authors to argue that nondeliberately managed learning and/or incidental learning was part of the concept of education (Januszewski, 1997).

However, the definitions of "technology" by Galbraith (1967), Hoban (1962), and Finn (1960a, 1965), which were used by the authors of the 1977 definition to discuss the term *technology* as it related to the concept of educational technology, all included the ideas of organization, management, and control (AECT, 1977). The writers of the 1977 definition considered organization, management, and control critical characteristics of technology; but these ideas were contrary to the idea of "incidental learning" and "learning that was not deliberately managed." Education, at least as it was distinguished from instruction included in the rationale of the 1977 definition, did not seem compatible with technology. It is difficult to conceive of a technology of the incidental, unmanaged, and unintended. The gains made in the organization of the framework of the concept of educational technology by distinguishing between education and instruction were lost when education was paired with technology (Januszewski, Butler, & Yeaman, 1996).

Theory or theoretical construct. The relationship of educational technology to "theory" presented another problem in the discussion of educational technology presented in the 1977 definition and rationale. There are three ways in which the concept of theory is related to educational technology in the 1977 definition statement: (1) the thought that educational technology was a "theoretical construct" (AECT, 1977, pp. 18, 20, 24); (2) the notion that educational technology itself was "a theory" (AECT, 1977, pp. 2, 135, 138); and (3) that the "definition of educational technology was a theory" (AECT, 1977, pp. 4, 20, 134). To some degree, all three of these discussions of theory and educational technology are accurate, but they cannot be used interchangeably as they are in the 1977 definition. A theoretical construct is not the same as a theory; nor is it the case, that because a definition of a concept is a theory, the concept itself a theory.

The word *theory* has been used in at least four ways in the literature of the field of education: (1) the "law like" theory of the hard sciences; (2) theories that are supported by statistical evidence; (3) theories that identify variables that influence the field of study; and (4) theory as a systematic analysis of a set of related concepts (Kliebard, 1977).

The fourth sense of theory is of interest to this analysis of the 1977 definition of educational technology. Systematic analyses of any abstract concept can be said to be theories of that concept. Referring to educational technology as a theoretical construct, or a theory, or calling the definition of educational technology a theory may be accurate if the construct or theory includes a systematic analysis of the concept of educational technology.

The writers of the 1977 definition provided criteria for "theory" that was not theory as a systematic analysis of related concepts. The 1977 view of theory was an attempt to establish general principles and predict outcomes (AECT, 1977). This approach was substantially different from the usage of the word *theory* in the 1963 definition statement. Further confusion arises because of the writers' claim that educational technology did indeed meet the criteria for being a predictive theory (Januszewski, 1995, 2001).

Certainly "educational technology" is a theoretical construct. "Educational technology" may also be considered a theory depending on what exactly is intended by the word *theory*. The 1977 definition of educational technology is a theory about the abstract concept of "educational technology." But because the definition of the concept of educational technology may be a theory of educational technology, it does not necessarily follow that the concept of educational technology is itself a theory. This is similar to saying that a definition of the concept of democracy may be a theory of democracy but that the concept of democracy itself is not a theory.

Few involved in the field of educational technology adopted this systematic treatment of the concepts provided in the 1977 definition. Many in the field adopted only portions of the definition (e.g., Gustafson, 1981). Certain parts of the definition and the supporting statements were cited by scholars in order to make erudite points about the field of educational technology (e.g., Romiszowski, 1981), but a reading of the literature of the field during this era reveals that the whole of the conceptual framework provided in the 1977 definition, specifically the part intended to distinguish educational technology from instructional technology, was not widely accepted by the professionals in the field of educational technology (Seels & Richey, 1994). This lack of acceptance led to the label changes in the 1994 definition.

Distinguishing between educational and instructional.   The effort to revise the 1977 definition addressed some of the conceptual incongruencies of previous definitions. The first of these was the difference between educational and instructional technology. Unlike the writers of the 1977 definition, who sought to distinguish between educational technology and instructional technology, the authors of the 1994 definition acknowledged that this problem had no easy answer. They admitted, "At present the terms 'Educational Technology' and 'Instructional Technology' are used interchangeably by most professionals in the field" (p. 5). But they argued,

> Because the term 'Instructional Technology' (a) is more commonly used today in the United States, (b) encompasses many practice settings, (c) describes more precisely the function of technology in education, and (d) allows for an emphasis on both instruction and learning in the same definitional sentence, the term 'Instructional Technology' is used in the 1994 definition, but the two terms are considered synonymous. (Seels & Richey, 1994, p. 5)

With that, the official label of the field was changed from "educational technology" to "instructional technology," although it was quite acceptable to continue to use the term *educational technology.*

Underlying Assumptions

Seels and Richey (1994) did differentiate the 1994 definition from previous definitions by identifying and analyzing some of the assumptions that underlie this definition. Identified assumptions included

- Instructional technology has evolved from a movement to a field and profession. Since a profession is concerned with a knowledge base, the 1994 definition must identify and emphasize instructional technology as a field of study as well as practice (p. 2).
- A revised definition of the field should encompass those areas of concern to practitioners and scholars. These areas are the domains of the field (p. 2).
- Both process and product are of vital importance to the field and need to be reflected in the definition (p. 2).
- Subtleties not clearly understood or recognized by the typical Instructional Technology professional should be removed from the definition and its more extended explanation (p. 3).
- It is assumed that both research and practice in the field are carried out in conformity with ethical norms of the profession (p. 3).
- Instructional technology is characterized by effectiveness and efficiency (p. 3).
- The concept of systematic is implicit in the 1994 definition because the domains are equivalent to the systematic process for developing instruction (p. 8).

The inclusion of these assumptions in the analysis and explanation accompanying the 1994 definition allowed for the publication of a definition that was much more "economical" than were previous definition efforts.

Theory and Practice

The authors of the 1994 definition stated that the definition was composed of four components: (a) theory and practice; (b) design, development, utilization, management and evaluation; (c) processes and resources; and (d) learning. These components were not necessarily new; but in this definition, they were reorganized, simplified, and connected, in a way making the 1994 definition unique.

The 1994 definition used the phrasing included in the 1963 definition when it called instructional technology "the theory and practice of." And the authors argued, "A profession must have a knowledge base that supports practice" (Seels & Richey, 1994, p. 9). The authors used a simple but rather clear notion that "theory consists of the concepts, constructs, principles, and propositions that contribute to the body of knowledge" and that "practice is the application of the knowledge" (p. 11). In so doing, the authors cleared up the problem of the meaning of theory that they had inherited from the writers of the 1977 definition, a definition of theory that had been too precise.

Domains

The concepts (or "domains" of the 1994 definition) of design, development, utilization, management, and evaluation comprise the accepted knowledge base of the field today as evidenced by the *Standards for the Accreditation of School Media Specialist and Educational Technology Specialist Programs* (AECT, 2000). When these concepts are taken together and conducted in sequential order, they are the same as the stages of "development" described in the 1977 definition. These concepts are directly traceable to the idea of educational engineering developed by W. W. Charters (1945). It is important to realize that the authors of the 1994 definition did not intend that practitioners of educational technology perform all of these tasks in the sequential order. Specializing in or focusing on one of these tasks would include broad practitioners in the field (Seels & Richey, 1994).

Seels and Richey (1994) provided definitions of processes and resources: "A process is a series of operations or activities directed towards a particular end" (p. 12). "Resources are sources of support for learning, including support systems and instructional materials and environments" (p. 12). These descriptions allowed the authors to (a) use process to reinforce notions of engineering and science in instruction; (b) maintain the distinction between resources as things and processes; and (c) be consistent with terminology used in all three previous definitions.

The concept of learning was not new to the 1994 definition; however, the definition of learning intended by the authors was new. In previous definitions, the term *learning* was intended to connote a change in behavior such as advocated by Tyler (1950). But the authors of the 1994 definition wanted to move away from a strong behaviorist orientation. They argued, "In this definition learning refers to the 'relatively permanent change in a person's knowledge or behavior due to experience'" (Mayer, 1982, as cited in Seels & Richey, 1994, p. 12). Including the phrase "due to experience" also aided in moving away from causal connections and allowed for incidental learning.

This interpretation signaled the acceptance of a different kind of science in education: one less grounded on prediction and control and more interested in applying other theoretical and research principles to the instructional process.

Criticism of the 1994 Definition

The primary criticism of the 1994 definition is that instructional technology appeared to look too much like the systems approach to instructional development while changes in the practice of the field (e.g., constructivist-based initiatives and the general acceptance of computer innovations in classroom methodologies) made the 1994 definition too restrictive for mainstream teachers and school administrators as well as researchers and scholars. These criticisms and further evolution of the research and practice in the field led to a need for reconsideration and revision of this definition after more than a decade of use.

The Current Definition

The task force empanelled by AECT to review the 1994 definition wrestled with the historical issues presented here and with other issues of perception, changing employment and training expectations, semantics, and a strong desire to develop a definition that both served to include the broad variety of practitioners in this field and one which would prompt renewed attention to the theory and research so critical to our continued contributions to learning.

In a sense, we are not so far removed in this century from the professional goal stated in the 1963 definition:

It is the responsibility of educational leaders to respond intelligently to technological change . . . If the DAVI membership is to support the leadership in such bold steps, definition and terminology as a basis for direction of professional growth is a prime prerequisite . . . Now that the field of audiovisual communications, the largest single segment of the growing technology of instruction, has reached the point of decision making, we find ourselves in the same quandary other fields have discovered when they have attempted to define their fields: i.e., definition exists at various levels of understanding but no *one* definition can be *the* definition. (Ely, 1963, pp. 16–18)

And so, the latest in the line of definitions of educational technology: "Educational technology is the study and ethical practice of facilitating learning and improving performance by creating, using, and managing appropriate technological processes and resources."

References

Association for Educational Communications and Technology. (1972). The field of educational technology: A statement of definition. *Audiovisual Instruction, 17,* 36–43.

Association for Educational Communications and Technology. (1977). *The definition of educational technology.* Washington, DC: Author.

Association for Educational Communications and Technology. (2000). *Standards for the accreditation of school media specialist and educational technology specialist programs.* Bloomington, IN: Author.

Berlo, D. (1960). *The process of communication.* New York: Holt, Rinehart and Winston.

Charters, W. W. (1945). Is there a field of educational engineering? *Educational Research Bulletin, 24*(2), 29–37, 53.

Commission on Instructional Technology. (1970). *To improve learning: A report to the President and the Congress of the United States.* Washington, DC: U.S. Government Printing Office.

Ely, D. P. (1963). The changing role of the audiovisual process: A definition and glossary of related terms. *Audiovisual Communication Review, 11*(1), Supplement 6.

Ely, D. P. (1972). The field of educational technology: A statement of definition. *Audiovisual Instruction, 17,* 36–43.

Ely, D. P. (1973). Defining the field of educational technology. *Audiovisual Instruction, 18*(3), 52–53.

Ely, D. P. (1982). The definition of educational technology: An emerging stability. *Educational Considerations, 10*(2), 24.

Ely, D. P. (1994). *Personal conversations.* Syracuse, NY: Syracuse University.

Ely, D. P., Funderburk, E., Briggs, L., Engler, D., Dietrich, J., Davis, R., et al. (1970). Comments on the report of the Commission on Instructional Technology. *Audiovisual Communications Review, 18*(3), 306–326.

Finn, J. D. (1953). Professionalizing the audiovisual field. *Audiovisual Communications Review, 1*(1), 617.

Finn, J. D. (1957). Automation and education: General aspects. *Audiovisual Communications Review, 5*(1), 343–360.

Finn, J. D. (1960a). Automation and education: A new theory for instructional technology. *Audiovisual Communications Review, 8*(1), 526.

Finn, J. D. (1960b). Teaching machines: Auto instructional devices for the teacher. *NEA Journal, 49*(8), 41–44.

Finn, J. D. (1965). Instructional technology. *Audiovisual Instruction, 10*(3), 192–194.

Finn, J. D. (1967, August). *Dialog in search of relevance.* Paper presented at the Audiovisual Communication Leadership Conference, Lake Okoboji, Iowa.

Galbraith, J. K. (1967). *The new industrial state.* Boston: Houghton Mifflin.

Gerbner, G. (1956). Toward a general model of communication. *Audiovisual Communications Review, 4,* 171–199.

Gustafson, K. (1981). *Survey of instructional development models.* Syracuse, NY: ERIC Clearinghouse on Information Resources. (ERIC Document Reproduction Service No. ED 211 097)

Heinich, R. (1970). *Technology and the management of instruction.* Washington, DC: Association for Educational Communications and Technology.

Hoban, C. F. (1962, March). *Implications of theory for research and implementation in the new media.* Paper presented at the Conference on Theory for the New Media in Education, Michigan State University, Lansing, Michigan.

Hoban, C. F. (1968). Man, ritual, the establishment and instructional technology. *Educational Technology, 10*(5), 11.

Januszewski, A. (1995). *The definition of educational technology: An intellectual and historical account.* Ann Arbor, MI: Microfilms International.

Januszewski, A. (1997, February). *Considerations for intellectual history in instructional design and technology.* Paper presented at the Annual Meeting of the Association for Educational Communications and Technology, Albuquerque, New Mexico.

Januszewski, A. (2001). *Educational technology: The development of a concept.* Libraries Unlimited: Englewood, CO.

Januszewski, A., Butler, R., & Yeaman, A. (1996, October). *Writing histories of visual literacy and educational technology.* Paper presented at the Annual Meeting of the International Visual Literacy Association, Cheyenne, Wyoming.

Kliebard, H. M. (1977). Curriculum theory: Give me a "for instance." *Curriculum Inquiry, 6*(4), 257–269.

Knowlton, J. Q. (1964). A conceptual scheme for the audiovisual field. *Bulletin of the School of Education, Indiana University, 40*(3).

Myers, D. C., & Cochran, L. M. (1973). Statement of definition: A response. *Audiovisual Instruction, 18*(5), 11–13.

Romiszowski, A. J. (1981). *Designing instructional systems.* London: Kogan Page.

Saettler, P. (1990). *The evolution of American educational technology.* Englewood, CO: Libraries Unlimited, Inc.

Seels, B., & Richey, R. (1994). *Instructional technology: The definition and domains of the field.* Washington, DC: AECT Press.

Silber, K. (1970). What field are we in, anyhow? *Audiovisual Instruction, 15*(5), 21–24.

Silber, K. (1972). The learning system. *Audiovisual Instruction, 17*(7), 10–27.

Twelker, P. A., Urbach, F. D., & Buck, J. E. (1972). *The systematic development of instruction: An overview and basic guide to the literature.* Stanford, CA: Stanford University. (ERIC Document Reproduction Service No. ED 059 629)

Tyler, R. (1950). *Basic principles of curriculum and instruction.* Chicago: University of Chicago Press.

Wallington, C. J. (1977). Preface. *The definition of educational technology.* Washington, DC: Association for Educational Communications and Technology.

PROFESSIONAL ETHICS AND EDUCATIONAL TECHNOLOGY

Andrew R. J. Yeaman
Detroit, Michigan

J. Nicholls Eastmond, Jr.
Utah State University

Vicki S. Napper
Weber State University

Introduction

Educational technology is the study and *ethical practice* of facilitating learning and improving performance by creating, using, and managing appropriate technological processes and resources.

*T*HE EPIGRAPHIC SENTENCE BY the Definition and Terminology Committee shows the successful promotion of professional ethics to the point where it deserves a full chapter (Januszewski, 2006). Previously, professional ethics was a topic only briefly mentioned. The earlier definitional works recognized it by each assigning two pages to reprint the AECT *Code of Ethics*. The AECT (1977) Task Force on Definition and Terminology welcomed the newly created code (pp. 116, 118–119). Seels and Richey (1994) recognized the codification of the profession's ethical standards, especially in light of ongoing and current concerns (pp. 106–107, 152–153). Inclusion in *this* book is an achievement based on the educational efforts of the AECT Professional Ethics Committee. Credit is due to the AECT leaders who have chaired the Committee and those members who have served on it productively.

Anything has potential for being assessed along ethical dimensions and that consideration applies to professional activities in educational technology. This chapter looks at the professional ethics of technologists first sociologically and historically, then by examining the current situation, and finally in predicting areas of progress and growth. The conclusion emphasizes what may be most valuable for readers: that the development of professional ethics offers a rather different way of defining educational technology.

Professional ethics should not be confused with the branch of philosophy known as *ethics*. Nor should "professionally unethical" be accepted as a euphemism for conduct which is illegal, immoral, sinful, or a violation of workplace rules. Similarly, judging a colleague as professionally ethical or unethical is not the same as judging a colleague as polite or rude, agreeable or disagreeable, or professional or unprofessional. Those things mean nothing more than how the person measures up against an unsaid, invisible criterion. It is recommended that, when speaking or writing about professional ethics, the full phrase is employed in order to avoid misunderstanding. It is also recommended that reference be made to upholding specific principles given in the AECT *Code*.

Professional Ethics: A Sociohistorical Perspective

Professional ethics was staked out for sociology as an intellectual territory by Durkheim (1957/1992) in three lectures on professional ethics that were first delivered in the 1890s (pp. xxx, xxxvii). Durkheim wanted to identify how society's rules of conduct were set up and understand what was achieved through them (p. 1). Freidson (2001) contended "that the resources for morality and ethics lie within occupations and are not available in other modern groupings" (p. 53). Although idealistic hopes for social change tainted those first analyses, and they were misunderstood, the methodological contributions remain foundational to social study (Freidson, 2001, pp. 52–54).

Today, it is understood that professions gain, maintain, and lose power through competition (Abbott, 1988, 1998, 2001) and by alliances with social institutions (Freidson, 1986, 2001). Having a code of professional ethics formalizes occupational territory aside from the requirements of government, law, institutional regulations, religion, and so on. What is "proper" reflects all of these authorities because professions work with them to hold social things in place. Professional ethics for educational technology are like any other profession's ethics because they are cultural standards. Professional ethics are politically negotiated and maintained as traditions.

Things, as they are thought of conventionally, are rarely based on understanding causal processes. In particular, social things tend not to be isomorphic with how they are regarded by conventional wisdom, which is extremely subjective. When Becker (2006) investigated judgments about quality and conventions for determining goodness or badness, similarities were detected in sociology, art, music, science, and engineering. Not surprisingly, the purpose of their "common sense" is not to aid the investigations of social scientists but to support the established order.

That is to say, "All social groups make rules and attempt, at some time and under some circumstances, to enforce them" (Becker, 1963/1997, p. 1). What is at issue is deviance, usually justified via a medical analogy as behavior that is regarded as sick. Complete agreement between everyone on what is "socially sick" is unusual and it is more enlightening to accept that *social groups create deviance by making the rules whose infraction constitutes deviance*, and by applying those rules to particular people and labeling them as outsiders" (Becker, 1963/1997, p. 9). Much as jazz musicians must endure the paying audience whom they disdain as squares (Becker, 1963/1997, pp. 85–91), the activities of education professors who regard themselves as researchers are justified by the existence of those whom they call practitioners. By extension, within the broad culture of education and training, "true" educational technologists have likewise been tolerated, for the most part, as members of a deviant subculture, albeit relatively harmless. This is the way things are, even more now than ever before, since with mass computerization almost everyone is—to at least some degree—a technologist. The dividing line is moving over and readers may have heard words like these: "I started teaching one of my classes online this term so now I know all about educational technology." A documented example is the contemporary newsletter headline announcing a conference program: "Jazz up your teaching with technology" (Schuetz, 2006).

Investigating a Professional Mystery

Statements on professional ethics should be reconsidered where there is a leaning toward syllogistic fallacies. Consider how the professional ethics discourse is widely accepted as a social convention. It is expected to be working in society but without much verification, either rationally or empirically: An example is the supposition that professions without codes of ethics are doomed to lose their niches (Gardner, Csiksentmihalyi, & Damon, 2001, pp. 22–24). This is a superficial belief. It gives too much weight to mind determining social function. Each profession selects and negotiates its ethical difficulties with the same code of ethics that polices the profession. In other words, codes of ethics establish the very thing they aim to prevent: Ethical violation is expected because it is named, and then blame can be assigned to individuals or groups. It is not the actuality but the possibility that is potent.

Professional ethics do not directly control and cannot force good behavior. Inclination toward being good is likely to be distributed throughout any population, much like intelligence. Belonging to an organization having an approved and enforceable code serves as a sign of holding professional status. Certainly, for educational technologists, this sign contributes to setting standards related to education, credentialing, and, at least on the face of things, in keeping out charlatans, impostors, and rival professional groups. Only a small percentage of technologists belong to the AECT, but it seems to be the leading established professional society in the technology field with an enforceable code and continual, serious efforts to educate its members about professional ethics.

Archaeology

Professional ethics are a mysterious aspect of educational technology. Many factors have influenced and will continue to influence what is considered professionally ethical for technology (see Table 11.1). It is based on Foucault's (1970/1972) social archaeology (pp. 21–76), which is not such a strange thing, as it may seem. An introductory reading of Foucault is part of the curriculum for many undergraduates, and the American Educational Research Association has a Foucault special interest group. On an episode of *The West Wing,* the cover of a Foucault (1997/2003) book is meaningfully shown twice in a montage (Noah & McCormick, 2006). Foucault rejects and goes beyond Sartre's existentialism (Paras, 2006). In doing so, Foucault wrote his archaeological method, which has subsequently inspired many writers. For instance, Rosemann (1999) found "a kind of manual of archaeological research, providing a comprehensive list of all the factors that need to be taken into consideration in the analysis of an episteme" (p. 40). Kendall and Wickham's (1999) textbook explains that, in opposition to reductive, total history, Foucault's historiography has the continuity of general history (p. 24). On putting this approach into action: "Archaeology helps us to explore the networks of what is said, and what can be seen in a set of social arrangements: in the conduct of an archaeology, one finds out

Table 11.1. An archaeology of discursive factors affecting professional ethics for educational technology. © A.R.J. Yeaman, 2007. Used with permission.

Public awareness and mass media influence	Traditions and customary expectations	Professional knowledge defining good practice
Neighborhood morality	Endless inventions, innovations, and new products, as well as relentless uncertainty	Academic base for preparing licensed professionals and continuing their education
Local laws	Economic forces	Utopian ideals
Regional laws	Technological practicality and probability versus possibility and desirability	Institutional philosophies, policies, and procedures
National laws	Rival professions	
Religious beliefs		Philosophical ethics
	Political interests and government	Individual interests and personal ethics

something about the visible in 'opening up' statements and something about the statement in 'opening up visibilities'" (Kendall & Wickham, 1999, p. 25). Archaeological "tasks" are described, such as "chart the relation between the sayable and the visible" (p. 26), and these are demonstrated through an archaeological analysis of schooling (pp. 27–28).

This is the origin of Table 11.1. Some things appear solid and lasting, such as laws, traditions, institutional rules, and professional licensing. Other factors in the map may be more transitory, such as economic pressures, government support, and the cycle of successful inventions becoming successful products and then falling into obsolescence as new inventions arrive. Religious convictions contribute as philosophy does. Do not expect consistency regarding results (consequentialism) or rights (deontology) because professional ethics is sociological.

Alterations in the factors named in Table 11.1 take place over time. At any moment, the people involved are barely conscious of the ongoing social constructions. It is as if demands for conformity and compliance are nothing more than perennial flowers. Technologists not only accept these requirements but also employ them for professional ends and come to expect them in their season. The purpose of these controlling devices may be no more than to assert control.

A Merton View

In structural functional sociology, the normative control of people is maintained by threats that those who do not conform to social norms will be labeled as "deviant" and, in the frame of this chapter, "unethical." This comes from Merton (1973), best known for identifying the Puritan norms of science. Social norms have several dimensions (Merton, 1982, p. 75). The spectrum runs from prescribing behavior to proscribing and from what is preferred to what is permitted. Consensus over norms varies. Norms receive unequal amounts of support. Control structures range from formal sanctions to informal reactions. Certain norms demand obvious compliance whereas others do not. Similarly, some norms are more or less flexible in this regard. The individual is subject to the norms imposed by the collective. From among the great mass of individuals, token deviants are labeled and stigmatized. They may be individually deviant or members of an out-group that is considered deviant by an in-group. The quantity of "bad people" thus selected, punished, and processed corresponds to public perceptions of there being a "real" social problem rather than an objective measure. Merton found it interesting that not all unacceptable or unethical actions are detected, and if they are detected, some do not undergo investigation.

Professional associations play a part in this control mechanism not only for their members but also in avoiding "the atomization of society into a sand heap of individuals intent on pursuing their own private interests" (Merton, 1982, p. 206). Professional affiliation is one of the ways of making connections between people possible. It enables the freer, less oppressive forms of society to flourish:

> The professional association is one of those intermediate organizations that furnish the social bonds through which society coheres. It provides unity in action and social cohesion without contiguity of its members. It mediates between practitioner and profession, on the one hand, and, on the other, between practitioner and social environment, of which the most important elements, apart from clientele, are allied occupations and professions, the universities, the local community, and the government. (Merton, 1982, p. 207)

A sobering perspective on those high thoughts is desirable: "Professional associations tend more to provide services to their members than to exercise control over their ethical or technical work behavior" (Freidson, 1986, p. 187). Professionalism itself comes from the professionals who gather, see themselves as being in a profession, and are seen that way by others. Professional ethics express professionalism by being normative and are definable through observable beliefs and behaviors.

Correspondingly, there are already strong external pressures on professional educational technologists. The cultural mapping in Table 11.1 shows the inhibiting presence of norming factors. Without being frozen in time, the components on display are likely to constrain educational technologists, to contribute to their professional ethics, and to define their professional ethics. The chart is illustrative, and these social things are not ordered by either hierarchy or structure, something that would be rather un-Mertonian (Gieryn, 2004, p. 93). The cell divisions are figurative, just as in everyday life where there is no risk of tripping over lines of latitude and longitude. Blanks have been left deliberately because it is probable that not all controlling mechanisms could ever be known and understood. Further inquiries are possible and readers can fill in the blanks for themselves. Obvious candidates may be the allied ideas of myth, discourse, and common sense (Mosco, 2004, pp. 12–13, 28–29).

The Pragmatist Philosophical Rationale

> One of the more lasting clichés of American culture is that of the redemptive benefits of education. (Stanley, 1978, p. 188)

While Stanley goes on to seek a "nontechnicist" philosophy of education—a discussion beyond the ambit of this chapter, the lasting cliché connects with John Dewey, the great American philosopher. Over 100 years have passed since Dewey's *Ethical Principles Underlying Education* (1897/1903) appeared. Dewey, the pragmatist, tells readers that ethics, inside and outside of school, are not different because "The moral responsibility of the school, and those who conduct it, is to society" (p. 10). There is a useful convergence here between education as a philosophical ideal and as a social practice requiring teachers to meet professional standards and function with professional autonomy.

As a psychologically inclined philosopher, Dewey emphasizes learners as individuals and emphasizes a need for building character toward social intelligence. Dewey's conclusion is not particularly optimistic: "We need to translate the moral into the actual conditions and working forces of our community life, and into the impulses and habits which make up the doing of the individual" (1897/1903, pp. 32–33).

Dewey's synthesis provided a desirable direction for schooling in the new century and supported progressivism as the ethical approach to mass education. It was not easy to disagree with the pragmatics of pulling these values together: traditional and progressive, academic and vocational, and religious and secular. During a career crossing six decades, Dewey sponsored obligatory consensus and professional unity. Although never Dewey's only message, it was always present between the lines: *Education is both a good thing and a necessity.*

Dewey's reputation as a 20th-century philosopher was built in part on this bridging between radical reform and established customs. Dewey's philosophical ethics ensconced education *comme il faut* (as it should be) on a foundational, taken for granted, ready-made Puritan ethic. This support of education as an ethical endeavor helped advance teaching as a profession, a role philosophers of education have played (argumentatively) ever since. A recent essay following in that tradition (Warnick & Waddington, 2004) draws on Martin Heidegger and Albert Borgmann to show that educational technology, when enacted *properly*, is philosophically ethical. Of course, any professional labor is ethical because that attribute is considered inherent to any occupation deemed worthy of professional status.

Historical Beginnings

The earlier sections of this chapter are prerequisite to comprehending the historical situation. The sociological explorations illustrate what was going on beneath the surface and what is always going on. The philosophical heritage is relevant, too. Whereas the theoretical insights of sociologists are largely neglected, philosophers are routinely called upon to justify the institution of education, especially the works of Dewey.

Walsh's (1926) book *Teaching as a Profession: Its Ethical Standards* provided a history of ethical codes in U.S. education. The first state education association to adopt a code of ethics was Georgia in 1896, followed by California in 1902, Alabama in 1908, and Arkansas in 1910 (pp. 376–377). By 1925, 15 more states had a code of professional ethics for teachers. Walsh commented, "These are usually of little influence," and "There are no universally recognized standards in most fields of teaching activity" (p. 6).

The claim of variance in what the public regards as appropriate educational conduct is illustrated by the story of Miss M.'s frustrating interviews for a high school teaching position (Walsh, 1926, pp. 7–8). The first school would not hire her because Miss M. said she danced and said she enjoyed it. The board informed her that was not thought proper. The second school would not hire Miss M. because "she replied she had danced but did not care for it and would gladly refrain from dancing." The superintendent "wanted someone who danced well" to chaperone student dances and "take part in the social life of the community" where dancing was considered popular. Walsh argued, "The time has come when many phases of teacher relationships must be standardized; when definite and generally accepted ethical principles must be established, if a profession of teaching is to be developed, or even if it is to hold its place among the respected vocations."

The "requirements of a profession" are listed with the eighth and last criterion being "each profession embodies the foregoing or similar principles in a code of ethics which is formally accepted and rigidly enforced by the members of the profession" (Walsh, 1926, pp. 22–23). Principles are explored in chapters ending with problems for discussion such as this one about teachers as colleagues:

> Case 6. Miss T., the fourth grade teacher of long experience, looked over the new third grade instructor, a bobbed-haired, short-skirted, hand-tinted Miss, and exclaimed in disgust to her friend: "Superintendent J. will have to give me another room. I can tell him right now that I shall not stay here next year and attempt to teach children whom that snip has had for a year." (p. 247)

Building on Walsh's (1926) book, Landis' (1927) dissertation examined the professional ethics of education and 10 other professions. It was found that *ethics codes developed around conflicts* with clients, employers, supervisors, and colleagues including competitors, material businesses, and service businesses (p. ix). Landis wrote, "Twenty-seven state education associations and some other groups of educators have adopted codes of ethics during the past thirty years" (p. 1).

A National Education Association (NEA) committee to develop a *national* code had begun work in 1924 (Rich, 1984, p. 128). Subsequently, the NEA adopted the first code of ethics for the education profession on July 1, 1929 (National Education Association of the United States, 1929, p. 69). This action was taken at the Representative Assembly meeting in Atlanta. Enforcement was through each state's teachers' organization having a professional ethics committee (p. 70).

There were three articles: Relations with Pupils and to the Community (with six sections), Relations to the Profession (with seven sections), and Relations to Members of the Profession (with eight sections). The justification for the NEA *Code* was

> that the aims of education may be realized more fully, that the welfare of the teaching profession may be promoted, that teachers may know what is considered proper procedure, and may bring to their professional relations high standards of conduct. (National Education Association of the United States, 1929, p. 69)

In the mid-1930s, an effort to produce a uniform national code of ethics for teachers did not gain the support of the American Federation of Teachers (AFT) (Davidson, 1936). There were several objections by the AFT (p. 34). The strongest was that a national code ignored other influences on the quality of schools. It also held the danger of being used against teachers in unforeseen ways.

The NEA *Code* was revised in 1941, 1944, 1952, 1963, 1968, and 1975 (Rich, 1984, p. 128). Immense efforts to increase acceptance and awareness were made in the 1950s and 1960s, such as distributing over half a million copies of the *Code* in 1964 (p. 129). The NEA informed its members of how the NEA code was being interpreted by publishing its collected *Opinions* (National Education Association of the United States, 1964). These accreted over time and the fourth edition records 44 opinions. Several editions came out, usually in runs of 5,000. Rich (1984) commented that coverage in teacher preparation programs continued to be neglected (p. 129). Books suitable for use in colleges and schools of education were published in recent years, such as Strike and Soltis' (1998) textbook and Wagner's (1996) fastback.

A Different Direction

The organization that became AECT started in 1923 as the Department of Visual Instruction (DVI) of the NEA (Januszewski, 2001). The Department became tied to the NEA *Code* approved six years later, but enforcement was not departmental at that time. References to the NEA *Code* have been sought but not located in either the DVI Board minutes and proceedings housed in the National Public Broadcasting archives at the University of Maryland or in the *Educational Screen*, which was the professional periodical of that era.

A discourse existed whereby the newest educational communication technology was accepted as beneficial. If motion picture projection could be picked up by educators of all types at all levels, it would be certain to bring improvements. The advance toward solving the latest education crisis was balanced by a fear of misuse (Painter, 1926; Yeaman, 2004b). New technology connected with supposedly new problems formed part of the base for professionalization. A related fear was that theatrical motion pictures would corrupt youth (Short, 1928).

The DVI was aligned with the Motion Picture Research Council, whose concerns for moral reform not only diverged from those of the NEA but also were secretly funded by the wealthy wife of a Republican senator. Massive and strenuous research began in the mid-1920s, under W. W. Charters' (1935) management, to scientifically assess the impact of theatrical films on the moral development of youth and to connect that with classroom instruction

(Jacobs, 1990). Mortimer Adler (1937) defended the Motion Picture Producers and Distributors Association (MPPDA) by asserting that, although parts of the research had been conducted brilliantly, much of the science was mediocre, and some was delirium. Edgar Dale started at Ohio State University in 1929 by working on Charters' project, and one of Dale's students from the 1930s and 1940s recalled that Adler's reputation was diminished by becoming thought of as "an industry flack" (R. W. Wagner, personal communication, May 18, 2003). Adler's criticisms were summarized in *Are We Movie Made?* (Moley, 1938), the publication of which was sponsored by the MPPDA (Jowett, Jarvie, & Fuller, 1996, p. 117). Becker (2002) gave a contemporary assessment with implications from a sociological point of view in regard to investigating new media.

The DVI became the Department of Audio-Visual Instruction (DAVI) at the Atlantic City conference held in March 1947. In the new constitution documents, there was no mention whatsoever of professional ethics (Dameron, 1947). Anna Hyer (1969), DAVI's long time executive secretary, recollected, "DAVI had less than 500 members in 1951; perhaps 300 persons attended our first independent convention held in Boston in 1952" (p. 108). Within a few years, the membership increased tenfold.

Growth

It was apparent considerably earlier that DVI was progressing away from NEA. In the early 1930s, Howard McClusky (1934) noted that anthropologists had characterized peoples in relation to their tools and, with a suspiciously mystical sense of certainty, wrote of the coming technology of education:

> It is not the province of this paper to venture a prediction of the exact date which will mark the advent of what might be termed the new technology of education. But the writer opines that just as the railroad, steamboat and automobile have transformed the mode of life in the last century, and just as the extension of electric power and the development of aviation will reshape the life of the coming century, just as certainly will the radio, motion picture, television, sound recording (musical and nonmusical), cheap printing, and similar devices transmogrify the cultural life of the masses and the technique of their instruction. The change is inevitable. Its arrival is only a matter of time and ingenuity. (p. 84)

It looks as if the first mention of professional ethics in the audiovisual literature came from Finn (1953). This article on the professionalization of the field has been selected as a classic reading (Finn, 1996). Landis (1927) is briefly cited because "many codes are window dressing" (Finn, 1953, p. 8).

While professional ethics are thought necessary for qualifying as a profession, no examples are given of either existing or anticipated ethical conflicts, issues, or duties.

By the mid-1960s, a Professional Standards Committee had been appointed with reporting subcommittees, sometimes also referred to as Commissions or as committees themselves. The Professional Standards Committee contained a subcommittee on professional ethics (P. W. Welliver, personal communication, January 20, 2005). That group grappled with applying the NEA *Code* to ethical problems specific to technology work, such as copyrights. It also addressed the status of members of the profession who are employed outside of NEA jurisdiction.

Brief descriptions of the activities of the Professional Ethics Committee began to appear in *Audiovisual Instruction*. These appear to be the earliest signs of attention being given to professional ethics as a specific topic of any organizational concern. Positions were taken on copyright compliance and the responsible use of advances in copying technology.

This news comes from the 1966 DAVI Convention in San Diego, California where a particular focus was the DAVI opinion on teacher strikes:

> The Commission on Professional Ethics continues to keep in close contact with the NEA's Committee on Professional Ethics in interpreting the NEA Code of ethics as it applies to the specific problems of the audiovisual educator. (Highlights of DAVI, 1966, p. 514)

John A. "Jack" Davis became Chair of the Professional Ethics Committee at the Atlantic City DAVI Convention (Schwartz & Davis, 1967), and next year reported the meeting of the Committee on Professional Ethics, a Professional Standards Subcommittee (Highlights of DAVI, 1968, p. 688). After reviewing "the disposition of cases considered during the year" the Professional Ethics Committee produced a Resolution on Professional Ethics for the DAVI Board of Directors. This was in regard to proposed changes in copyright law (1968 DAVI Resolutions, 1968, p. 679). The Committee on Professional Ethics planned to "study examples and specific cases of ethics problems encountered by the audiovisual profession to determine the desirability and feasibility of adopting or drafting a Code of Ethics for DAVI" (Davis, 1969).

Following the separation from NEA, the AECT Professional Ethics Committee came into existence as a governance committee in 1970. The new executive director wrote that it is easier to standardize hardware than human conduct, and he quoted from the report of the DAVI Professional Ethics Committee (Hitchens, 1970). The report recommended adoption of the NEA

Code because it had previously been accepted by the DAVI Board of Directors 18 months earlier. However, "We have yet to put teeth in the enforcement of that professional code—but that will evolve in due time."

The editorial's themes had been drawn from Finn (1953) and concluded by echoing that essay:

> Does our field, then, satisfy the criterion of having a standard and a code of ethics to which we can subscribe? I believe we do, even though the job is still unfinished and will perhaps ever remain so. However, each of us should direct his efforts toward the further definition of our professional standards and our professional code of ethics. This is perhaps more appropriate for our field than for any other sector of education, for we have a unique interest in change. (Hitchens, 1970)

A response was printed the next year. The jurisdiction of the NEA was limited and there ought to be a frame for ethical actions by professional educators who work for noneducation organizations. It will help them avoid being "caught in a professional and ethical abyss where normal professional channels for inquiry, employment security, professional rights, and protection against professional and personal abuse are closed" (Welliver, 1971, p. 43).

A public correspondence about professional ethics followed (P. W. Welliver, personal communication, February 27, 2005). The letters are by Davis (1972) and Welliver (1972).

AECT's Code of Ethics

Jack Davis remained the Chair until 1975 and then continued as a member. A new code of professional ethics was drawn up, partly based on the NEA Code, and approved in 1974 when Gerald M. Torkelson was president (J. A. Davis, personal communication, June 7, 2005). The bylaws continued to acknowledge NEA's *Code of Ethics of the Education Profession* for another 10 years (Association for Educational Communications and Technology, 1984, p. 12). The current version of the AECT *Code* was approved by the Board of Directors on November 6, 2001, and it is displayed in Table 11.2. It can also be viewed via a link on the AECT Web site (Association for Educational Communications and Technology, n.d.a.).

Although AECT had its own code of ethics by the mid-1970s, supported intellectual freedom, affirmative action, and "humane" technology, and opposed stereotyping, it did not "enforce its ethical and value positions, and professionals in educational technology do not show a widespread concern for the importance of these positions" (Silber, 1978, p. 179).

Table 11.2. AECT *Code of Professional Ethics.* This version was approved by the AECT Board of Directors on November 6, 2001. Used with permission of AECT.

Preamble

1. The Code of Ethics contained herein shall be considered to be principles of ethics. These principles are intended to aid members individually and collectively in maintaining a high level of professional conduct.
2. The Professional Ethics Committee will build documentation of opinion (interpretive briefs or ramifications of intent) relating to specific ethical statements enumerated herein.
3. Opinions may be generated in response to specific cases brought before the Professional Ethics Committee.
4. Amplification and/or clarification of the ethical principles may be generated by the Committee in response to a request submitted by a member.

Section 1—Commitment to the Individual

In fulfilling obligations to the individual, the members:

1. Shall encourage independent action in an individual's pursuit of learning and shall provide open access to knowledge regardless of delivery medium or varying points of view on the knowledge.
2. Shall protect the individual rights of access to materials of varying points of view.
3. Shall guarantee to each individual the opportunity to participate in any appropriate program.
4. Shall conduct professional business so as to protect the privacy and maintain the personal integrity of the individual.
5. Shall follow sound professional procedures for evaluation and selection of materials, equipment, and furniture/carts used to create educational work areas.
6. Shall make reasonable efforts to protect the individual from conditions harmful to health and safety, including harmful conditions caused by technology itself.
7. Shall promote current and sound professional practices in the appropriate use of technology in education.
8. Shall in the design and selection of any educational program or media seek to avoid content that reinforces or promotes gender, ethnic, racial, or religious stereotypes. Shall seek to encourage the development of programs and media that emphasize the diversity of our society as a multicultural community.
9. Shall refrain from any behavior that would be judged to be discriminatory, harassing, insensitive, or offensive and, thus, is in conflict with valuing and promoting each individual's integrity, rights, and opportunity within a diverse profession and society.

Section 2—Commitment to Society

In fulfilling obligations to society, the member:

1. Shall honestly represent the institution or organization with which that person is affiliated, and shall take adequate precautions to distinguish between personal and institutional or organizational views.

Table 11.2 (continued)

2. Shall represent accurately and truthfully the facts concerning educational matters in direct and indirect public expressions.
3. Shall not use institutional or Associational privileges for private gain.
4. Shall accept no gratuities, gifts, or favors that might impair or appear to impair professional judgment, or offer any favor, service, or thing of value to obtain special advantage.
5. Shall engage in fair and equitable practices with those rendering service to the profession.
6. Shall promote positive and minimize negative environmental impacts of educational technologies.

Section 3—Commitment to the Profession
In fulfilling obligations to the profession, the member:
 I. Shall accord just and equitable treatment to all members of the profession in terms of professional rights and responsibilities, including being actively committed to providing opportunities for culturally and intellectually diverse points of view in publications and conferences.
 2. Shall not use coercive means or promise special treatment in order to influence professional decisions or colleagues.
 3. Shall avoid commercial exploitation of that person's membership in the Association.
 4. Shall strive continually to improve professional knowledge and skill and to make available to patrons and colleagues the benefit of that person's professional attainments.
 5. Shall present honestly personal professional qualifications and the professional qualifications and evaluations of colleagues, including giving accurate credit to those whose work and ideas are associated with publishing in any form.
 6. Shall conduct professional business through proper channels.
 7. Shall delegate assigned tasks to qualified personnel. Qualified personnel are those who have appropriate training or credentials and/or who can demonstrate competency in performing the task.
 8. Shall inform users of the stipulations and interpretations of the copyright law and other laws affecting the profession and encourage compliance.
 9. Shall observe all laws relating to or affecting the profession; shall report, without hesitation, illegal or unethical conduct of fellow members of the profession to the AECT Professional Ethics Committee; shall participate in professional inquiry when requested by the Association.
10. Shall conduct research and practice using professionally accepted and Institutional Review Board guidelines and procedures, especially as they apply to protecting human participants and other animals from harm. Humans and other animals shall not be used in any procedure that is physically invasive to them.

During the presidency of Paul Welliver, Section 19 of the bylaws was changed so that, in place of the NEA *Code*, "Adherence to the AECT Code of Ethics shall be a condition of membership" (Association for Educational Communications and Technology, 1985, p. 10). That requirement of members is now in Section 16 of the bylaws. Last amended and ratified on August 2, 1999, the bylaws can be accessed online via the AECT Web site (Association for Educational Communications and Technology, n.d.b.).

Discipline.   Paul Welliver became chair of the Professional Ethics Committee in 1987. At the Committee's meeting during the Atlanta Convention that year, it was "agreed that some assessment of the current thinking of AECT members toward ethics and educational technology was needed since this sort of information was not available and since ethical issues around educational technology are likely to become more pronounced as technological capacity increases" (Nichols, Martin, & Welliver, 1988). A survey was constructed and administered at the May meeting of Professors of Instructional Design and Technology (PIDT). Important issues needing attention were elicited. Completed questionnaires were returned by 43 of the 80 attendees.

In analyzing these data, responses were merged into 11 categories: ownership as in copyright and plagiarism; issues in work with noneducation clientele, such as "needs assessments conflicting with agencies' expectations;" treatment of students; misrepresentation of technological capabilities: the differences between the "promise and the reality" of instructional technology and the "exaggeration of benefits—especially computers;" instructional development/design issues; sociocultural issues; academic/political issues; research issues; ethical issues *per se*; values transmission; and honesty in personal relationships. Respondents were generally favorable to professional ethics and wanted an increased emphasis on professional ethics although slightly more opposed monitoring practicing professionals than supported the idea. The report concluded with five questions and comments generated from the data. The first one was "What does only moderate (Is it *only* moderate?) interest of the professors about raising issues and monitoring imply about AECT membership as a whole?"

Educating the membership.   By the time Paul Welliver took on chairing the Committee, the *Code* may have become mostly decorative. It looked like a code, but the meanings of the principles in how they applied to working with technology were seldom obvious. For example, Section 1, Principle 8 prohibited sexual stereotypes. This phrase from the 1960s was replaced with gender stereotypes. However, there was no concrete information at hand on what those who drafted the *Code* intended.

Therefore, as chair, Welliver wisely directed his efforts at raising the consciousness of AECT members about their professional ethics. Consequently, the Committee was put to work. Each member was requested to write one scenario to show a principle helping illuminate an ethics problem. The immediate goal was to have each principle in the *Code* introduced and explained at least once. The Committee's opinions on each principle were published in *TechTrends* and then as a book (Welliver, 2001).

The thoroughness of the Association's illustrated casebook on professional ethics (Welliver, 2001) stands favorably alongside publications from much larger professions. Welliver provided succinct, plain English annotations of each principle based on what, in his informed opinion, was the core meaning of each principle (pp. 11–17).

The biggest issue for any profession in educating members about their profession's ethics is in getting away from textbook answers to textbook problems: The end goal is to prepare and support members for engagement with the imprecise difficulties of professional life. This is where AECT's *A Code of Professional Ethics* (Welliver, 2001) excels in being instructionally effective. The scenario problems appear on right-hand pages. The analysis is not immediately available because it follows overleaf. The pages depicting the ethical dilemmas can be photocopied and handed out to facilitate small group discussions. Here, the text format matches up well with an instructional design calling for *thinking* and for *dialog* before encountering the words of authority. This instructional tactic can work well in promoting the flexibility needed for making ethical assessments.

Reconceptualization

After some years, the AECT Professional Ethics Committee saw that its activities had produced noticeably good results and the Committee's own consciousness had been raised, too. Self-awareness reached the point where certain logical difficulties surrounding the Committee's business could be perceived. These were the unforeseen consequences of having professional ethics for technology.

At the 1992 meeting in Washington, DC, a draft code for information processing professionals had been brought to the Committee's attention and reconceptualization of the AECT *Code* was suggested. It was noted that codes of ethics could be thought of as developing perpetually, especially since they provide guidance not only for individuals but also for their professional societies and the relevant social institutions.

A subcommittee formed and Nick Eastmond, Dennis Fields, and Andrew Yeaman met the next year. Randy Nichols and Paul Welliver were unable to attend, but they contributed later through extensive phone conversations. Subsequently, the subcommittee reported,

> From a process of rereading and rethinking, a need has been perceived for affirming the Association's ethical commitments to our society and the world. . . . It is not so much necessary to threaten members with investigation as it is necessary to recognize our Association as a group of professionals who are active in the real world. There is a political problem in reducing the membership to individuals without collective power. The stereotype individual is disconnected, ineffectual, passive, and shallow. Our Code threatens to police our members as individuals but no one has ever been punished. Instead, there are ethical areas where our Association should be effective in showing social concern and in taking action.

The subcommittee identified, discussed, and listed new ethical topics for AECT. Here are 4 of the 21 items from the subcommittee report:

- We are in a business that is careless about environmental pollution. We need to negate environmental pollution, especially that caused by our technologically based profession. Pollutants include worn out videocassettes, used toner cartridges, discarded paper, and obsolete hardware.
- Physical and psychological aspects of learning environments are not only less than ideal for effective instruction but they are detrimental to learners' well being. Our field frequently overlooks these human factors and ergonomics considerations.
- Information technology such as word processing putatively increases efficiency but also deskills the task of the writer who now has to spend more time on clerical chores. Authors who are required to distinguish between three types of hyphen in their manuscripts, for instance, are drowning in the minutiae of details.
- Of all the research and development that is done in our field, how much consideration is given to the world's major problems?

The report concluded,

> The work of this subcommittee is grounded in our mutual certainty there are ethical issues that are the responsibility of our profession. We believe, therefore, these ethical issues are the responsibility of the Association. The subcommittee is looking at possible avenues for putting our vision of a socially ethical Association into effect, including revising and adding to the Code. Does educational technology have a conscience?

Thinking otherwise.   The forward movement in AECT's professional ethics was convergent with another social development in educational technology. Furthermore, some of the same people were involved. In the late 1980s, it became apparent there was an invisible college of educational technology scholars who upheld values somewhat distinct from the wholehearted support of the technostructure (as defined in Galbraith, 1967; Yeaman, Nichols, & Koetting, 1994). Although strongly connected to previous generations (Yeaman et al., 1994, pp. 8–9), they were alternate from their own time in speaking of ethics, social responsibility, and conscience. See chapter 9 for a contrasting view of values in educational technology, one more aligned with the technostructure.

Galbraith (1967) was cited and quoted first by Heinich (1968) and later by others, especially the AECT Task Force on Definition and Terminology (AECT, 1977, p. 57) but there are alternate readings. Galbraith wrote about education but not educational technology and his enthusiasm was unrealistic:

> Colleges and universities can serve the needs of the technostructure and reinforce the goals of the industrial system. They can train the people and cultivate the attitudes which insure technological advance, allow of effective planning and insure acquiescence in the management of consumer and public demand. . . . Or colleges and universities can strongly assert the values and goals of educated men—those that serve not the production of goods and associated planning but the intellectual and artistic development of man. It is hard to believe there is a choice. (pp. 375–376)

Winner (1977) noticed that these ideas did not add up to indicate any reasonable hope for the either the present or the future (p. 169). Lyotard's (1984) moral and aesthetic response was to identify the postmodern condition. Fuller (2000) objected to the mixing of political interests with the generation of scientific knowledge in universities. That work on social epistemology is close to the alternative points of view in educational technology where critical theory, postmodern thought, and poststructural thought are engaged. An example of thinking otherwise with theory follows in the next paragraph, a project that the successful book by Hlynka and Belland (1991) inspired.

Based on the voices of members heard at national conferences, as well as read in the educational technology literature, the possibility of rethinking educational technology was summarized in 26 points (Yeaman, 1994a, pp. 20–22). They were described collectively as a draft agenda for a postmodern educational technology (Yeaman, 1994b, p. 61; 1996b, p. 285). However, "The contribution of postmodern and poststructural theory appears not in a new social theory but as a sensibility modulating existing theories" (Yeaman, 1996a, p. 293). For example, under the heading *Technoscience,* technologists were asked to "Acknowledge that technology has art and craft as the foundation of its design creativity rather than science" (Yeaman, 1994a, p. 21). Under *Cultural Aspects,* technologists were asked to "Realize that all educational communications are non-neutral and exist in a sociopolitical context." While these are uncomfortable observations today, just as they were in the 1990s, such critical demands no longer seem alien.

Nine issues were raised around the practice of instructional design, many of which are now diffusing into the production of educational technology, although probably not from being collected together and printed in a scholarly professional journal:

- Accept there are probably several workable solutions to every instructional design problem, not just one ideal solution.
- Examine and learn from instruction that supposedly fails as well as instruction that succeeds as predicted.
- Be cautious: All media are metaphorical and never mean exactly what they seem to convey. It is not possible to escape from language, but metaphors, symbols, and models should be used with care.
- Look for self-contradictions in your own messages and in other peoples' messages.
- Expect diversity in the way students and trainees understand and what they understand. This increasingly comes from the teaching of English, and related subjects, where common sense understandings of media are being replaced by analysis and interpretation. Advocate this way of understanding as superior to the myth of the linear, pipeline transmission of knowledge.
- Break away from the tradition of communication that assigns power to the creators of instructional messages and denies it to learners. Its authoritarian approach is its failing.
- Avoid idealism suggesting there is a perfect meeting of minds. Although people are engaged in communication all their lives, there is seldom an absolute correspondence in understandings.
- Evaluate technological fixes, not only to see if the original problem has been solved but also to see what else has been changed. Have new problems been created?
- Plan by considering needs—not just technologies. Your task is to solve real world problems and not to advocate mythical solutions such as computers.

Sorry, not an instant winner. The Board of Directors asked for an addition to the *Code* to cover harassment. The result was Section 1, Principle 9 (Welliver,

1995a). Following this precedent, a decision to further revise and improve the *Code* was made by the Committee at its meeting in February 1995:

> The Committee has come to recognize the fact that the existing principles, set forth by the code, deal exclusively with the ethical obligations of individual members of AECT. After considerable discussion, the conclusion was reached that perhaps we also hold important ethical responsibilities, collectively, as an association representing our profession. (Welliver, 1995b, p. 9)

The Committee members wrote out their ideas for new principles and exchanged these drafts in 1996. They were also discussed in two workshop sessions on professional ethics at the leadership conference later that year. Following the usual sort of negotiations, the first two principles were passed on February 12, 1997. After an explanatory memo was composed, these were mailed to each member of the Board of Directors on April 27, 1997.

At the next annual convention, the improvements were announced to the AECT membership. This took place in St. Louis on February 20, 1998, at the session on *Professional Ethics in Practice* presented by Nick Eastmond, Vicki Napper, Randy Nichols, Annette Sherry, and Paul Welliver (Heebner, 1998).

The Committee remained active in following its strategic plan (Welliver, 1995c) and provided a professional development workshop on research ethics with the cosponsorship of the Research and Theory Division at AECT's national convention in 2000. Presentations were made by Rob Branch, Frank Dwyer, Leslie Hall, Steve Ross, and David Shutkin. Randy Nichols and Al Januszewski led the discussion and commentary. In general, the Committee kept up its productivity with panel sessions at the annual conference, annual reports were issued, and budget requests submitted in support of the strategic plan (though neither funded nor acknowledged). Committee members were recruited, appointed, and oriented. Committee members organized workshops at state and local conferences and initiated the teaching of experimental classes on professional ethics.

Following a turnover in AECT staff and the move of the national office to Indiana, the "ethics book" (Welliver, 2001) went into publication with explicit mention of a new section on the Commitment of the Profession to Society. Eventually, it became known from browsing official minutes posted on the AECT Web site that Section 4 and its new principles had been discussed by the Board in Scottsdale in August of 1998 but not fully understood. Forward movement had been anticipated but there was no change. Nothing had happened. Even the textual bloopers in the *Code* continued like the scribal errors and inspired distortions found in sacred scriptures.

It seemed a lack of understanding existed in regard to the status of the Professional Ethics Committee, which, in the bylaws, is one of the required governance committees. What was in play here may have been the confusion of power with knowledge in the hierarchy of credibility, factors that have been delineated to some extent by sociologists (Becker, 1998, pp. 90–91).

Please try again.　The situation was disappointing but the Committee soon moved on. Change in the Board was inevitable and organizational memory is short so persistence was likely to pay off. The new goal was to achieve the desired results another way and go even further in the intended direction. Above all else, the Committee kept on going and more convention sessions were organized. The International Council became involved in addressing professional ethics in different nations and across borders (Sherry, 2000). Other sessions looked at philosophy, methodology, and research ethics (Januszewski, Nichols, & Yeaman, 2001) as well as further model cases.

Moving the chairing of the Committee around so that more members had leadership experience would strengthen the Committee and vary its public face. Randy Nichols was appointed as the incoming Chair and went about organizing another revision of the *Code*. Randy made certain this was going to be completed. By the Denver meeting in 2000, it was largely acceptable to the Committee although there was no section showing Commitment to Society.

Assimilation.　The next year it was Annette Sherry who took office as the Chair and the Board of Directors adopted the latest revised *Code*. The updated and expanded AECT *Code of Professional Ethics* was submitted to the AECT Board of Directors and approved on Tuesday, November 6, 2001. The principles with alterations are Section 1, Principles 1, 5, 6, and 7 and Section 3, Principles 1, 2, and 5. The new principles are Section 2, Principle 6 and Section 3, Principle 10.

Leadership

The Professional Ethics Committee has been chaired by 10 members as recorded in the DAVI and AECT membership directories, as shown in Table 11.3. This information is particularly relevant to generating sociohistorical understandings. There have been as many chairs since the turn of the century as there had been in the previous 25 years. The rapid turnover may or may not be healthy. Some assessment of this shift is necessary.

The ongoing role of the Committee has been educational rather than as a mechanism for disciplining the errant (Eastmond, 2001). Support for Paul Welliver's emphasis on informing members is also provided by answers to a

Table 11.3. Leaders who chaired the AECT Professional Ethics Committee, 1965–2006. The year each person began service as the chair is to the left. Predecessors continued their leadership into that year. © A.R.J. Yeaman, 2006. Used by permission.

Year	Chair
1965	John C. Schwartz
1967	John A. "Jack" Davis
1975	Margaret E. Chisholm
1985	Richard "Dick" Hubbard
1987	Paul W. Welliver
1995	Andrew R. J. Yeaman
2000	G. Randall "Randy" Nichols
2002	Annette C. Sherry
2004	J. Nicholls "Nick" Eastmond, Jr.
2006	Vicki S. Napper

few "frequently asked questions" (Yeaman, 2001). An ongoing series of essays in *TechTrends* aims at further increasing awareness and understanding of the Committee's procedures and the *Code's* principles and social context (Yeaman, 2004a, 2004b, 2004c, 2004d, 2004e, 2004f, 2006a, 2006b). These articles drew upon the author's continuous experience as a Committee member since 1987, five years service as Chair, and initiating and ensuring that the first major revision of the *Code* was carried through.

Professional Ethics: The Present

Becoming Informed

The major thrust of the AECT Ethics Committee at the present is based upon the collective belief that maintaining ethical awareness is vitally important both for individual members and for the Association. Before any AECT member can be expected to act ethically, that person must become informed as to exactly what are the Association's professional ethics. Exposure to principles and cases, and then discussing and reasoning about them, preferably in a group setting with other professionals, is part of socialization into the profession.

A column in the professional journal, *TechTrends* was first printed in October 1989, with the main aim of educating AECT members about professional ethics. The column's name changed with the editors: Paul Welliver's column was "Ethics Today," Vicki Napper's was "Ethically Speaking," and Andrew Yeaman's is "Professional Ethics." There was a gap from the mid- to late 1990s. Due to Vicki Napper's successful tenure, the status was changed from column to section, which means there may be more than one item in an issue. All of the editors have diligently attempted to educate members of the profession in navigating the intricacies of professionally ethical behaviors and expressions. Many of the articles in past years have been presented as case studies to illustrate a given principle from the *Code*. In addition, the articles sometimes provided a tentative solution to the dilemma or situation. They are composed in such a way that the scenario problems may function as general purpose instructional materials for teaching and learning about professional ethics.

The articles in the original *TechTrends* series were revised and expanded with additional essays as *A Code of Professional Ethics: A Guide to Professional Conduct in the Field of Educational Communications and Technology* (Welliver, 2001). Besides being available as a book, this publication is also available to AECT members with a required password at the AECT Web site (http://www.aect.org). In a few instances, the cases have been presented as trigger videos to stimulate classroom discussion.

What Is Needed

In sum, technology colleagues and students can be alert and informed about what to anticipate. A reading of a scenario that applies a relevant principle could prepare AECT members for how to think and how to proceed when there are suspicions of unethical occurrences. Rather than practice at judging and blaming, practice at seeking understanding and resolution are going to be the activities that are the most beneficial. The articles should be documenting the profession's jurisdictions, areas of expertise, and the performance of tasks and techniques in relation to the profession's ethical principles. Each professional ethics article currently has three parts: a scenario, a principle, and an analysis.

Each article starts with a scenario in which a realistic but hypothetical problem is presented for readers to think through. When the situation is described, it is written as a brief instructional fiction. One of the sites of work where the profession holds jurisdiction, such as a school library media center, is identified in the scenario. It includes a network of actors such as a media specialist, a computer system, a software corporation, and a student's parent. One of the domains of the field is implied: management, utilization, development, evaluation, or design.

The dilemma progresses through factual statements, basic descriptions, and short dialog exchanges but is not overly complicated. The narrative takes readers quickly to the central problem and with sufficient details to be credible. It is readily comprehensible but the sketch is not colored in, which makes it more evocative. Readers may notice absent information and generate "what if" questions.

In the second section, the relevant principle is quoted from the *Code*. The author's goal in writing the article was to augment general understanding about this principle. At this place, readers may stop to consider the situation for themselves before continuing on.

In the third section, a concise critique that looks at the scenario in light of the principle is provided. The writing style is different. Analytic prose explores the layers of reality found in the scenario. As in Paul Welliver's memo to the Professional Ethics Committee, dated March 2, 1990, this should aim at "an open discussion which examines a variety of perspectives on the issues and avoids an arbitrarily simplistic answer."

The analysis does not cover all eventualities and may best provoke insight by suggesting varying interpretations and a range of possible resolutions. It can help defeat the fallacy that professional ethics could ever be 100% effective in being preventive. Being optimistic, there is a strong possibility of becoming ready to handle difficulties. One of the domains of the field is likely to figure explicitly.

The following disclaimer applies: Illustrative articles appearing in the Professional Ethics section of *TechTrends* are either fictionalized or completely made up. The scenario format allows authors to be imaginative in engaging readers with lifelike characters, dramatic events, and realistic details. Although an author's inspiration for writing could possibly come from something witnessed, experienced, or heard about from another source, there is never any intended resemblance to specific individuals or specific institutions. The instructional purpose is to raise consciousness about AECT's professional ethics.

How May AECT's *Code* Actually Operate Among Members?

Beyond subscribing to the *Code*, members have various frameworks for ethics and interpretations of them. Brewer, Eastmond and Geertsen (2003) took a wide view of conduct ranging from acceptable to unethical, to illegal, and to heinous. They described one possible framework for identifying and for acting on ethical infringements. The rationale for building an ethical awareness is that people can analyze their action along a scale ranging from basic consideration for others, professional judgment, moral considerations, ethical standards, or legal action (or threat). While different considerations may apply under different circumstances, these mechanisms are seen to act as a "series of fences" designed to protect the individual and society from acting inappropriately.

In the case of ethical concerns, the weight of the person's membership in the Association is proposed as a deterrent to inappropriate action. A better motive than possible expulsion from membership in the Association, *if educational efforts are effective*, is that of individuals wanting to be consistently ethical. The increase of that kind of thinking among the members is what this particular chapter attempts to bring about.

In drawing on the concept of the role set (Merton, 1957), Brewer et al. (2003) converged with the Modern Language Association (MLA) in the acknowledgment that differing levels of intervention are possible. A contemporary discussion of practical ideas for coping with bigotry and suspicions of bigotry was provided by the MLA Committee on Academic Freedom (2003). It identified who may intervene and described how to intervene.

Finding Examples in Daily Life

The point of the written cases and the trigger videos is to encourage colleagues to see ethical situations and trade-offs in everyday life. There are choices to be made in these situations that are experienced by professionals at all career stages. Sensitization to the existence of unethical possibilities should improve ethical performance.

A small change can alter ethical acceptability. For example, if a librarian were asked to supply the names of those who had checked out a certain kind of book, the librarian would be unethical in providing such a list. Even if the reasoning was considered beneficial, such as to enable placement in a community-based literacy program, compliance would be wrong under American Library Association guidelines (Foerstel, 2004) or under the AECT ethics, Section 1, Principle 4 requirement "to conduct professional business so as to protect the privacy and maintain the personal integrity of the individual."

However, if the librarian were to make available a stack of brochures for the literacy program and mention to patrons about the program's availability at the time of checkout, that action would be ethically acceptable. The one action is unethical, while the other is not.

As AECT members become more adept at identifying potential ethical violations, they can take action to avoid ethical problems. Sometimes people can only learn from making mistakes, but others can spot an ethical violation ahead of time. Knowing the potential and practicing avoidance is the preferable course of action. An approach to self-monitoring ethical behavior is the practice of keeping a personal "ethical encounters journal" where ethics issues that surface in daily life can be analyzed privately and the lessons recorded for one's own use.

Update Knowledge as Conditions Change

New technologies may create opportunities for learning, but they also create ways to run afoul of ethics. For example, someone acting out of maliciousness from the anonymity of the Internet can quickly step over ethical lines in electronic mail or listserv correspondence (Eastmond, 2002). The ease of duplicating digital material available on the Internet continues to have ethical and legal cases and controversies. The point is that, in addition to learning to use new technologies effectively, Association members need to learn how to use these technologies ethically. R. W. Burniske (2004) developed a Web-based program to teach proper Internet behavior, a program called the Cyber-Pilot's License. It received sustained interest and was tested in cross-cultural settings in Hawaii and Brazil.

Pushing on

Several developments promise to change the practice of ethics within AECT. Some are well under way, while others have only been mentioned and are still being considered.

That technology is having negative effects is an area for attention (Yeaman, 2004f). The human factors and ergonomic aspects of computers in educational settings needs further research into the position of keyboard, chair, and screen for healthy long-term usage. The political, economic, and philosophical aspects of technology and learners need to be comprehended (Nichols, 1991, 2002). The redesign and replacement of tall, unstable equipment carts so they can work more safely needs to be implemented (Sherry, 1998; Sherry & Strojny, 1993). Each of these areas represents an avenue of potential inquiry.

Cultural and cross-cultural aspects need attention. Cultural diversity and cultural pluralism need to be included in instruction (Branch, Brigham, Chang, & Stout, 1991). The *Code* itself is read and used not only within the United States but also internationally; see the special section of the International Review section of *ETR&D* on cross-cultural issues (Sherry, 2000). That theme was taken up in *TechTrends* with more voices from different cultures (Bradshaw, Keller, & Chen, 2003). The steps to have the AECT *Code* translated into other languages have been described (Sherry et al., 2003). In some instances, the meaning conveyed in the *Code* can only partially be transmitted in another language. It is questionable how much any single code of professional ethics can function across varied cultural settings.

Professional Ethics: Building Into the Future

Like the section of this chapter on the present, this section on the future is presented here because it illustrates the ongoing discourse. There definitely are real fears in people of youth being corrupted by technological media, fears of the dangers of physical media, and fears of victimization and criminality.

The idea that professional behavior should be based on a defined system of ethics is not new. As previously described, the AECT *Code* was adapted from an existing code of the NEA. The current AECT *Code* promotes ethical behavior through sections on the individual, society, and the profession.

Just as the professions within the community of educational communications and technology have developed, so have the definitions of ethical behavior. Based on the contexts of contemporary issues, defining ethical behavior is a developing process. Codes of ethics help define responses to modern issues and give guidance to the individuals who acknowledge those codes.

There are many troubling issues unique to this time in history. New generations encounter the confusions of a computer-emulated world. There have been numerous scandals involving leaders of major corporations whose unethical behavior has caused the loss of millions of dollars and in some cases reduced retirement pension funds to nothing or next to nothing. Just as the actions of individual leaders now effect more than the employees of their companies, the actions of the individuals of those companies effect more than the stockholders. Computer-based technology and its impact on the associated educational and work environments are changing the parameters of ethical action in work and education.

Professional Ethics for Individuals

Many people believe the growth of computer-based technologies has accelerated globalization, a complex process that embraces the rapid transfer of information, funds, and goods from one part of the world to another. Globalization amplifies the consequences of the actions of individuals by transforming them within the larger framework of the worldwide arena. Globalization also amplifies the consequences of actions of entities such as corporations and nations. The behavior of the single unit of the individual is no longer the only factor involved in ethical behavior. One person can change the world, but a group of people may be able to change the world much faster.

The Internet is an example of a globalized technology unique to the 21st-century world. The linking of people in many lands through the mechanism of the World Wide Web has led to an explosion of information and knowledge and a whole new set of ethical dilemmas. Although some people might argue that the information age is not truly the model of the entire world, it certainly is shaping the economy and ecology of the nations leading the world in gross national production and technological innovation. The rapid change occurring today can be threatening to third world nations through an ever-increasing digital divide. Separation between the electronic haves and the electronic have nots is one of the current ethical concerns that the world will continue to face in coming decades.

The AECT *Code* (Table 11.2) was recently revised to reflect an increased awareness of the complex role of the individual in educational settings or workplaces. Individuals who ostensibly work alone but in reality work in the community of the World Wide Web are frequently confronted with issues related to privacy in a public world. This web of information surrounding the life, actions, and identity of employers, workers, students, and citizens in general does not allow technology-based consumers to assume their actions are isolated or independent of consequences affecting others.

Professional decision making now demands an understanding of issues beyond the scope of the local production facility. For example, workers in a multinational educational corporation may reside in different communities of the world with different cultures and rules for behavior in homes, schools, or the workplace. This type of globalization may drive future issues of ethical behavior for all professions connected by technology for knowledge distribution.

Another example of new ethical concerns can be linked from the *Code* to the Internet (see Section 1, Principle 1 in Table 11.2). The *Code* declared,

> In fulfilling obligations to the individual, the members shall encourage independent action in an individual's pursuit of learning and shall provide open access to knowledge regardless of delivery medium or varying points of view on the knowledge.

Certainly, the Internet provides this type of multidimensional social environment, and this freedom of information is both desirable and problematic.

Similarly, Section 1, Principle 8 tells members to avoid either designing or selecting educational programs or media with "content that reinforces or promotes gender, ethnic, racial, or religious stereotypes." Again, the Internet provides an environment that may be filtered to eliminate content that reinforces or promotes gender, ethnic, racial, or religious stereotypes. Without those filters, however, some mechanical and some personally imposed, the individual faces an onslaught of ethical dilemmas. Further, members "shall seek to encourage the development of programs and media that emphasize the diversity of our society as a multicultural community."

Protecting Children

In most cases, this ethical dichotomy lies in the boundary between individual freedom and societal concern. The estimate for pornography sales on U.S. Internet Web sites is $12 billion a year, and the worldwide estimate is $57 billion (Family Safe Media, 2004). Thus, the total expenditure for pornography exceeds the "combined revenues of all professional football, baseball and basketball franchises" (Family Safe Media, 2004).

Societal concern has given rise to an entirely new category of employment to provide child-safe environments, free of inappropriate media. Current federal laws have been enacted to balance the safety of children and the rights of adults. The Children's Online Privacy Protection Act (COPPA) went into effect on April 21, 2000 and the Children's Internet Protection Act (CIPA) on April 20, 2001 (Carroll & Witherspoon, 2002). Both of these laws protect children from the free ranging environment of the Internet but create situations limiting adult access to information.

The American Library Association (2004) balanced the rights of free speech with the need to protect children. Libraries are the places that offer the wealth of programs and information resources for children and parents in safe and supportive environments in schools as well as after school. School library media specialists need both to support intellectual freedom and to maintain concern about content controls. The ethical considerations of the library media specialist must balance freedom of speech with freedom from unethical uses of information.

Digital Insecurity

The rights of the individual are at stake because of threats to the security of personal information. The Federal Trade Commission (2004) addressed the growth of "identity theft" where one person masquerades as another for purposes of obtaining financial resources or other benefits by fraudulent means. Previously, the United States reported 161,819 identity-theft cases with 214,905 cases reported a year later—a 33% increase. Categories of identity theft included falsification of government documents, employment related fraud, and Internet and e-mail fraud. The category of highest percentage of identity theft was ages 18–29 (28%). Children less than 18 years of age constituted 3% of the victims. Intel Corporation (2004) estimated 800 megabytes of data is recorded yearly on every person on the planet. Never before has it been so easy to use tools that give freedom to take freedom away.

The processes of information access and identity protection are now primary concerns whenever a piece of individually identifiable information passes a checkpoint of access. Technology support professionals are now held accountable for all of the bytes of information flowing through their systems. Filtering for accuracy, content, and privacy are requirements of information environments. The issue of who filters for accuracy of information, what is acceptable content, and the level of privacy of information access are growing ethical issues. The ethical issues of a decade ago have become legal issues of today. Ethical behavior will continue to develop the individual's commitment to society and to protect privacy while promoting freedom of access.

Professional Ethics and Research

An online publication from the Association of Internet Researchers described the ethics involved in Internet research (Wes & AoIR Ethics Working Committee, 2002). The document described ways to go about protecting the privacy of human subjects and obtaining informed consent in Internet related situations. While it is couched more in terms of questions that the researcher should ask while beginning to do Internet research, the group supplied enough guidelines to help researchers stay clear of ethical violations. Particularly interesting are the differences noted for ethical decisions given in the United States—generally more utilitarian and oriented toward outcomes—and those provided in the European Union—more deontological and oriented toward moral content of an action, right, and wrong, according to these researchers. The world of research ethics has been permanently altered by the new means available through the Internet.

The AECT *Code* does not supply guidelines for conducting research. Principle 10 was added to Section 3 but only enjoins the member to

> Conduct research and practice using professionally accepted and Institutional Review Board guidelines and procedures, especially as they apply to protecting human participants and other animals from harm. Humans and other animals shall not be used in any procedure that is physically invasive to them.

The *Code* defaults to the guidelines of the institutional review boards and to the ethical codes of other professional societies, like the American Psychological Association (see Nagy, 2000) or the American Anthropological Association (see Fluehr-Lobban, 2003). Now that Section 3, Principle 10 is in the *Code*, perhaps the Professional Ethics Committee should try outlining what is specifically expected of members while conducting research on educational communications and technology. The topic can be anchored in relation to foundations (Januszewski et al., 2001).

Technology and Health and Safety

This topic may seem beyond the scope of the chapter but it fits in with the discursive heritage of professional ethics. Responsibility emerges when protecting fellow employees or students from conditions harmful to health and safety, including harmful conditions caused by technology itself. Section 1 states the commitment of the members to

- Following "sound professional procedures for evaluation and selection of materials, equipment, and furniture/carts used to create educational work areas" (Principle 5)
- Making "reasonable efforts to protect the individual from conditions harmful to health and safety, including harmful conditions caused by technology itself" (Principle 6)

An epidemic of repetitive-motion injuries in the workplace beginning in the late 1970s raised an alarm for the potential for harm in seemingly safe computerized school environments. An ethical issue has arisen from the fact that, in the United States, the National Institute of Occupational Safety and Health (NIOSH) regulates the workplace environment but not classrooms or other areas used in schools for learning. At the present time, there is no national or state organization responsible for safety and health in classroom environments like there currently is for workers.

Although the idea of safety is not a new issue for professions involved with the daily use of hardware, the idea of creating safe educational study and learning areas is different than in bygone years. The chalkboard for display and sharing of ideas within a classroom may be permanently shifting to computer projection systems and multimedia computers. These machines are not only a production resource and an instrument of learning but also a potential cause of injury.

The evidence is growing that there are health concerns for students related to computer usage. These issues include but are not limited to

- The weight of backpacks for small children
- The amount of time students spend keying information into a computers

- Development of eyesight and the impact of video display terminals on that process
- The decibel level of sound in educational environments and resultant disruption to learning and hearing
- Safe handling of heavy equipment by children and teachers (Ergonomics for Children in Educational Environments, 2004)

As individual members of professions that place us in educational environments, we become legally and perhaps morally responsible for the students placed in our charge. These issues are both imminent and professionally ethical. When budgetary realities conflict with perceptions of safety, the ethical issues arise. For example, the long-term effects of technology usage are easy to ignore because they may not materialize for years. They may not create a sense of danger just as the routine use of televisions did not suggest the heavy carts would fall on small children, causing injuries and deaths (Sherry, 1998; Sherry & Strojny, 1993). The ethics of safety need to become the rules of the classrooms. Future ethical concerns will undoubtedly embrace concerns emerging from the changing physical and psychosocial environments of education.

The *Code* contains sections devoted to the commitment of the member to society and the profession. The idea that the individual is an important shaper of society may grow in importance as technology industries increasingly globalize their operations. Section 2, Principle 1 of the *Code* states that the individual's commitment to society "shall promote positive and minimize negative environmental impacts of educational technologies." This commitment takes on new meaning when educational technologies are being tried out in the jungles of Brazil, a centuries-old village of India, or a classroom on Mars.

Future ethical issues reside in the present as seeds of future actions. These ethical seeds arise from the actions of today but the consequences arrive in the future. The current AECT *Code* was developed with the idea of a culture primarily within the bounds of the United States. Those boundaries have grown considerably since the early years of AECT and now include members from a variety of countries around the world. The commitment of the individual to self, society, and profession will need to expand to include communities of students and workers as defined by globalization and by exploration of space. The future holds the promise of AECT members who live beyond the confines of their own countries and potentially beyond earth.

Conclusions

Professional Ethics Makes Educational Technology Visible

An important way to assess and define the profession of educational technology is to think seriously about the stories that are told inside the profession. It is an old question in cultural anthropology: How do a people see themselves? The most insight-provoking tales are seldom straightforward. Being held together by congruent elements in a logical progression is frequently balanced by mystery. Similarly, the stories technologists tell about what is and what is not professionally ethical can be informative.

Readers are probably already familiar with the foundational concept that "technology makes instruction visible" (Heinich, 1970, pp. 157–163). Heinich had discovered a generalizable principle about educational technology and this visibility is distinct from instructing with visual materials (p. 159). It means actions are made manifest, concrete, and perhaps empirically measurable. This is the investigative technique of using a thing, about which many aspects are known, to make an estimation of something else. Likewise, technologists may examine what their professional ethics mean and how their professional ethics are applied, and consequently, they may better comprehend technology itself.

Heinich's (1971) insight also had a corollary: "Technology can only be effective when we pull apart the elements of a process and step by step devise technical means to achieve our goals in a systematic way" (p. 80). This is sensible as greater awareness of any process should result in improvements. The symmetry applies to professional ethics, too, and "Ethics does not solve problems, it structures them" (Harpham, 1995, p. 404). Therefore, it can be said, "Professional ethics makes educational technology visible" because it is possible to analyze educational technology processes and products to see whether ethical principles are carried out or abrogated. It would be possible for technologists to follow professionally ethical principles in relating to learners and colleagues, and in maintaining social order. While this is associated with the many influential forces and continuing tensions (shown in Table 11.1), it is reasonable to expect that the level of sophistication of AECT members about professional ethics ought to grow over time. In this fashion, professional ethics provides a way of knowing the responsibilities of educational technologists.

Obstacles

From an instructional point of view, not only has a code been established but also a functioning set of illustrative scenarios serve to encourage discussion and generalization in the site of work. Nevertheless, a concerning and substantial question is whether technologists have professional ethics only because they are supposed to have them. As with the NEA, the founders of AECT believed a code of professional ethics is one of the attributes which causes an occupational grouping to merit and gain classification as a profession.

Possibly, these things will be better understood in the future, but at this time, there possibly appears to be a deficiency regarding the *Code*'s foundational concepts and their origins. Unlike Heinich's (1970, 1971) *technology*, there seems no overarching structure to the *Code* to help it be understood as part of a process. It seems there are neither reminiscences nor written records explaining the decision to have three sections indicating commitment to the individual, society, and the profession, respectively. It is not obvious from the principles in each section what those abstractly titled sections signify. Readers may wish to refer to Table 11.2 and reflect on this for themselves. Is the first section in the *Code* referring to "the individual" as an abstraction meaning respect for each person as unique or to "the individual learner?" In the latter case, the object of Section 1 would be the person who is the immediate client and supposed beneficiary of the personal transformations promised by educational technology.

Nor is it known why Section 1 states the obligations of the members and the others state the obligations of the member. The distinction between the members and the member is plural compared to singular. This could be a typing error, a stylistic mistake in producing sentence stems that are not parallel, or something meaningful on which accurate interpretation of the *Code* may hinge. Differing from Table 11.2, one printing of the *Code* gave the obligation of members in the plural form in each section (Welliver, 1989, p. 53).

At least two principles are matched like bookends: Compare Section 2, Principle 4 with Section 3, Principle 2, and note their similarities. Furthermore, several principles cover forms of corruption but corruption may be unlikely as a major concern regarding educational technology. Note that Section 2 on society is the shortest and Section 1, Principles 8 and 9 refer to society. The new principles are left to stand at face value. What does the jargon "other animals" really mean in Section 3, Principle 10? Perhaps this inscrutable phrase humorously refers to the hazards of home schooling without technology (see *My Family and Other Animals;* Durrell, 2000).

Developments

Regardless of what has been lost from living memory, the explication of the *Code* continues. Fresh ethics scenarios and analyses are being written and published for the professional ethics section of *TechTrends*. As before, they are constructed as general-purpose instructional materials with just enough detail to be ambiguous yet evocative of a specific principle.

In the first published collection of cases, one third of the colleagues with a professional ethics problem were media managers and one half were professors (Welliver, 2001). It is not known if this is either accurately representative or generalizable. The extent of any influence on what technologists do is probably unknowable, just as people deviate from norms without being detected, but it will be interesting to see which technologists are afflicted with ethical dilemmas in coming issues of *TechTrends*.

There also exists potential for comparing the AECT *Code* with other codes in educational technology's area of professional jurisdiction. These are the code of the American Society for Training and Development (n.d.), Johnson's (2004) human resources booklet *Ethics for Trainers*, the code of the International Society for Performance Improvement (2006), and the code of a smaller group, the International Board of Standards for Training, Performance, and Instruction, which published (with J. Michael Spector) the *Code of Ethical Standards for Instructional Designers* (Richey, Fields, & Foxon, 2001, pp. 201–202).

No Science Necessary

Preparation in professional ethics is being incorporated as a requirement in the graduate curriculum. An important aspect is that awareness of professional ethics requires thinking through how we want our profession to be positioned in society in the future. Having formal professional ethics helps make that goal both tangible and knowable.

The search for conscience continues in regard to what technologists do, what they say they do, and how things are done because the potential for ethical violations is foundational to technology (Januszewski et al., 2001). The most necessary question regarding technology being ethical is a sign of our profession "coming of age" and an indication that professionals with a conscience are needed (Yeaman, 2000). Nevertheless, it would be misleading to suggest there is (or should be) a *scientific* basis for linking the *Code* to what choices are made by professional technologists (as we may be calling ourselves this decade) and for deciding whether or not technologists' actions (in particular or in general) are in fact professionally ethical. The ethical standards of a profession do not require scientific justification and it would be an opening for excess utilitarianism.

This chapter's engagement with technology through its professional ethics should make technology more visible. It should help technologists do technology well and increase the likelihood of doing well. "How are we to be ethical professionals?" ought to be the root question in defining educational technology.

Authors' Note

While some pieces of the text have appeared in a different form in *Tech Trends,* many of the authors' ideas given this chapter were first presented by them at AECT conferences.

Acknowledgment

This is to sincerely thank all of AECT's members who, over the years, have generously given their time, wisdom, and resources by sitting on the Professional Ethics Committee. As the profession continues to emerge and mature, their conscientious efforts have decreased the superstitions and increased the rationality of professional ethics for technology.

References

1968 DAVI Resolutions. (1968). *Audiovisual Instruction 13*(6), 676–679.

Abbott, A. (1988). *The system of professions: An essay on the division of expert labor.* Chicago: University of Chicago Press.

Abbott, A. (1998). Professionalism and the future of librarianship [Special issue]. *Library Trends, 46*(3).

Abbott, A. (2001). *Chaos of disciplines.* Chicago: University of Chicago Press.

Adler, M. J. (1937). *Art and prudence: A study in practical philosophy.* New York: Longmans, Green and Co.

American Library Association. (2004). *Civil liberties, intellectual freedom, and privacy.* Retrieved May 11, 2004, from http://www.ala.org/ala/washoff/WOissues/civilliberties/civilliberties.htm

American Society for Training and Development (ASTD). (n.d.). *Code of ethics.* Retrieved June 27, 2006, from http://www.org/NR/rdonlyres/5DBEF5A3-EC0E-4C5C-9FA5-4DD47C19A4A8/8544/CodeofEthics.pdf

Association for Educational Communications and Technology. (1977). *Educational technology: Definition and glossary of terms: Volume I.* Washington, DC: Author.

Association for Educational Communications and Technology. (1984). *Human resources: The Association for Educational Communications and Technology: Membership Directory: 1984–1985.* Washington, DC: Author.

Association for Educational Communications and Technology. (1985). *Human resources: The Association for Educational Communications and Technology: Membership Directory: 1985–1986*. Washington, DC: Author.

Association for Educational Communications and Technology. (n.d.a). *A code of professional ethics*. Retrieved July 28, 2006, from http://www.aect.org/Intranet/Publications/ethics/index.html

Association for Educational Communications and Technology. (n.d.b). *AECT: Bylaws*. Retrieved July 28, 2006, from http://www.aect.org/About/Governance/Bylaws.asp.

Becker, H. S. (1997). *Outsiders: Studies in the sociology of deviance*. New York: The Free Press. (Original work published 1963)

Becker, H. S. (1998). *Tricks of the trade: How to think about your research while you're doing it*. Chicago: University of Chicago Press.

Becker, H. S. (2002). Studying the new media. *Qualitative sociology, 25*(3), 337–343.

Becker, H. S. (2006). *What about Mozart? What about murder?* Retrieved June 9, 2006, from http://home.earthlink.net/~hsbecker/mozart.htm

Bradshaw, A. C., Keller, C. O., & Chen, C. (2003). Reflecting on ethics, ethical codes, and relevance in an international instructional technology community. *TechTrends, 47*(6), 12–18.

Branch, R. C., Brigham, D., Chang, E., & Stout, P. (1991). Incorporating cultural diversity into instruction. *Community Education Journal, 18*(4), 20–21, 30.

Brewer, E., Eastmond, N., & Geertsen, R. (2003). Considerations for addressing ethical issues. In M. A. Fitzgerald, M. Orey, & R. M. Branch (Eds.), *Educational media and technology yearbook* (Vol. 28, pp. 67–76). Westport, CT: Libraries Unlimited.

Burniske, R. W. (2004). *The CyberPilot's license*. Retrieved May 5, 2005, from http://etec.hawaii.edu/cpl/home.html

Charters, W. W. (1935). *Motion pictures and youth: A summary*. New York: Macmillan.

Carroll, J. A., & Witherspoon, T. L. (2002). *Linking technology and curriculum: Integrating the ISTE NETS standards into teaching and learning* (2nd ed.). Lebanon, IN: Merrill Prentice Hall.

Dameron, V. G. (1947). An explanation—the new DAVI Constitution. *Educational Screen, 26*, 442, 448.

Davidson, B. (1936). Code of ethics or statement of objectives. *The American teacher, 20*(4), 33–34.

Davis, J. (1969). Committee on Professional Ethics. *Audiovisual Instruction, 14*(6), 82.

Davis, J. A. (1972). Dear editor [Letter to the editor]. *Audiovisual Instruction, 17*(1), 94–95.

Dewey, J. (1903). *Ethical principles underlying education.* Chicago: University of Chicago Press. (Original work published 1897)

Durkheim, E. (1992). *Professional ethics and civic morals* (C. Brookfield, Trans.). London: Routledge. (Original work published 1957)

Durrell, G. M. (2000). *My family and other animals.* New York: Penguin.

Eastmond, J. N. (2001). A historical perspective. In P. W. Welliver (Ed.), *A code of professional ethics: A guide to professional conduct in the field of educational communications and technology* (pp. 5–6). Bloomington, IN: Association for Educational Communications and Technology.

Eastmond, N. (2002). What is a cyberpredator? *TechTrends, 46*(4), 8–10.

Ergonomics for Children in Educational Environments. (2004). Retrieved May 15, 2004, from http://education.umn.edu/kls/ecee/default.html

Family Safe Media. (2004). *Pornography statistics 2003.* Retrieved May 11, 2004, from http://www.familysafemedia.com/pornography_statistics.html

Federal Trade Commission. (2004). *Sentinel complaints by calendar year.* Retrieved May 11, 2004, from http://www.consumer.gov/sentinel/trends.htm

Finn, J. D. (1953). Professionalizing the audio-visual field. *AVCR, 1,* 6–17.

Finn, J. D. (1996). Professionalizing the audio-visual field. In D. P. Ely, & T. Plomp (Eds.), *Classic writings on instructional technology* (pp. 231–241). Englewood, CO: Libraries Unlimited.

Fluehr-Lobban, C. (Ed.). (2003). *Ethics and the profession of anthropology: Dialogue for ethically conscious practice* (2nd ed.). Walnut Creek, CA: Alta Mira Press.

Foerstel, H. N. (2004). *Refuge of a scoundrel: The Patriot Act in libraries.* Westport, CT: Libraries Unlimited.

Foucault, M. (1972). *The archaeology of knowledge and the discourse on language* (A. M. Sheridan Smith, Trans.). New York: Pantheon Books. (Original work published 1970)

Foucault, M. (2003). *"Society must be defended": Lectures at the College de France, 1975–76* (D. Macey, Trans.). New York: Picador. (Original work published 1997)

Freidson, E. (1986). *Professional powers: A study of the institutionalization of formal knowledge.* Chicago: University of Chicago Press.

Freidson, E. (2001). *Professionalism: The third logic.* Chicago: University of Chicago Press.

Fuller, S. (2000). *The governance of science.* Buckingham, UK: Open University Press.

Galbraith, J. K. (1967). *The new industrial state.* Boston: Houghton Mifflin.

Gardner, H., Csiksentmihalyi, M., & Damon, W. (2001). *Good work: Where excellence and ethics meet.* New York: Basic Books.

Gieryn, T. F. (2004). Eloge: Robert K. Merton, 1910–2003. *Isis, 95*(1), 91–94.

Harpham, G. G. (1995). Ethics. In F. Lentricchia, & T. McLaughlin (Eds.), *Critical terms for literary study* (2nd ed., pp. 387–405). Chicago: University of Chicago Press.

Heebner, A. L. (1998). Professional ethics in practice. *TechTrends, 43*(3), 58.

Heinich, R. (1968). Educational technology as technology. *Educational Technology, 7*(1), 4.

Heinich, R. (1970). *Technology and the management of instruction* (Monograph No. 4). Washington, DC: Association for Educational Communications and Technology.

Heinich, R. (1971). Technology and teacher productivity. *Audiovisual Instruction, 16*(1), 79–82.

Highlights of DAVI committee and commission activities. (1966). *Audiovisual Instruction, 11*(6), 513–515.

Highlights of DAVI committee and commission meetings. (1968). *Audiovisual Instruction, 13*(6), 650–671.

Hitchens, H., Jr. (1970). Six characteristics in search of a profession: Two. *Audiovisual Instruction, 15*(4), 120.

Hlynka, D., & Belland, J. C. (Eds.). (1991). *Paradigms regained: The uses of illuminative, semiotic and post-modern criticism as modes of inquiry in educational technology: A book of readings.* Englewood Cliffs, NJ: Educational Technology.

Hyer, A. L. (1969). A time to keep and a time to cast away. *Audiovisual Instruction, 14*(4), 108–109.

Intel Corporation. (2004). *Digital transformation fuels new opportunities.* Retrieved May 15, 2004, from http://www.intel.com/labs/teraera/index.htm

International Society for Performance Improvement. (2006). *Code of ethics.* Retrieved June 27, 2006, from http://www.certifiedpt.org

Jacobs, L. (1990, January). Reformers and spectators: The film education movement in the thirties. *Camera obscura: A journal of feminism and film theory, 22,* 29–49.

Januszewski, A. (2001). *Educational technology: The development of a concept.* Englewood, CO: Libraries Unlimited.

Januszewski, A. (2006). Definition and Terminology Committee. *TechTrends, 50*(1), 10.

Januszewski, A., Nichols, R. G., & Yeaman, A. R. J. (2001). Philosophy, methodology, and research ethics. *TechTrends, 45*(1), 24–27.

Johnson, J. (2004). *Ethics for trainers.* Alexandria, VA: American Society for Training & Development.

Jowett, G. S., Jarvie, I. C., & Fuller, K. H. (1996). *Children and the movies: Media influence and the Payne Fund controversy.* Cambridge, UK: Cambridge University Press.

Kendall, G., & Wickham, G. (1999). *Using Foucault's methods.* London: Sage Publications.

Landis, B. Y. (1927). *Professional codes: A sociological analysis to determine applications to the educational profession.* New York: Teachers College, Columbia University.

Lyotard, J. (1984). *The postmodern condition: A report on knowledge.* Minneapolis, MN: University of Minnesota Press.

McClusky, H. Y. (1934, October). Mechanical aids to education and the new teacher—A prophecy. *Education, 55*, 83–88.

Merton, R. K. (1957). *Social theory and social structure* (Rev. ed.). Glencoe, IL: The Free Press of Glencoe/Macmillan.

Merton, R. K. (1973). *The sociology of science: Theoretical and empirical investigations* (N. W. Storer, Ed.). Chicago: University of Chicago Press.

Merton, R. K. (1982). *Social research and the practicing professions* (A. Rosenblatt, & T. F. Gieryn, Eds.). Cambridge, MA: Abt Books.

Modern Language Association Committee on Academic Freedom and Professional Rights and Responsibilities. (2003). Advice for combating campus bigotry and fostering respect in the academic community. *Profession*, 158–172.

Moley, R. (1938). *Are we movie-made?* New York: Macy-Masius.

Mosco, V. (2004). *The digital sublime: Myth, power, and cyberspace.* Cambridge, MA: MIT Press.

Nagy, T. F. (2000). *Ethics in plain English: An illustrative casebook for psychologists.* Washington, DC: American Psychological Association.

National Education Association of the United States. (1929, October). Code of Ethics of the National Education Association of the United States. *Phi Delta Kappan, 12*(3), 69–71.

National Education Association of the United States. (1964). *Opinions of the Committee on Professional Ethics: 1964 edition with reference to the code of ethics of the education profession* (4th ed.). Washington, DC: Author.

Nichols, R. G. (1991). Toward a conscience: Negative aspects of educational technology. In D. Hlynka, & J. C. Belland (Eds.), *Paradigms regained: The uses of illuminative, semiotic and post-modern criticism as modes of inquiry in educational technology: A book of readings* (pp. 121–150). Englewood Cliffs, NJ: Educational Technology.

Nichols, R. G. (2002). Meeting our ethical obligations in educational technology. *TechTrends, 46*(1), 52–53.

Nichols, R., Martin, B., & Welliver, P. (1988). *Concern about ethics and ethical issues among professors of instructional systems design and technology.* Syracuse, NY: Syracuse University, ERIC Clearinghouse on Information & Technology. (ERIC Document Reproduction Service No. ED 304099)

Noah, P. (Writer), & McCormick, N. (Director). (2006). Transition [Television series episode]. In A. Sorkin (Producer), *The West Wing.* Burbank, CA: Warner Bros. Entertainment Inc.

Painter, G. S. (1926, January). Psychological background of visual instruction. *American Education,* 210–215.

Paras, E. (2006). *Foucault 2.0: Beyond power and knowledge.* New York: Other Press.

Rich, J. M. (1984). *Professional ethics in education.* Springfield, IL: Charles C Thomas.

Richey, R. C., Fields, D. C., & Foxon, M. (2001). *Instructional design competencies: The standards* (3rd ed.). Syracuse, NY: ERIC Clearinghouse on Information & Technology.

Rosemann, P. W. (1999). *Understanding scholastic thought with Foucault.* New York: St. Martin's Press.

Schuetz, C. (2006). From the president. *Library Instruction Round Table News, 28*(4), 1.

Schwartz, J., & Davis, J. (1967). Professional ethics. *Audiovisual Instruction, 12,* 659.

Seels, B. B., & Richey, R. C. (1994). *Instructional technology: The definition and domains of the field.* Washington, DC: Association for Educational Communications and Technology.

Sherry, A. C. (1998). Is a safety policy enough? *TechTrends, 43*(3), 17–18.

Sherry, A. C. (2000). Introduction. *ETR&D, 48*(4), 100–101.

Sherry, A. C., & Strojny, A. (1993). Design for safety: The audiovisual cart hazard revisited. *Educational Technology, 33*(12), 42–47.

Sherry, A. C., Burniske, R. W., de Freitas, C. M., Varela de Rabago, J. D., Johari, A., Chu, C., & Marchessou, F. (2003). Traductore, traditore: Can the AECT Code of Ethics "speak" across cultures? *TechTrends, 47*(6), 19–25.

Short, W. H. (1928). *A generation of motion pictures. A review of social values in recreational films.* New York: National Committee for Study of Social Values in Motion Pictures.

Silber, K. H. (1978). Problems and needed directions in the profession of educational technology. *Educational Communications and Technology Journal, 26,* 174–185.

Stanley, M. (1978). *The technological conscience: Survival in an age of expertise.* New York: The Free Press.

Strike, K. A., & Soltis, J. F. (1998). *The ethics of teaching* (3rd ed.). New York: Teachers College Press.

Wagner, P. A. (1996). *Understanding professional ethics.* Bloomington, IN: Phi Delta Kappa Educational Foundation.

Walsh, M. J. (1926). *Teaching as a profession: Its ethical standards.* New York: Henry Holt.

Warnick, B. R., & Waddington, D. (2004). *The Gathering*: An ethical and educational criterion for educational technology. *Educational Technology, 44*(5), 24–32.

Welliver, P. W. (1971). Consumerism in education: Ethical implications. *Audiovisual Instruction, 16*(7), 43–44.

Welliver, P. W. (1972). Dear editor [Letter to the editor]. *Audiovisual Instruction, 17*(10), 53.

Welliver, P. W. (1989). The AECT code of professional ethics: A guide to professional conduct in the field. *TechTrends, 34*(5), 52–53.

Welliver, P. W. (1995a). Harassment, bias, and discrimination. *TechTrends, 40*(1), 13.

Welliver, P. W. (1995b). Looking to the future. *TechTrends, 40*(4), 9.

Welliver, P. W. (1995c, June 15). The role of the AECT Committee on Professional Ethics in the AECT Vision 2000. June 15, 1995. (Response to Dr. Lynn Milet, President, AECT, regarding the Self Evaluation and Review and the Strategic Plan)

Welliver, P. W. (2001). *A code of professional ethics: A guide to professional conduct in the field of educational communications and technology.* Bloomington, IN: Association for Educational Communications and Technology.

Wes, C., & AoIR Ethics Working Committee. (2002). Ethical decision making and Internet research. Retrieved November 27, 2002, from http://www.aoir.org/reports/ethics.pdf

Winner, L. (1977). *Autonomous technology: Technics-out-of-control as a theme in political thought.* Cambridge, MA: MIT Press.

Yeaman, A. R. J. (1983). Microcomputer learning stations and student health and safety: Planning, evaluation, and revision of physical arrangements. *Educational Technology, 23*(12), 16–22.

Yeaman, A. R. J. (1989). Resources for improving computerized learning environments: An annotated review of human factors literature. *TechTrends, 34*(6), 30–33.

Yeaman, A. R. J. (1994a). Deconstructing modern educational technology. *Educational Technology, 34*(2), 15–24.

Yeaman, A. R. J. (1994b). Where in the world is Jacques Derrida? A true fiction with an annotated bibliography. *Educational Technology, 34*(2), 57–64.

Yeaman, A. R. J. (1996a). Envoi. In D. H. Jonassen (Ed.), *The handbook of research for educational communications and technology* (pp. 292–295). New York: Simon & Schuster/Macmillan.

Yeaman, A. R. J. (1996b). Postmodern and poststructural theory: Version 1.0. In D. H. Jonassen (Ed.), *The handbook of research for educational communications and technology* (pp. 275–292). New York: Simon & Schuster/Macmillan.

Yeaman, A. R. J. (2000). Coming of age in cyberspace. *ETR&D, 48*(4), 102–106.

Yeaman, A. R. J. (2001). Foreword. In P. W. Welliver (Ed.), *A code of professional ethics: A guide to professional conduct in the field of educational communications and technology* (pp. v–viii). Bloomington, IN: Association for Educational Communications and Technology.

Yeaman, A. R. J. (2004a). Enforcement for victims and villains, complainants and respondents. *TechTrends, 48*(3), 9–11.

Yeaman, A. R. J. (2004b). The misuse of technology. *TechTrends, 48*(5), 14–16, 83.

Yeaman, A. R. J. (2004c). The origins of educational technology's professional ethics: Part one. *TechTrends, 48*(6), 13–14.

Yeaman, A. R. J. (2004d). Professional ethics for technology. *TechTrends, 48*(2), 11–15.

Yeaman, A. R. J. (2004e). Professionalism, ethics, and social practice. *TechTrends, 48*(4), 7–11.

Yeaman, A. R. J. (2004f). Technologist by day, technologist by night. *TechTrends, 48*(1), 7–10.

Yeaman, A. R. J. (2006a). Protecting learners from technology. *TechTrends, 50*(2), 11.

Yeaman, A. R. J. (2006b). Scenarios and principles. *TechTrends, 50*(2), 10–11.

Yeaman, A. R. J., Nichols, R. G., & Koetting, J. R. (1994). Critical theory, cultural analysis, and the ethics of educational technology as social responsibility. *Educational Technology, 34*(2), 5–13.

12

IMPLICATIONS FOR ACADEMIC PROGRAMS

Kay A. Persichitte
University of Wyoming

Introduction

*I*N MANY RESPECTS, THE Association for Educational Communications and Technology (AECT) faces unique challenges for a professional organization: from the diversity of professional practice that we attempt to include under the organizational umbrella to our repeated efforts to adopt a definition of our field which represents that breadth of philosophy, practice, and research. In many respects, our field is a moving target and most of us, practitioners and academics, embrace this environment of continuous change and challenge on some level.

The Evolution of Names

As evidence of this perpetual evolution, one cannot ignore the fact that we have not been able to reach consensus about the name that we use to identify our area of work. Over a period of several decades, we have referred to ourselves and our programs under many titles including, but not limited to, visual instruction, audiovisual specialists, instructional media, instructional design and technology, educational media, school library media specialists, instructional systems development, instructional systems design, instructional systems technology, instructional technology, and educational technology. These program titles are one indication of the impact of the evolution of our practice, but they are also indicative of the evolution of the definitions described in chapter 10. One might argue that these titles represent both a response to changes in the field and an influence on changes in the field. It is perhaps impossible for us to separate the applications of our field from the definitions or from the reality that systemic change is inherent to the field. It is reasonable, however, to consider the implications of these evolutionary factors on the academic programs which prepare the next generation for the next iteration of a definition for the field.

Audiovisual Education

Saettler's (1990) *The Evolution of American Educational Technology* traces the historical development of our practice and the contextual events that have influenced the definitions described in chapter 10. In the first half of the 20th century, Thorndike, Dewey, Skinner, Piaget, and others offered psychological bases for a science of teaching that incorporated contemporary works related to individualized and learner-centered instruction. As discussed in chapter 8, by the 1920s, the use of silent films, slides, and other pictorial materials in education stimulated the growth of a field of visual instruction, and university courses began to be offered on this subject. The addition of phonograph records, radio, sound films, and other auditory media expanded this concept to audiovisual instruction by the 1950s.

The flowering of media research during and after World War II sparked new enthusiasm and raised expectations for combining media and instructional method to improve teaching. Practitioners in our field were commonly referred to as "AV specialists" and our work was largely relegated to supporting instruction. Audiovisual (AV) courses were common in teacher preparation programs, and several higher education institutions offered doctoral programs in audiovisual education as early as the 1940s. For example, Indiana University approved master's and doctoral degree programs in Audio-Visual Instruction for the 1946–1947 academic year, and graduate-level summer courses in AV were also offered at the University of Chicago and University of Wisconsin in the same time frame (Cook, 1980). As another example, from 1947 to 1962 the academic program at Syracuse University was titled Audio-Visual Education.

Communications Era

The post–World War II period, 1945 to the early 1960s, was heavily influenced by the development of communication theories, and many curricula (high school through graduate school) included courses and degrees in mass communications. McLuhan and Fiore's (1967) work, *The medium is the massage*, was influential for this evolution of our field as we turned our attention to two other "new" technologies that were destined to leave an indelible imprint on our research and practice: television and computers. These authors open their commentary with this paragraph that is relevant yet today:

> The medium, or process, of our time—electric technology—is reshaping and restructuring patterns of social interdependence and every aspect of our personal life. It is forcing us to reconsider and re-evaluate practically every thought, every action, and every institution formerly taken for granted. Everything is changing—you, your family, your neighborhood, your education, your job, your government, your relation to "the others." And they're changing dramatically. (McLuhan & Fiore, 1967, p. 8)

McLuhan (1969) wrote, "In the Age of Information, media . . . are in themselves new natural resources increasing the wealth of the community" (p. 37). Many in our field would argue that the influence of media on our economy and our culture continues to be significant and mostly positive.

As the tools of teaching began to morph, so, too, did our thinking about the internal processes associated with learning. Claude Shannon's (1949) communication theory, supplemented by the Shannon–Weaver visual model, introduced a mathematical definition of information, which became the foundation of information science. For audiovisual education, it offered a way to deconstruct the complexities of the communication process, making it more amenable to research and theorizing.

As academic theories of mass communication grew, radio broadcasting and later television broadcasting, also grew in scope and societal influence. Societies around the world found it necessary to regulate these mass communications media. The British and most other European countries, for example, chose to put broadcasting directly under public control, while the United States developed a hybrid system of commercial ownership of stations, regulated by government licensing. However, in 1938 and 1940 the Federal Communications Commission (FCC) reserved certain radio channels for educational use, and then repeated this pattern in 1952 for television through the "Sixth Report and Order." Those engaged in educational broadcasting affiliated themselves primarily with the National Association of Educational Broadcasters (NAEB), but there was a considerable overlap of membership and of professional concerns with the Department of Audio-Visual Instruction (DAVI) of the National Education Association (NEA), the predecessor of AECT. Consequently, the communications perspective was reflected in the audiovisual education field as well. This is indicated, for example, by the name change of the Syracuse University program from Audio-Visual Education to Instructional Communications in the mid-1960s.

University program names, such as Instructional Systems at Florida State University, reflect the growing saliency of theory and research related to systems theory and its application to instructional development that gained traction in the 1960s and continued through the current era. This intellectual movement grew out of practical research in systems analysis during World War II and the pioneering books and lectures of Ludwig von Bertalanffy on general systems theory in biology in the 1960s.

The transformative impact of the psycho-technology of programmed instruction in the 1960s, plus the emergence of digital technology broadened the horizons of the field once again, leading to widespread adoption of the term *technology.* The University of Southern California was the first to use the program name Instructional Technology in the late 1960s, followed shortly thereafter by Syracuse University. Indiana University chose Instructional Systems Technology in 1968, thus subsuming both "systems" and "technology."

This period was also characterized by much new research and experimentation in education. The launching of Sputnik and Sputnik II in 1957 and a host of other international challenges brought a surge to the federal funding of education-related research in the United States, including media applications. The influence of the National Defense Education Act (NDEA) of 1958, which used the term *new educational media,* was profound for all levels of education programs, from kindergarten through postgraduate. Then the Elementary and Secondary Education Act (ESEA) of 1965 added further fuel to the fire. Local schools were able to increase their holdings of audiovisual equipment and materials many-fold, and they were supported in launching innovative instructional activities. Membership in DAVI and attendance at its conventions grew to all-time highs by 1970. On the research front, programmed instruction and computer-assisted instruction (CAI) were fertile areas for exploration given the impact on public schools of the baby boom generation.

Outcomes-based education and standards-based education caused us to reflect on the systematic and iterative principles that ground much of our practice. In the 1970s and 1980s, our vocabulary expanded to include terms such as *instructional systems design, metacognition, public television,* and *microcomputers.* Graduate programs in our field shifted focus from AV to instructional design (ID). Some academic program titles documented this shift as well. For example, Syracuse University's program chose as its fourth name Instructional Design, Development, and Evaluation.

Recent Transitions and Transformations

Since the 1980s, the transitions and transformations of our field have been phenomenal. We have turned our attention to the contextualized nature of teaching and learning with research on motivation, media selection, feedback, assessment, learning styles, rapid application prototyping, and a host of other complexities that influence teaching and/or learning in any setting. The influence of computers as the "hot new technology" is indelible on our field. The emphasis in instruction turned from "teacher centered" to "learner centered" as the trend toward constructivist learning theories emerged out of the cognitivist theories of Vygotsky, Piaget, Bruner, Atkinson and Schiffrin, Ausubel, Paivio, and others still working from the "teaching as a science" perspective (e.g., Ausubel, 1977; Bruner, 2004; Paivio & Walsh, 1994; Vygotsky, 2004).

The Web Era

After 1994, the World Wide Web became a foundational resource for teachers and students at all levels of education (National Center for Educational Statistics, NCES, 2003) and in our daily lives. The dominance of online learning activities made learning objects (Recker & Wiley, 2001), online learning environments (e.g., Lowell & Persichitte, 2000; McConnell, 2005), digital content protection (Nathans, 2002), Web casting (Gasaway, 2003), and blogs (e.g., Johnson & Kaye, 2004; Kirkpatrick, Roth, & Ryan, 2005) part of the everyday terminology of the field. Constructivism and postmodernism have influenced our research and our practice (Hannafin, M. J., Hannafin, K. M., Land, & Oliver, 1997; Hannafin, M., & Rieber, 1989; Jonassen, Peck, & Wilson, 1999; Peters, 2003).

In the most recent decade, there has been some movement to expand academic programs to include *performance technology* or *performance improvement* and to address problems other than those of an instructional nature, as discussed in chapter 3. While this seems a natural progression for our field, the focus of this chapter, and indeed this book, is on the concept—and definition—of *educational* technology. Just as we saw some academic programs in the post–World War II era shift their emphasis to AV, technological advances and external pressures of the 21st century may result in further shifts in academic programs as well.

As a matter of professional reflection, AECT has taken a proactive approach to reflecting on who we are, what we do, where we do it, and what tools we employ in our work as educational technologists. Interactive multimedia systems, artificial intelligence, brain research, the Internet, informatics, virtual spaces, podcasting, open courseware architectures, and hardware/software technologies never imagined make these reflections difficult, but necessary to our future as we prepare the next generation of professionals for our field.

Programs, Curricula, and the Next Generation

As we review the historical context of this chapter, the multiple versions of our definitions do not seem so far-fetched. Indeed, each definition that AECT has adopted has been both an influence on and influenced by research, practice, theory, and technological innovations of that time. Most programs that prepare graduate students to work in this field are unabashed in admitting that we rely on the research and theory bases of many disciplines (e.g., educational psychology, communications, organizational management and development, adult learning theory, change theory, psychology, computer science, etc.) to prepare new professionals for work in settings that will challenge even the most competent if one does not continue to monitor changes in our field and others that contribute to our efficacy.

New definitions should cause us to reflect on the extent to which the academic programs reflect both the vocabulary and the application of the definition. Some have even attempted to differentiate programs by distinguishing between instructional technology and educational technology (see Seels & Richey, 1994, pp. 3–5). Definitions and vocabulary are not just exercises in semantics; rather they provide us with a broad structure for the applications of our profession. The important point is not whether the definition (or the program title) is *educational technology* or *instructional technology* or any other combination of relevant terms. The important consideration is whether the curricula and experiences required in the preparation of our next generation of practitioners is aligned with contemporary interpretations of the explication of the accepted definition for the field.

As a profession, we must acknowledge and act upon an expectation that the knowledge and skills embedded in today's preparation of new educational technologists provide a foundation for unknown professional demands in the future. Wiley (2004) offered a powerful challenge to professional educational technologists in a parable whose message is to think differently about both process and product as media continue to morph. The AECT definition of *educational technology* is intentionally broad in the interpretation of the vocabulary to allow for response to unanticipated consequences of future process and product evolutions. The key concepts and their interrelationships are shown in Figure 12.1.

As the number of practitioners in our field has grown, the types of settings in which we work have expanded, and the expectations for our performance have broadened, the current AECT definition specifies "... the study and ethical practice of facilitating learning and improving performance. ..." These words reflect both the process and the assessment perspective of our work. The tension in some programs between competing (and sometimes overlapping) directions for the preparation of practitioners of educational technology is implicit in the choice of these words. Examples of such competing/overlapping directions for preparing educational technologists include

- Educational settings versus BIG (business, industry, government) settings
- For-profit versus not-for-profit
- Government R&D versus contract R&D
- International versus local/regional concerns

Many institutions of higher education have deep historical and/or institutional mandates to emphasize one or the other, while other programs struggle

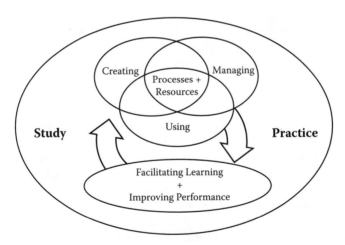

Figure 12.1. A visual summary of key elements of the current definition. The definition intentionally uses broad terms to communicate essential ideas, recognizing that the names for these ideas may change over time.

to find the balance and overlap between competing directions where both may be real options for graduates of the program.

Ethical Component

Inclusion of the word *ethical* was purposeful in this definition to turn our attention to issues of access, program delivery, product development, and equity in our graduate programs. Copyright, educational fair use, and issues of access to all forms of digital content are ethical and legal concerns for practitioners today regardless of setting. As discussed in chapter 11, ethical considerations go beyond just technological concerns.

Accountability Component

The term *improving performance* is a dual-edged sword that cuts across evaluation of our programs and assessment of the preprofessionals in those programs. Political pressures and public expectations for increased accountability have led to an environment where performance measurement and outcome assessments are becoming the norm (Fitzpatrick, Sanders, & Worthern, 2004). Performance is measured and reported for programs, graduates, organizations, and learners. Education has become a data-driven, decision-making enterprise where all share accountability for the outcome and the stinging impression of a person not well-prepared costs us individually and collectively. *Performance* also carries applied meaning in this field often distinguished by the context of *where* and *for whom* we apply our skills and knowledge. This portion of the AECT definition highlights the more dispositional elements of the definition and our practice.

Skill Component

In contrast, this definition continues to focus on some traditional aspects of preparation associated with instructional design and systems development: ". . . creating, using, and managing. . . ." A skill component to our practice has been evident in every definition of the field. Practitioners must know what the toolbox has in it, and we must have at least rudimentary skills for the use and application of those tools for a variety of audiences. Many of the current academic programs have a history of emphasis in this "media" side of our field, or they have such a reputation from graduates and/or program promotion. This new definition offers the opportunity for faculty to consider mission and vision statements and reflect as to whether this is the appropriate window to revise the conceptual base of the program or to move forward on an established base. These reflections are the root of any program evaluation effort.

Knowledge Component

Finally, there is a knowledge component to the AECT definition: ". . . appropriate technological processes and resources." Processes, conceptual bases, research, and access to new innovations require systematic review of academic programs to assure that curricula and experiences allow for development of such knowledge by our graduates. Review of the curricula should also include a review of recent research and theory for inclusion in the program to allow new professionals the knowledge base that will be necessary to continue to develop their skill base once they are practicing in the field.

Program Review and Professional Standards

The mission and vision of AECT have influenced the current definition and this definition will inevitably have an influence on the standards that AECT has established for evaluation and accreditation of programs that prepare new professionals in school media and educational technology. In 2000, the National Council for Accreditation of Teacher Education (NCATE) adopted these standards (AECT, 2000), as this national accrediting body recognized AECT as a Specialty Professional Association (SPA) responsible for providing program review expertise for programs in educational technology and those that prepare school media specialists. In the same way, one expects the mission and vision of all academic programs to influence decisions about the curricula and learning experiences, which consequently impact the knowledge, skills, and dispositions of graduates.

Program review may be a formal (as in the case of an NCATE/AECT review) or an informal process (as in the case of an annual report compiled for a department) that documents how data are utilized to drive decisions about program revisions. Regardless, the focus of the effort is on program relevance and opportunities for improvement to achieve a net positive effect, then, on the next generation of professionals for our field. Performance assessment data and standards-based curricula are not only historical foundations of our field; these elements are also required for any level of program review. Fitzpatrick et al. (2004) makes a strong case for using program evaluation as a mechanism for "informing the appropriate audience(s) about the findings and conclusions resulting from the collection, analysis, and interpretation of evaluation information" (p. 377).

Current Standards

The AECT (2000) standards are directly aligned with the knowledge base and skill set articulated by the Seels and Richey (1994) definition: design, development, utilization, management, and evaluation. Within each standard are multiple indicators that provide examples of the types of knowledge or skills that might be assessed among students in our graduate programs to document acceptable performance in each domain. The complete AECT definition provides a context for the standards and a base for the setting of mission and vision within our graduate programs. The assessment cycle is complete when program curricula, experiences, and assessments are aligned with standards and the program-specific mission and vision.

Future Standards

The adoption of a new definition for our field is timely from the academic program perspective, as well. The AECT (2000) standards are due for review and revision. The Seels and Richey (1994) definition had a significant impact on the AECT (2000) standards, and we should expect that a similar revisioning will occur as the AECT 2009 standards are developed. Similarly, the AECT (2000) standards were purposely broad in identifying performance indicators that encompass a variety of contexts for the application of our practice (e.g., preservice K–12 teacher preparation, higher education faculty, school media specialists, corporate and government trainers and developers). Given this AECT definition, one would expect the AECT 2009 standards to incorporate a broad view of the contexts for application of our field, as well. Academic programs that seek national recognition by AECT and those that seek to continue their national recognition should take note of this evolution of the definition for our field and the next evolution of AECT standards.

While one cannot predict with any certainty the revisions that will emerge in the next iteration of AECT standards, we should expect that the new definition will result in revisions to the AECT (2000) standards and the performance indicators. The inclusion of "ethical practice" in the new definition will most likely stimulate much conversation among AECT members/practitioners around the academic program indicators that we should expect and the measurement of those behaviors, attitudes, and dispositions. To date, academic programs have not been expected to provide any documentation of ethical practice among their faculty or their students. Many graduate programs have used the AECT code of ethics as a basis for some curricula, but the standards have not required evidence of the teaching of ethical practice at the program level or the student level. It is likely that this will be a significant revision to the AECT (2000) standards.

Each of the current five standards, and their indicators, may continue to provide the structure for documenting proficiency in the new knowledge and skills language of the definition, ". . . facilitating learning and improving performance by creating, using, and managing appropriate technological processes and resources." Design might be interpreted to encompass "facilitating learning" and "creating." Development might be interpreted to encompass "creating . . . appropriate technological processes and resources." Utilization clearly maps to "using" and management maps to "managing." Evaluation is implied in the systems approach to design/development and is also part of the meaning of "improving performance."

The process of iterative review of professional standards and indicators for relevance to contemporary applications within our practice is an important result of the impact of the definition evolutions in our field. One should expect significant revisions in the indicators to keep pace with developments in the research, theory, and best practices as recognized today. On the other hand, as these conversations among AECT members occur, we may take advantage of the opportunity to recast the AECT 2009 standards and indicators to align directly with the new definition so that academic programs have a clear path for program changes to align with the AECT professional knowledge base and definition.

Conclusion

This AECT definition provides us with opportunities and, perhaps, motivation, to reflect on our programs, our graduates, and our professional vision for the future of the field. A new definition should also challenge us to wrestle with newly emerging questions that impact new professionals directly and the profession, indirectly. For instance,

- Should we use the AECT standards as a measure of our academic and professional accountability?
- Do our professional preparation programs reflect the AECT code of ethics, and how are we measuring that among our students and graduates?
- Does the new AECT definition of the field alter the vision, mission, or emphases of our professional preparation programs?

- What is the impact of the new AECT definition on our program review, accreditation processes, curricula audits, course development, student assessment, and external credibility across diverse audiences?

Program review in the context of current AECT standards, regardless of the version, provides us with a process and a structure to monitor the evolution of our practice and to engage in some level of quality control of the products of these academic programs that prepare the next generation of educational technology practitioners. To the extent that we recognize and embrace the opportunity to join in the peer review of our field, the AECT definition grounds our academic programs and sets a course for the curricula and experiences that we value and expect from future practitioners.

References

Association for Educational Communications and Technology (2000). *Standards for the accreditation of school media specialist and educational technology specialist programs.* Bloomington, IN: Author.

Ausubel, D. P. (1977). The facilitation of meaningful verbal learning in the classroom. *Educational Psychologist, 12*(2), 162–179.

Bruner, J. (2004). A short history of psychological theories of learning. *Daedalus, 133*(1), 13–20.

Cook, A. W. (1980). A history of the Indiana University Audio-Visual Center: 1913–1975. Unpublished doctoral dissertation, Indiana University, Bloomington, Indiana.

Fitzpatrick, J. L., Sanders, J. R., & Worthen, B. R. (2004). *Program evaluation: Alternative approaches and practical guidelines* (3rd ed.). Boston: Pearson Education.

Gasaway, L. (2003). Webcasting and copyright. *Information Outlook, 7*(2), 38–40.

Hannafin, M. J., Hannafin, K. M., Land, S., & Oliver, K. (1997). Grounded practice and the design of constructivist learning environments. *Educational Technology Research and Development, 45*(3), 101–117.

Hannafin, M., & Rieber, L. (1989). Psychological foundations of instructional design for emerging computer-based instructional technologies: Parts I and II. *Educational Technology Research and Development, 37*(2), 91–114.

Johnson, T. J., & Kaye, B. K. (2004). Wag the blog: How reliance on traditional media and the Internet influence credibility perceptions of weblogs among blog users. *Journalism & Mass Communication Quarterly, 81*(3), 622–643.

Jonassen, D., Peck, K., & Wilson, B. (1999). *Learning with technology: A constructivist perspective.* Upper Saddle River, NJ: Merrill.

Kirkpatrick, D., Roth, D., & Ryan, O. (2005). Why there's no escaping the blog. *Fortune, 151*(1), 44–50.

Lowell, N. O., & Persichitte, K. A. (2000). Virtual ropes course: Creating online community [Electronic version]. *Asynchronous Learning Networks, 4*(1).

McConnell, D. (2005). Examining the dynamics of networked e-learning groups and communities. *Studies in Higher Education, 30*(1), 25–43.

McLuhan, M. (1969). *Counterblast.* New York: Harcourt, Brace & World.

McLuhan, M., & Fiore, Q. (1967). *The medium is the massage.* New York: Bantam Books.

Nathans, S. F. (2002, July). Government, copyright issues step to fore at DVD 2002. *EMedia Magazine,* 12–13.

National Center for Educational Statistics. (2003). *Distance education at degree-granting postsecondary institutions: 2000–2001* (NCES 2003-017). Washington, DC: U.S. Department of Education.

Paivio, A., & Walsh, M. (1994). Concreteness effects on memory: When and why? *Journal of Experimental Psychology/Learning, Memory & Cognition, 20*(5), 1196–2005.

Peters, M. A. (2003). Technologising pedagogy: The Internet, nihilism, and the phenomenology of learning [Electronic version]. *Simile, 3*(1).

Recker, M. M., & Wiley, D. A. (2001). A non-authoritative educational metadata ontology for filtering and recommending learning objects. *Interactive Learning Environments, 9*(3), 255–271.

Saettler, P. (1990). *The evolution of American educational technology.* Englewood, CO: Libraries Unlimited.

Seels, B. B., & Richey, R. C. (1994). *Instructional technology: The definition and domains of the field.* Washington, DC: Association for Educational Communications and Technology.

Shannon, C. E. (1949). *The mathematical theory of communication.* Urbana, IL: University of Illinois Press.

Vygotsky, L. S. (2004). Imagination and creativity in childhood. *Journal of Russian and East European Psychology, 42*(1), 7–97.

Wiley, D. (2004). The polo parable. *TechTrends, 48*(3), 76.

AFTERWORD

Alan Januszewski

State University of New York at Potsdam

Introduction

*A*s a way to "wrap up" this document (likely the project of defining educational technology is never ending) the committee suggested that I, as chair, prepare an afterword that discusses, at least, (a) the timetable that the committee followed and some of the committee's activities, (b) a few of the more overarching considerations involved in defining educational technology, and (c) a look at the committee's thinking and reasoning as it decided upon a definition. Clearly, this afterword will not address all of the factors that the committee considered as it set about its task. It may even appear to lack a unified theme. There are some things that I am certain I must have forgotten. I have, using my best judgment, omitted other things by intent.

Obviously, this particular piece is my observation of the committee's activities and has not undergone the stringent review process that the other chapters in this volume have. I do owe a debt to Mike Molenda for trying to ensure the factual accuracy of this chapter as well as ensure some measure of grammatical and stylistic integrity in my writing. Actually, this is something that he did through all of the chapters of this manuscript. He has an eye for detail and a command of ideas, which greatly improved this volume.

Timetable

The AECT Board of Directors charged the Definition and Terminology Committee to prepare a new statement of definition at the meeting of the Board in the summer of 2002. In the time period between that summer meeting and the first Board meeting of the 2002 convention in Dallas, members of the AECT Executive Committee and I addressed several ideas including determining the primary audience and purpose of the definition statement (book); intellectual property rights for the definition statement and supporting materials, including possible book format; a target date for completion based on AECT's Accreditation Committee's need to use this definition in the writing of the new NCATE standards, and the make-up of the committee.

By the conclusion of the meetings of the Definition and Terminology Committee in Dallas, the committee agreed that the primary intended audience for this book was students entering the graduate programs of our field, although it was recognized that there would indeed be others who would read and use the book. All of the members of the committee had experience using either the 1977 or 1994 definition books in their teaching so there was confidence in the committee's direction.

The Board and the Committee came to agree that the book should be an edited volume with authorship acknowledgement granted to authors who wrote explanatory chapters. The decision was made to give authorship credit to the association itself for the one-sentence definition and the first chapter that provides a basic explanation of the one-sentence definition because it is believed that since this definition is the product of a committee of the association and because it is approved by its Board of Directors, no individual should own it or be credited with it. Although, to be fair, I would like to recognize Mike Molenda and Rhonda Robinson as having done the lion's share of work in documenting the Committee's words and intentions and then converting them into a coherent narrative, with other Committee members editing and elaborating as appropriate. Citations for the first chapter in this book should read Association for Educational Communications and Technology (2007). Definition. In A. Januszewski, & M. Molenda (Eds.), *Educational technology: A definition with commentary* (pp. 1–14). New York: Lawrence Erlbaum Associates.

The committee held an informational session at the 2002 Dallas convention where we announced our project and invited comments from the organization's membership. At the conclusion of the Dallas convention the Committee agreed to complete a "one-sentence definition" while communicating by telephone and e-mail, since we had set a deadline for completion of drafts of all manuscripts by the convention in Chicago in October 2004. During the time between the Dallas (2002) and Anaheim (2003) conventions, the Committee agreed that we should use "nontechnical language" in writing the one-sentence definition. The Committee also agreed on some of the terminology that should be included in that definition. But we could not meet the original goal of completing a draft of a one-sentence definition before the 2003 convention in Anaheim.

At the Anaheim convention, the Committee met at length, and, after substantial deliberation, agreed to a tentative one-sentence definition that it shared with the membership at an informational session during that convention. Again, we accepted comments and considered these as we struggled to finalize our wording. Also at the committee meetings in Anaheim, we determined which members of the Committee would take the lead on writing the different chapters that would explain the major ideas, if not specific terms, used in the definition. In the course of time, some of the Committee members were unable to complete their writing assignments, so other Committee members volunteered to fill the gaps.

After the Anaheim convention, the Committee tried to negotiate the final wording of the one-sentence definition, once again, by using e-mail and telephone. Again, we were unable to come to agreement and conclude this piece that was the key to the overall definition project. At the May 2004 meeting of the Professors of Instructional Development and Technology (PIDT), a quorum of the Committee was able to meet face to face with other members of the Committee participating via conference calls. It was during this series of Definition and Terminology Committee meetings, informed by excellent constructive feedback provided by PIDT attendees, that the committee came to a final consensus on this one-sentence definition:

> Educational technology is the study and ethical practice of facilitating learning and improving performance by creating, using, and managing appropriate technological processes and resources.

It was this definition that AECT's Board of Directors approved at their summer meeting in Chicago in 2004. Once this definition was formally approved, authors began work on the explanatory chapters for this volume in earnest.

Overarching Considerations Involved in Defining Educational Technology

Heinich (1971) argued, "A definition can be viewed as an attempt to establish a power base" (p. 9). It follows from his statement that the act of establishing a definition is a political act in that one is intending to influence the establishment of a power base. As individuals, whether they were members of the Definitions and Terminology Committee or not, tried to influence the content of the definitions of educational technology, they were engaged in a type of political action. In past efforts to define educational technology, some tried to influence the definitions to include activities that they performed or felt ought to be performed in the field. Undoubtedly, this is true for the work of this AECT Definitions and Terminology Committee as well. Creating such a definition of educational technology is not a value neutral act.

The meanings of abstract concepts are often contested based on different interpretations of those concepts. Abstract concepts do not represent physical things, which can simply be pointed to, and their meanings easily agreed upon. Therefore, it is difficult to isolate and gain specific agreement about the "is" of an abstract concept. It is difficult to gain specific agreement about the "is" in the practice of educational technology. It is even more difficult to establish specific agreement on the "oughts" of the field. Are definitions of educational technology merely descriptions of what *is* occurring in practice? Or are they based on individual theories of what *ought* to be involved in educational technology? Whose practice is being analyzed to gain insight to the "is" related questions? Whose theories and ideas are being included in the analyses of any "ought" related questions?

Different individuals have different understandings of educational technology; both what it *is* and what it *ought* to be. This fact helps to explain why there were differences of opinion on what the definition of educational technology should be and why it has changed over time. These definitions are far from being precise and objective statements typically associated with definitions. Rather, they are the products of contemporary context and political activity, most specifically, negotiation and compromise. The definitions were products of a group process. Each of the members of the group brought their own ideas to the process. But, since at least some of the members of the group disagreed on specific points to be included in the definition, negotiation, and compromise were required in order to maintain consensus among the members of the group on the final definition statement. The resulting statement of definition is a "political" document by both intent and process (Januszewski, 2001). Political is not necessarily negative, as seems to be implied in common parlance. But it is not something that we often think of ourselves as doing. We often view others as being political, while we tend to view ourselves as simply working through things.

The Committee's Reasoning

As you might imagine, during the many hours that the members of the Definition and Terminology Committee spent in group deliberation, they had friendly but sometimes spirited, even intense, debates about issues framing the definition as well as about the terms of the definition itself. Why else would it have taken years to come to agreement? I would like to cover some of the key points, briefly summarizing the issues and the reasoning behind the consensus reached.

Educational Versus Instructional Technology

This choice provoked the first and longest discussion. Everyone agreed that the term representing the core of the field should be the one that was broadest in scope. Some argued that *instructional* is broader because instruction takes place in all settings, formal and informal, aimed at children or adults, while *education* seems to imply formal school and college settings. Others felt that *education* subsumes *instruction,* covering both purposive and spontaneous learning, both teacher led and learner initiated, while *instruction* connotes teacher-led education and training. It was acknowledged that this distinction between instructional technology and educational technology that was a part of the 1977 definition statement was unhelpful and even confusing. For some, the clincher was that the dictionary definition of *instruction* focuses on *imparting* knowledge or giving directions—a connotation that does not fit well with the current view of how most learning takes place. In the end, *educational* was conceded to have broader scope.

For others on the committee, there was no real distinction between instructional technology and educational technology when these are viewed as whole or unified constructs, as seems to appear in most of our literature, and not as types of technology with different preceding adjectives.

Domains of the Field

The 1994 definition established the notion that educational technology encompassed a number of *domains*: design, development, utilization, management, and evaluation. Earlier definitions had viewed these elements as *functions* carried out by practitioners. The 1994 Committee chose to call them domains, following the pattern of Bloom's taxonomy of educational objectives with its domains of learning. There was broad agreement that the five domains provided a handy, useful classification scheme, but disagreement as to whether this label could stand up to logical scrutiny.

In the end, the choice was simplified because the Committee rejected the five subdivisions for other reasons. The categories themselves were reconstrued and "creating, using, and managing" were now noticeably more like functions than constituent parts of the field.

Specialized Technical Terms Versus Common Lay Terms

Several members of the committee argued that "design, development, utilization, management, and evaluation" were most obviously interpreted as stages in the systems approach to instructional development. That is, insiders would most likely interpret them as "terms of art" in the profession, not as broad and generic functions, regardless of the intentions of the 1994 Committee. One of the first criteria agreed upon for the new definition was that it be intelligible to outsiders, avoiding specialized "insider" jargon. The adoption of broader terms (e.g., *creating* and *using*) served two purposes—to avoid implying a sole commitment to the systems approach to instructional development and to appear more approachable to nonspecialists.

Value Terms

Most of the committee members came into the discussion without the assumption that the definition should contain explicit references to the *values* embraced in the field. Previous definition books acknowledged that educational technology professionals tended to hold some common values, but they were not considered to be defining traits. Some members, however, argued that you cannot distinguish "what educational technologists do" from "what anyone can do" without bringing in value terms, such as *efficiency* and *effectiveness*. Nowadays, anyone with computer hardware and software can concoct a slide show, handout, or job aid. What distinguishes such amateur concoctions from "professionally designed" products? Surely educational technologists must claim to do something "better than" amateurs.

There was fairly quick agreement to include *ethical* as a required trait. Other value terms were not so easy to reach consensus. No one could disagree with *appropriate*, but what did this broad label encompass? It was decided to leave that interpretation to the authors of each of the relevant chapters.

The term *technological* was also hardly debatable since it was the core term of the overall definition, but lexicographers argue that you cannot use a form of a word to define the word itself. In the end, the committee decided that the word *technological* was acceptable as a convenient shorthand for a whole set of core values—technology as a way of thinking, commitment to decisions based on scientific and other organized types of knowledge, and a commitment to dealing with practical tasks. Again, it was left to the individual chapter authors to decide how to interpret these broad values with regard to the concepts and functions discussed in each chapter.

Any discussion as to whether or not to include value terms in the one-sentence definition did not really involve disagreement on the importance of the values connoted in those words to our field. Rather, it was a difference of opinion over the necessity of including those terms in the one-sentence definition. It was believed by some that those terms could be articulated at length in the first explanatory chapter. You could view this disagreement as a primary need for the inclusion of value terminology versus an economy of words, or perhaps an elegance that might be lost due to wordiness. In the end, the decision was made to accept the inclusion of several value terms to maintain an overt clarity tied directly to the desire to have this definition "provide direction."

Including these value terms also opened the door to some substantial philosophical considerations. Were these necessary or sufficient or either to a definition of educational technology? Could certain terms be eliminated and still connote an accurate definition? This form of philosophical analysis was called "the A without B procedure" by philosophy professor Tom Green at Syracuse University, my alma mater. Looking at an abstract-contested concept like educational technology in this way seemed to pose an insurmountable problem.

The solution to this problem came from two realizations: first, as Fischer (1971) argued there were at least 28 different kinds of definitions used in conceptual studies and different rules are applied to the use of each of them. This definition of educational technology is not an analytical definition. Rather, it is more of a theoretical definition. It both describes the current state and prescribes for the future at the same time. Second, when formulating this definition the Committee followed the model used by Ely (1963) and the writers of the association's first definition. That definition was intended to be forward looking and to give direction to the field. It is in that spirit that this work was done.

This "giving direction" is also a different mindset than "defining educational technology by what educational technologists do," like the authors of the 1972 and 1977 definitions attempted to do. That is clearly a behavioral approach. It is conservative in that it looks at the current state of practice and does not look to the future. By the way, Kliebard (2004) provided an excellent explanation of the conservative nature of job and activity analysis in conceptualizing and developing a curriculum. Clearly, this sort of approach to defining invites the question, "If you define educational technology by what educational technologists do, who's an educational technologist in the first place?" "How do you determine who to look at to see what they are doing?" It was an easy decision to avoid this mindset as we began our process.

As you might have surmised, the committee dealt with ideas, factors, influences, and conceptions as it worked through the process of defining educational technology. It was certainly not, at all times, a linear process, most likely because members could not identify and agree on all of the necessary factors prior to beginning our actual task. Members would advance particular ideas and these would be scrutinized for possible alternatives along with considerations of logic and consistency. We would move on, back up, and start again. We agreed to the inclusion of some words much more easily than others. In-depth discussion of the practice of philosophical and conceptual analysis went hand in hand with the discussions and decisions involved in writing the one-sentence definition.

Conclusion

Did you ever wonder why it was, and still seems to be, the case that educational technologists spend so much energy trying to formally and officially define educational technology? Obviously, I have or I would not have bothered to raise the question at this point. To my knowledge, no other academic discipline, area of study, or field of professional practice—call it what you will—invests as much time and effort trying to define itself as does educational technology, at least the individuals who comprise AECT. Do a search of other professional organizations and see what you come up with. You will see many *conceptual analyses* of those professions conducted by *individual* scholars. I am not at all certain that AECT members are better or smarter than other professionals are because they engage in this "defining" activity. Neither do I believe the opposite. I do believe that, on the whole, educational technologists are preoccupied with trying to systematize and bring order to things.

I had the good fortune to study with Donald Ely at Syracuse University 20-plus years ago. At one point, he explained that the decision to use the term *definition* was originally suggested by Jim Finn. Finn believed that certainty in the meaning and use of terminology in our field was necessary to the field being recognized as a profession. Further, he saw the role of the then Commission on Definition and Terminology as being the functional equivalent of L'Académie Française (French Academy). Such a commission would arbitrate and determine the meaning of particular terms used in the field, also acting in an enforcement role. Needless to say, this has not happened— mercifully, in my view. I believe that this would have stifled the growth and development of the area of study and practice that is educational technology. Linguists and philosophers have been able to show that there are strong connections between how we talk, how we think, and how we act.

I think that the desire to define is part of the need to attain certainty and order that is rooted in the field's past. Historically, the concept of technology and the developing and applying of a science of education were focused on gaining certainty and order. Defining is a natural outgrowth of that need for certainty, order, and consistency. I believe that is why we define rather than do conceptual analyses. Of course, this invites the question, "Why define educational technology in the first place?" What practical value does this have?

As someone whose undergraduate and graduate studies included a substantial amount of work in philosophy and the social sciences I would answer, with no small degree of smugness, "What could be more practical than having some idea of what you are talking about?" I say "some idea" because I want to recognize that many conceptualizations of educational technology and many interpretations of the supporting concepts help flesh out those conceptualizations. I think we should all recognize the potential and possibilities that exist outside of our own conceptualizations of educational technology. It is constantly changing. The change may be characterized as evolutionary or revolutionary but it is constantly changing. When looked at from this perspective, when it comes to educational technology, we may never really know what we are talking about. All we get is a snapshot at a moment in time. And this picture can be looked at from a variety of angles.

Emerson said, "A foolish consistency is the hobgoblin of little minds." I think that there is a lot to that observation. It is then fair to ask me, "If you believe all of what you've just said, then why were you involved in this project?" I have no real answer to give you for that question . . . except, perhaps, to try to quote Burgess Meredith's character from the movie *Grumpy Old Men*. It goes something like this: "When you get old and you look back over your life all that you are going to have left is the experiences. It's all about the experiences."

Shepherding this project has certainly been an experience. I would like to thank Mike, the other members of the Committee, the authors, and the rest of those who have been involved for the experience.

References

Ely, D. P. (1963). The changing role of the audiovisual process: A definition and glossary of terms. *AV Communication Review, 11*(1), Supplement 6.

Fischer, D. H. (1971). *Historians' fallacies: Toward a logic of historical thought.* New York: Harper & Row.

Heinich, R. (1971). Toward a definition of instructional development: An eclectic approach. In I. K. Davies, & T. M. Schwen (Eds.), *Toward a definition of instructional development.* Washington, DC: Association for Educational Communications and Technology.

Januszewski, A. (2001). *Educational technology: The development of a concept.* Englewood, CO: Libraries Unlimited.

Kliebard, H. M. (2004). *The struggle for the American curriculum: 1893–1958* (3rd ed.). New York: Routledge and Falmer.

CONTRIBUTORS

Anthony Karl Betrus
State University of New York at Potsdam
Dr. Betrus (PhD in Instructional Systems Technology, Indiana University) is an associate professor and chair of the Department of Information and Communication Technology at the State University of New York at Potsdam. He has served as a member of the AECT History and Archives Committee and the Definitions and Terminology Committee, and as the president of the Multimedia Production Division. His research interests focus on design and use of simulations and games in education.

Elizabeth Boling
Indiana University
Prof. Boling (MFA in Printmaking, Indiana University) is an associate professor and chairperson of the Department of Instructional Systems Technology at Indiana University. She previously worked as an interface designer and production manager for educational software development and as a graphics and animation manager for instructional products at Apple Computer, Inc. She is a past president of AECT's Design and Development Division. She served as editor-in-chief of *TechTrends* from 2004 to 2006. Her research interests include visual design for interactive information and instruction, and design process and methods.

Robert Maribe Branch
University of Georgia
Dr. Branch (EdD in Instructional Technology, Virginia Tech) is a professor in the Department of Educational Psychology and Instructional Technology at the University of Georgia. He began his career in education as a Peace Corps volunteer high school teacher in Botswana and then as a lecturer at the University of Botswana. Dr. Branch has coedited the *Educational Media and Technology Yearbook* since 1997, and he coauthored the two most recent editions of *Survey of Instructional Development Models*. He teaches courses related to instructional systems design and consults regularly with governments, businesses, and other educational institutions. His published research focuses on diagramming complex conceptual relationships and other flow processes.

Christa Harrelson Deissler
University of Georgia
Dr. Deissler (PhD in Instructional Technology, the University of Georgia) is a grants development specialist in the Office of Research and External Affairs of the College of Education at the University of Georgia, where she teaches graduate courses on grant development and supports grant development efforts. Dr. Deissler taught technology integration to public school educators prior to her current position. Her research on the educational beliefs of teachers who integrate technology has been reported at conferences of the AECT, the Society for Information Technology and Teacher Education (SITE), and the Eastern Educational Research Association (EERA). Her current research focuses on faculty attitudes toward seeking external funding.

J. Ana Donaldson
University of Northern Iowa
Dr. Donaldson (EdD in Instructional Design, Northern Illinois University) is an associate professor in the Department of Curriculum and Instruction at the University of Northern Iowa. She serves as lead instructor for the performance and training technology emphasis within the Instructional Technology Division. She is coauthor with Rita-Marie Conrad of *Engaging the Online Learner: Activities for Creative Instruction*. Dr. Donaldson has written many journal articles and book chapters on topics related to educational technology. She has been active in AECT leadership, having served as a member of the board of directors, as an officer of the Training and Performance Division, and as chair of the Leadership Development Committee. She is currently a member of the Ethics Committee. Her research and teaching interests include performance and training technology, visual literacy, and learner engagement through technology integration.

J. Nicholls Eastmond, Jr.
Utah State University
Dr. Eastmond (PhD in Educational Psychology, University of Utah) is a professor in the Department of Instructional Technology at Utah State University. He is coauthor with Robert Brien of the book *Cognitive Science and Instruction*. He has been a member of the AECT Professional Ethics Committee since 1985 and served two years as head. He edited the International Review section of *Educational Technology Research and Development* for ten years. He speaks French and has done overseas work in France, Canada, South Africa, Mauritius, and Hawaii. His interests include the use of educational technology for foreign language learning, for online learning at a distance, and for improving cross-cultural understanding.

Alan Januszewski, Coeditor
State University of New York at Potsdam
Dr. Januszewski (PhD in Instructional Design, Development, and Evaluation, Syracuse University) is a professor in the Department of Information and Communication Technology at the State University of New York at Potsdam. He is the author of *Educational Technology: The Development of a Concept*. He has also written a variety of journal articles and book chapters on topics related to foundations issues in educational technology. He has been a member of the AECT Definition and Terminology Committee since 1991 and has served as chair since 1996. His research and teaching interests include the history of educational technology, philosophy and educational technology, and cultural and social impact of educational technology.

Michael Molenda, Coeditor
Indiana University
Dr. Molenda (PhD in Instructional Technology, Syracuse University) is an associate professor emeritus in the Department of Instructional Systems Technology at Indiana University. He is coauthor of the first five editions of *Instructional Media and the New Technologies of Instruction*, an award-winning textbook for college courses in media utilization. He has written numerous journal articles, encyclopedia entries, and book chapters on topics related to educational technology. He has lectured and consulted in Spain, Indonesia, Korea, China, and several countries in Latin America and the Middle East. He has been an active member of AECT since 1963 and held numerous leadership positions.

Vicki Napper
Weber State University
Dr. Napper (PhD in Instructional Technology, Utah State University) is an associate professor and technology director in the Department of Teacher Education at Weber State University. She was a Cochran Intern at the 1996 AECT National Convention, and she presented a keynote session entitled *Can You Improve Your EQ? Designing Workstations* at the AECT National Convention in 1997. She has served in various capacities related to ethics, including editor of the Ethically Speaking column in *TechTrends*, for which she received AECT's Annual Achievement Award; as chair of AECT's Professional Ethics Committee; and as a member of the Technical Committee on Ergonomics for Children and Educational Environments of the International Ergonomics Association. Her past and present research interests include copyright and fair use in educational settings, ergonomic considerations in K–12 education, and human–computer interaction.

Robert Pearson
Performx Group
Dr. Pearson (PhD in Instructional Design, Development, and Evaluation, Syracuse University) is currently president of Performx Group, a Toronto-based group of learning services companies with operations in five cities across North America. Over the past 15 years, Dr. Pearson has worked as a learning solutions consultant with a wide range of Fortune 500 companies. He has complemented his work in business with active participation in a number of professional organizations including AECT, ASTD, and ISPI. He is a member of the board of directors of the Canadian Society for Training and Development. He has written a number of journal articles, and he is a frequent presenter at learning-related conferences around the world.

James A. Pershing
Indiana University
Dr. Pershing (PhD in Education, University of Missouri) is a professor in the Department of Instructional Systems Technology at Indiana University. He is the editor of the *Handbook of Human Performance Technology*, 3rd edition. In 1999, he served as interim executive director of AECT and as editor of *TechTrends*. From 1999 to 2005, he was editor-in-chief of *Performance Improvement*, the journal of the International Society for Performance Improvement (ISPI). He has written numerous journal articles on topics related to performance technology. He is a recipient of the President's Award from ISPI and the Distinguished Service Award from AECT, honoring his exceptional contributions to performance technology and educational technology. His teaching and research focus on research methods in education, performance improvement, evaluation, and change management.

Kay A. Persichitte
University of Wyoming
Dr. Persichitte (PhD in Educational Technology, University of Northern Colorado) is a professor and director of teacher education at the University of Wyoming. She is the author of numerous articles in the areas of technology integration and distance education. She has served AECT as chair and as a member of the Standards and Accreditation Committee, as president of the Teacher Education Division, and as a member of the Definition and Terminology Committee. She currently serves on the board of directors. She served for a decade on the National Council for Accreditation of Teacher Education Board of Examiners and is currently the AECT representative to the Specialty Areas Studies Board. Her current research interests include curriculum and

program development in teacher education, accreditation, standards-based instruction, and preservice teacher technology preparation.

Landra L. Rezabek
University of Wyoming
Dr. Rezabek (PhD in Educational Technology, University of Oklahoma) is an associate professor of instructional technology in the Department of Adult Learning and Technology, College of Education, University of Wyoming. Her scholarly publications in visual literacy, instructional design, and distance education underscore her interest in the study of human learning as a problem-solving activity. Dr. Rezabek has an extensive record of professional activity, including numerous leadership roles in AECT and the International Visual Literacy Association (IVLA). She believes one of her greatest accomplishments is reflected in the tribute from students who observe, "She practices what she teaches."

Rhonda S. Robinson
Northern Illinois University
Dr. Robinson (PhD in Educational Communications and Technology, University of Wisconsin) is a distinguished teaching professor in the Department of Educational Technology, Research, and Assessment at Northern Illinois University (NIU). She is the author of several book chapters and a variety of journal articles on topics related to visual literacy, technology integration in schools, and qualitative research methods. Her research and teaching interests include encouraging qualitative research in educational technology, the social and cultural impact of visual literacy and technology on teaching, and the integration of technology and visual literacy concepts and skills into teaching and learning. She has served as president of the Research and Theory Division of AECT and on the Definition and Terminology Committee. She is currently chair of the Publications Committee and is a member of the Teacher Education Division. She is a past president of the International Visual Literacy Association (IVLA) and currently serves on the board of directors and as membership cochair. She is an editorial board member of the *Journal of Visual Literacy*. She has been recognized by NIU as a Presidential Teaching Professor and by IVLA as an Outstanding Educator.

Sharon Smaldino
Northern Illinois University
Dr. Smaldino (PhD in Educational Technology, Southern Illinois University) holds the L.D. and Ruth G. Morgridge Chair in Teacher Education and

Preparation in the College of Education and serves as director of the College of Education Partnership Office at Northern Illinois University. She is coauthor of several widely used textbooks in educational technology, including *Teaching at a Distance: Foundations for Distance Education, Instructional Technologies and Media for Learning* (5th through 8th editions), and *Planning for Interactive Distance Education: A Handbook*. She has taught courses and conducted numerous workshops related to distance education, technology integration, and professional standards. Dr. Smaldino is past president of AECT and has served on the board of directors and several AECT committees. She is currently editor-in-chief of *TechTrends*.

Andrew R. J. Yeaman
Detroit, Michigan
Mr. Yeaman joined AECT in 1981 while a doctoral student at the University of Washington. He was appointed to the Professional Ethics Committee in 1987 and later served as chair for several years. He edited the 75th anniversary issue of *TechTrends* and an issue of *Educational Technology* on ethics and critical theory. He is currently the editor of the Professional Ethics section in *TechTrends*.

AUTHOR INDEX

SUBJECT INDEX

E